Praise

"Walk a mile in someone else's shoes—in fact, two pairs at once. In this sharply insightful and engaging work of creative fiction, a reader will certainly see himself (yes, him) or herself repeatedly in the situations, the thoughts, and the reactions and interactions that each of the authors reveal as they unfold, in parallel, this tumultuous period of their lives. At journey's end we emerge from the family's conflicts and comforts with a rare awareness—the value, indeed the triumph of seeing beyond ourselves."

—Nancy Clark, author and poet

"Exceptionally portrayed by mother and daughter authors, each gives her own perception of events challenging their family stability that occur after one of their many corporate moves. Each family member tackles the adjustment to their new community, but it is particularly difficult for daughter Krista. Inclusion of suspense, drama, humor, and interesting characters make this novel an excellent read."

—Lane Schroeder, author and poet

"*Depression Cookies* is a lively telling of the two very different experiences of mother and daughter as they face yet another uprooting of family life. Vividly written by a mother-daughter writing team, the dueling narrative spurs readers onward as Krista battles acne, jaw restructuring and junior high bullies at the same time mid-life matriarch Abby is struggling for control over aging parents, an often-absent husband and her own unrealized dreams. Grab this book for your mother or daughter—and be inspired to examine your own divergent viewpoints on family and relationships, past, present—and future."

—Stacey Hartmann, journalist

DEPRESSION COOKIES

❖

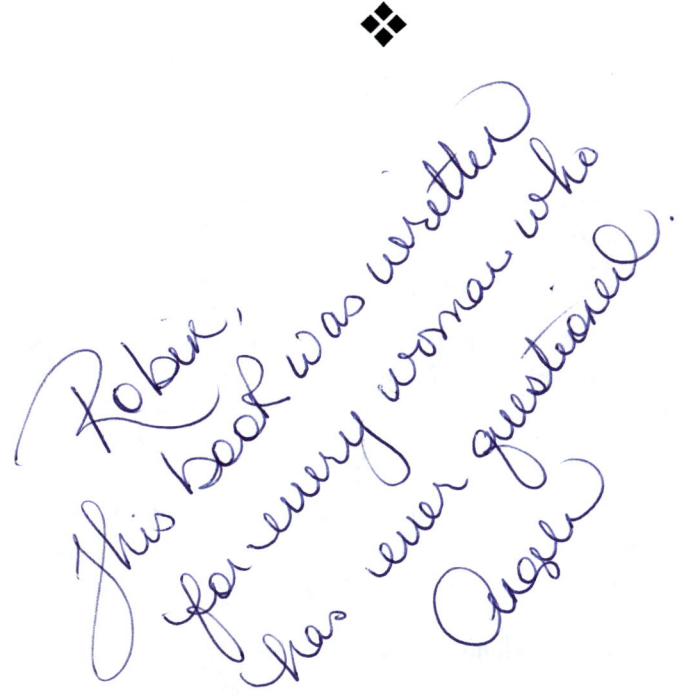

Robin,
This book was written for every woman who has ever questioned.
Angela

Tia Silverthorne Bach
&
Angela Beach Silverthorne

Copyright © 2010 by Tia Silverthorne Bach & Angela Beach Silverthorne.

Library of Congress Control Number:		2010912787
ISBN:	Hardcover	978-1-4535-6734-0
	Softcover	978-1-4535-6733-3
	Ebook	978-1-4535-6735-7

All rights reserved. No part of this book may be reproduced or transmitted in any form or by any means, electronic or mechanical, including photocopying, recording, or by any information storage and retrieval system, without permission in writing from the copyright owner.

This is a work of fiction. Names, characters, places and incidents either are the product of the author's imagination or are used fictitiously, and any resemblance to any actual persons, living or dead, events, or locales is entirely coincidental.

This book was printed in the United States of America.

To order additional copies of this book, contact:
Xlibris Corporation
1-888-795-4274
www.Xlibris.com
Orders@Xlibris.com
84405

ACKNOWLEDGMENTS

We would like to thank all the women in our lives. You teach us, support us, make us laugh, and bring enrichment to our lives and certainly this story.

Tia: I would like to especially thank my mom who has always been there for me and encouraged me every day throughout the ten-year process of writing this book. I would not be where I am without her. I thank my dad for always pushing me to do my best but ready to help me if I failed. Ed, thanks for always believing in me and being my rock. My three beautiful girls—Jackie, Reagan, and Maddie—inspire me every day to be a better person and certainly helped me remember what it was like to be a kid and preteen. Thanks to Tara, Dana, and Stacey H. for reading the manuscript and giving me your commentary and encouragement. Finally, to all my friends, you know who you are.

Angela: Dallas, thank you for sitting close and rarely complaining when I spent hours writing and editing. Without the support, listening, and push from my daughters, Tia, Tara, and Dana, this book would have gathered dust in a binder. There's no doubt without Tia's inspiration and determination, this book would have never been birthed. A special thanks to all the members of Pamlico Writers. You have taught me a lot about the writing process. And to Nancy and Lane who trudged through the manuscript, editing and adding commentary . . . a huge thank you!

A special thank you to the late Stephan Horvath, founder of Blue Moon Press and Tortuga Pines Publishing, who read our book and decided it was worth completing. We will be forever grateful to him as our initial editor and mentor.

Often intimacies between women go backwards, beginning with revelations and ending up in small talk without loss of esteem.

—Elizabeth Bowen, *The Death of the Heart*

PART I

Table Talk

Summer 2000

Abby

Sitting at the kitchen table, I watched my three daughters and was amazed at the ease with which they shared funny, anecdotal tidbits of their lives. Animated, they acted out various scenarios, ribbing one another in humorous play and laughing unabashedly. Resting my gaze on first one, and then another, I smiled, letting my mind wander back to pigtails, lace socks, mud pies, and hopscotch.

Pain seared through my right hip, snapping me back to the present. I gripped the table's edge, struggling to hoist myself to a standing position. That's when I spied its worn Formica edges. This had been the first piece of furniture Bob and I had bought. It had become the central coming-together place in our household for meals and to share, but it was more. Each of us had grown up around this piece of wood. Our hearts had been spilled over its solid surface; issues had been endorsed and tossed aside; lessons had been learned; friends and family had made restitutions and resolutions around its oval shape; and once, puppies had been delivered and cradled on its top.

What magical properties did it possess that drew the best and worst from us? I stretched my legs and eased myself back into the chair and glanced over at Janie, my six-month-old granddaughter. She was playing in her walker and completely oblivious to the cacophony of female voices.

"Mama, are you listening?" Krista asked, drawing me back into the conversation.

I shook my head, "No, I guess not. I was actually enjoying a moment of reflection."

"We were talking about the past," Katie said, bringing me up to speed. Then, added, "How did we ever do it?"

"We did what we had to," I replied, hoping my terse tone would defuse further discussion on the topic. I wanted to stay in this melancholy moment, but the spell had already been broken.

I was surprised Katie would be the one to open the family's Pandora's box. She usually avoided personal involvement. Being the middle child, caught between a vocal older sister and a younger one with health issues, she had somehow managed to escape most family traumas, while creating plenty of pandemonium in her own life to send us all to the funny farm.

"No, we did what Dad wanted," Krista hissed, pushing back in her chair, slinging her arm over its back, posturing for the next remark.

"Now, come on, guys, let's don't put all the blame on Dad," Chelsea said, defensively, "What did we ever do to change things?"

Chelsea, the youngest, was the peacemaker. In every who-done-it situation, Chelsea backed the one on the wrong side of attention. Even now, at twenty, she was still refereeing.

I sat quiet, listening. Our lives had not been easy. There were times when each of us had wanted to run and hide. We had never met another family who had experienced twelve corporate moves in twenty years, yet we were still together. That was the victory. The relationship we forged from turmoil had made us a strong family, but the sting of those upheavals was carved into our physical and mental makeup.

Bob entered the kitchen and went straight for the cookie drawer. We stopped and watched him pull out a Hostess Twinkie, unwrap it, and devour it in two successive bites. He turned and faced us, cookie cream outlining the edges of his mouth, "Whew, my ears are burning!

I certainly don't have to be a rocket scientist to figure out who's the brunt of this conversation. Let me guess. Huh, it must be the psych-evaluation-on-moves conversation." He cast a sarcastic, finger-pointing look and continued, "Get over it! Haven't you heard that beating yourself to death is sadistic?"

We sat watching as he licked his lips, savoring the cream as much as our soured expressions. Before we could respond, he rushed out of the kitchen, hoping to escape our reactionary comments.

Bob loved having our daughters home, but hated the resurrection séances on his career, especially the part that centered on the many moves we made to help advance it. In his mind, every decision he made, every opportunity given to him was for the family and completely void of any personal agenda. He claimed the estrogen levels in our household drowned out any clear-thinking directives in the murky waters of too much discussion. His mantra, which Krista entitled the "Fulfillment Prophecy," could not be disputed, "If I hadn't worked all those hours and saluted every time the company said move, you girls would be up shit creek without a paddle and sinking, trying to pay off all those college loans. It's a shame I have to pat myself on the back to get a little credit around here!"

Krista, the oldest and often most critical, jumped up from the table, flinging her last thoughts over her shoulder, "You know, that's easy for him to say. He didn't have to attend three high schools and get slam-dunked by obstinate teens! But he's right about one thing, this conversation is over!"

Stopping in midstride, she softened, "Mom, would you watch Janie? I'm going upstairs for a few minutes." She cast a glance at her daughter, adding, "I realize I can't change the past, but I hope I can change the future. I don't want the same mayhem for Janie. I'd like to offer her a life where she feels safe and secure outside her home and develop a sense of community belonging. Maybe she'll be able to brag about going from first grade through high school with her best friend.

It may be impossible to achieve, but I'm going to try." Krista turned, leaving without her usual huff.

Silence trapped us. No one moved until Chelsea sighed, flat-toned, "Mom, Krista's bitter."

I shifted in my chair, "I realize she has issues, but your dad's right. Sometimes we get together and make our lives sound miserable." My pitch belied my words. It was an unresolved, rhetorical question despite our thousand and one attempts to explain it or cover it up. When alone, I refused to think about those early years. It was nonproductive, leaving a bruised, hollow place I didn't know how to treat. Now, I typically resorted to prescriptive avoidance, my preferred strategy, to handle things I could not bundle into neat packages.

But these thoughts brought back the same disturbing questions I tried to forget. Had I failed my daughters by not pushing back on Bob's insane need to succeed, resulting in one move after another? Did they blame me for the inconsistency in their lives? Did I actually push his agenda by not standing firm? Had we all been snared, sucked into bittersweet promises?

Each move followed a pattern. Bob would come home with the grand announcement. We would sit, stone-faced, listening to his spill on the latest advancement opportunity. After awhile, the rhetoric mimicked a giant parade. While Bob talked, I imagined the scenario unfolding in distorted images. The executive presenting the deal, turned from an empty navy suit to an overly animated sepia-toned clown. As he artfully coaxed the contractual agreement, his rust-lined smile spread downward, exposing yellow ochre teeth, creating a horrid impression of a grin. His dull onyx eyes drew into a glaring fold. When Bob failed to be impressed, he would do somersaults and back-over handsprings to court Bob further with more never-never land promises.

Seduction was a craft well learned. If Bob faltered, a clown ushered in the heavy artillery: marching bands, beauty queens, money contracts, stocks, bonds, titles, benefits, and options—anything to seal the deal. It was always the same hoopla, the same fanfare, the same game. For our family, it was never a journey to a land of cotton candy and caramel apples, but a purging. An overdose of Sunkist prune juice resulting in severe cramping before we finally adjusted, spent, and exhausted.

Bob did not understand. When we gathered, it wasn't the physical moves we sat around talking about. It was the in-between lost periods when we didn't know anyone, couldn't find our way around, and mourned our losses.

Often these conversations dredged up our nastiness and insecurities. For me, it often ripped open the past, making me feel like a little girl again—out of control, fearful, afraid to make mistakes or disappoint anyone, especially myself.

I heard Chelsea get up from the table, followed by Katie. Janie threw a rattle on the floor to get my attention. It was sweet having a granddaughter; she reminded me so much of Krista. Watching her gave me a thrill, and yet made me sad, evoking more questions. How do we protect her? She was trusting, vulnerable, and smart, already noting and responding to discrepancies in emotions and beliefs. How would she process them, absorb and mesh them into her own unique being? Right now, I wanted to inoculate her from humanity, bypassing heartache and struggles.

I knew struggles. I wore them like armor attached to my skin. In an effort to keep them from erupting, I oiled the steel rivets to prevent rust and repaired it when necessary. Over the course of life, the armor had been ruptured, torn and scarred. It had been cast into fiery furnaces to refine. It had sat in damp pits absorbing moisture, drawing tighter. It had provided security and protection. Even now, equipped with the knowledge to allay outbursts, there were exposed areas that were weak, vulnerable targets.

I fought the past like a disease, not wanting to be like Mama, Grandmother, or any of my high school girlfriends. Few lived full lives or escaped the traps they chose or were set for them. At times, I saw them rear up to fight; but most succumbed, battle weary, lifeless, and bitter. I did not want my daughters or Janie to be a part of this legacy.

I glanced over, and Janie was still playing. Every once and awhile she'd check, making sure I hadn't left the room. In the background, Katie and Chelsea were arguing over the name of an actor they'd seen at the theater. Bob came in from the garage, whistling a George Jones tune, and rushed over to sit on the floor to play with Janie.

Seeing my husband with our granddaughter made me stop and hold my breath. He opened up fully. He was attentive, loving, and different from his involvement with his own daughters. Then again, I understood. Before retirement, he set the alarm for four thirty in the morning, five days a week, and rarely got home before seven at night. He was exhausted, having given away the best part of himself. Being an engineer with an engineer's mind-set, he had little patience for the flighty, unruly, and shrill giggles of little girls. But with Janie, it was different. Every squeal brought rounds of robust laughter free pouring from his love-stricken face.

Sensing everyone occupied, I slipped quietly upstairs. Overloaded, my mind rolled in uneasiness. I could feel the armor expanding, threatening to rupture. Tension threaded through its eyelets, running from one shoulder across the apex of my head to the other. It laced tight, knotting into deep plaits down the back of my neck.

When I entered my bedroom, I looked around and sighed. I loved this room; it was my haven. When I refurbished it two years ago, I drew from my soul's color palette: navy, olive, peach, and gold. Bob thought the large flowers on the duvet were hideous, and he swore the oversized picture of the Golden-capped Conure gawked at him. Every time he entered the room, he grunted. But I knew giving over the bedroom wasn't much of a sacrifice; his sacred sanctuary was the garage.

I walked over to my writing table, an eighteenth-century French piece. I smiled thinking about the thin young man who sold it to me. He wept, sharing his angst over selling his deceased grandmother's household goods. Leaving with my treasured piece, I was stopped by an elderly woman directing traffic. She inquired, "Did he cry?" Taken off-guard, I merely stared at her. She threw her head back and cackled, "He's a loser! He forages through dumpsters and landfills and rips off the poor and elderly. You've been duped!" I shrugged, deciding her agenda was menacing. How could that sweet young man put up with such a shrew?

When I drug the table into the house, I thought Bob would faint. The legs looked like a bulldog had gnawed them, and the veneer top was curled and blistered. But three hundred dollars later, I had a masterpiece. Bob groaned, saying while I was out spending his money, his hair was falling out in the sink.

I reached down, running my hand around the table's smooth, finished edge. It was a nervous habit that gave me time to pause, reflect, and figure out what I was supposed to do next. I stopped and glanced down when my hand snagged on a jagged corner. A note protruded from the end of my calendar:

Life doesn't have to be perfect to be wonderful.

I loved this quote from Annette Funicello. For someone who used to associate chaos with unhappiness and perfection with harmony, it was a gentle nudge for me to remain on my present course, letting go of ordered routine and taking more risks. But getting to this realization had not been easy. From experience, I learned habits and responses were broken by hard knocks or confirmed by successes. When life's storms overwhelmed me, the past kept me from overreacting. I drew comfort realizing it was cyclic, going from overflowing to barely a ripple as quickly as a pulse beat. Consequently, I weathered the next ones more expertly, batting down the hatches quicker, knowing fair skies were right above the blanket of threatening ones.

My head throbbed. I walked over to the bed and sank into the thick down comforter. Leaning back against the headboard, I glanced over and spotted my high school graduation picture on the dresser. I sighed, thinking, *Who was that girl?* Then I smiled, *Eighteen, full of hopes and dreams . . .*

* * *

My greatest pastime was dreaming. I couldn't wait to get my hands on a magazine, or better yet, the Sears, Roebuck catalog. I would steal away to my room and spend hours going from page to page imagining how other people lived, ate, and dressed. It was fascinating to tear pages out, to cut and paste, building collages of families, adding wardrobe ensembles and household furnishings. I would dream of their life, home, travel, and friends. When I finished one family, I would create another. I stored them under my bed in a large box I found in the trash behind David Moore's grocery store. These pictures became my surrogate family. I loved them.

By my junior year in high school, I was getting bored with pretend. The girls were grading one another, pitching camp into smaller, more defined groups. I had always stood on the fence between the losers and developers. I chose the developers. They were a lot more fun.

They taught me a new sense of self. The definition was easy to learn—copy the senior developers. They laid the path; we modified. I learned to live for the day. It was rare to think about tomorrow unless it was to plan a date or a new adventure.

Then, I met Drake McGregor, a top senior developer. He was a dream angel with a fast car and hungry lips, the perfect combination. Our dates were crazy wild. But it didn't last. Little things started to nag at me: how he blamed everyone else because he couldn't keep a job; how he touched every girl he passed; and how he was beginning to dictate how I was to think, speak, and act. Looking around I saw my girlfriends getting wired to punks who used the same strategies, and I bailed. But not before having my heart ripped out.

That's when I met Bob. He bumped into me in the lunch room, or maybe it was the other way around; and I yelled, "What do you think you're doing?" I had no patience or couth.

When I looked at him, he was smiling. It was the kind of smile that was one hundred percent genuine. Hook number one!

I recognized him as an insider, part of the studious group who walked around with slide rules attached to their belt loops. I often wondered if they saw it as a badge of honor. I saw it as the worst dress accessory, ever. Despite the bad taste, this guy looked interesting.

Giving him the once-over, I finally smiled back, readdressing my first comment, "Hey, I'm sorry. It was probably my fault." I leaned over to pick up the fork I had dropped, and he touched my arm, replying, "No, let me get it for you. By the way, my name is Bob Simmons. Have we met?"

Have we met? He must be kidding! I thought. His group did not mingle with mine, except by chance or by arrangement, whispered about, but never openly discussed.

I laid my tray down on the table, extended my hand, and said, "Not really. I'm Abby Bowers. It's nice to meet you, Bob."

He blushed. He actually blushed. A new excitement rushed through me. Images of the collages under my bed resurfaced. Our faces began to fill those pages.

We were established that day as a couple. Bob broadened my thinking and grounded me. He talked to me, not at me. Hook number two!

My girlfriends railed me, threatened to drop me from the group, but I kept walking straight toward Bob. They said he was changing me. I didn't know how to tell them I was evolving, not changing. They wouldn't understand. I barely understood myself.

It wasn't hard to adore Bob. He was quite the charmer. Even when he cracked jokes about my awkwardness, I couldn't be mad. He was only teasing. That's when I would run my palm over the top of his dark brown crew cut hair to chide him. I liked its spiky feel, which reminded me of the neighbor's short-haired Chihuahua. When he smiled, I would reach up and trace the soft lines that etched his eyes. I wanted to memorize these new feelings and put them in my memory bank for safekeeping.

The popular girls hated me. Who cared! Two months from our fateful meeting, I had his class ring. But it didn't stop the vultures from asking him to escort them on the football field when they won their Miss Whatever titles. From where I sat in the bleachers, Bob appeared to be in the center of their world. A chilling air would bank against me. Watching, I wondered if I would ever fit in.

Bob started talking to me about his future. He had big plans. As I kissed his neck and tried to wind myself around him for attention, he talked endlessly about what he wanted to do with his life and how he planned to get there. He was confident and bold. Hook number three!

It wasn't long before I sat riveted to the other side of the car, facing him, listening. Dreams cropped up in my alone time, causing me to pull out my pencil and paper and write down my options. For the first time in my life, I wanted something bigger than I'd ever dare dream.

My parents, like Bob's, could not afford college. Bob had a scholarship and community support. I had neither. When the school counselor handed out application forms for the North Carolina Teacher's Scholarship, I grabbed one, quickly shoving it into my pocketbook. That night, I filled it out, and the next day, mailed it off. My hope for the future was predicated on getting it. I didn't tell anyone what I had done. I figured dealing with my own disappointment was easier than having people feel sorry for me when I didn't get it.

In early May, over the intercom, the school's secretary, Patsy Leggett, announced I had won the teacher's scholarship. Tears welled, and I threw up my hand up to squelch a sudden gasp. I looked around the room wondering if anyone else had heard it; but no one said a word to me, not even my teacher. Well, not until I got called into the counselor's office.

Mr. Tiddle, the high school counselor, looked at me with his pasty smile, scratching at his toupee, pushing it from side to side. He didn't waste a lot of words on me. He wanted to know if I had a mind to hold Bob back from a successful career at State. When I shook my head, no, he pulled out my standardized test scores, saying college might not be for me. He suggested I look into a cosmetology program or go to work for the local Roses Department Store. I stared, never saying a word. I left his office, walked home to an empty house, and cried myself to sleep.

The cold hard facts hit me. Dreaming did not mesh with reality. They were in total opposition to one another. The pictures under my bed were make-believe. I had to create real pictures of a future life and pursue it.

The next day, I marched myself into Mrs. Langley's room after school and asked her to help me with the paperwork for college. I knew she could handle it; she had been voted teacher of the year. She was very impressed with my decision and set up an appointment the next day to advise me.

The first thing she told me was I had to take Algebra II during summer school. No one had ever mentioned this before. Then again, I don't believe anyone would have suggested it after I failed Algebra I, twice. In spite of my questions, I listened. Mrs. Langley helped with the final papers for college and registration for summer school.

Everything was in motion. Now I had to talk to Mom and Dad. Talking to them about going to college panicked me more than talking to them about sex, a taboo subject in our household. I took out my pen and paper and listed the challenges I faced. First, going to summer school meant cutting my job hours, which equaled less money. Second, cutting hours and receiving less money would not pay for all the additional expenses I would incur at college. Third, I had no transportation.

I didn't have to guess, I knew what Dad's first comment would be, "How in the hell are *you* going to pay for all this?" Second comment, "Well, *you* can leave me out, that's for damn sure."

Regardless, I had to try. I couldn't sit by and wait for the hand of luck. My hand was all I had.

At supper that night, everyone was quiet. Dad was bent over his food, never lifting his head as he crammed one fork full of food after another into his mouth. He was wearing an old paint-stained T-shirt. His dark brown hair, creased by his hat, flipped up over his ears. When he stopped and looked up, I noticed paint streaked across each cheek. His eyes glazed dull and nonexpressive.

Mama sat, one hand folded in her lap, eating her food in slow, controlled bites. Her new perm had left her dry, parched hair springing in untamable corkscrews. From the stretch of her neck, I knew she was defying any more comments on her new do. Teddy had already received wrath from his Frankenstein's wife comment.

Feeling I could contain myself no longer, I blurted out as fast as I could, "I've got a scholarship to go to college. It covers tuition and books, but not transportation, and I will have to live at home. Will you help me?"

Dad dropped his fork on his plate. Mom stared. The silence continued until she got up from the table, picked up her plate, and walked over to the sink. I lowered my head, hearing water running and dishes clanging. My eyes were still focused on the bright red and white checkered table cloth when Dad said, "You must be crazy. There's no extra money or car. Forget it!"

Mama kept doing dishes and never turned around. Knowing her, I knew she needed time to think; but the longer I sat there, the madder

I got. *I didn't need them,* I kept thinking. *I'm going to do this on my own.* When I lifted my head, I marched to my bedroom, determined to make plans to go to college. *They would all be sorry,* I steamed, picking up my writing tablet off the nightstand. My first entry on the top line was Plan A, next to it, Plan B.

During the twelve weeks of summer break, Plan A was accomplished. I worked two jobs, made an A in Algebra II, and packed away most of my salary.

In early August, I began Plan B. I made arrangements to bum a ride to East Carolina University to register for fall classes. I spent hours in line getting my scholarship acknowledged. At the end of the day, I had a student identification card and was registered for five freshmen classes. Excitement and tears rolled into one. Even my ride, an elderly neighbor, could not slash my happiness when she screamed at me on the way home for wasting her time.

At this point, I had three days to work out how I was going to get to and from classes, praying a glass bottle would fall into my lap, and my own personal genie would appear. I needed this opportunity. I feared being stuck in this hick town or bailing hay in Bear Grass with the likes of Joe Bailey.

Now, it was time to face Mama, again. My only hope was she had witnessed my determination during the summer and had softened. Convincing Mama to help me was my last hope.

I found her in the kitchen. She was attempting to wash my baby sister's hair when I slipped inside. I planned to strike up a casual conversation and ease my way into the school business, but Mandy kept screaming at the top of her lungs. I recoiled, remembering I had forgotten to wash her hair that morning. The Bantam rooster in Uncle John's backyard looked more merciful than Mama did right now.

Cautiously, I walked over to the sink and asked if I could help. Mama threw the towel at me, yelling, "I give up!"

I tried to forget my problems, concentrating on Mandy. Being with her was like comfort food, soothing and delicious. My frayed nerves unwired, and hers did, too.

Having Mandy had been hard on Mama. She had to go back to work when Mandy was ten-days-old. Feeling helpless, I offered to get up and help with her nightly feedings. Two days into her job, Mama brought Mandy to my room and said she couldn't work on no sleep. Dad moved her crib into my room the next day.

Sometimes it was hard getting up at night and going to school the next day, but I managed. Mama was working like a fiend and dragging home ragged looking. I tried not to whine; but when I did, she encouraged me, saying I could handle sleepless nights; I was a lot younger than her.

After I got Mandy settled down, I went looking for Mama. As usual, she had gone from one battle to another. Now she was begging my brother, Teddy, to stop playing with his stupid green army men and take a bath. I could tell he was trying to royally piss her off by pretending he couldn't hear her. He kept picking up one man after another, blowing them to smithereens with gunfire sounds, erupting from between his clenched teeth. Mama finally had to drag him into the bathroom. He ignored her completely and kept firing missiles the whole time Mama was pushing, shoving, straining, and cussing.

I decided to follow her until she stopped. It took awhile. She ran from the bathroom to her bedroom, balancing a clothes basket on each hip, talking the whole way, "If I don't get this wash folded, you all will be naked as a jaybird come Monday morning. Maybe that'll teach you. I'm only one person, and I'm doing the work of three!"

As she folded clothes and ranted to the wall, I sat on the other side of the bed and sorted them into neat piles. She never looked my way but continued to toss half-folded clothes in my direction.

Finally I drew in a deep breath and went for it, "Mama, how am I going to get to college on Monday?"

For a moment, I thought she had forgotten I was there. She stopped what she was doing, looked out the window, and kept expelling air like an infuriated, trapped bull. I chewed on my thumbnail, wishing I could raise a finger and test the wind direction.

I was about to ask my question again, thinking she hadn't heard me, when she looked straight into my eyes and said, "You just won't leave well enough alone, will you? I work nine hours a day, rush home to two children and a squalling baby, cook dinner, clean house, wash clothes, and then scream at kids to take a bath. I hardly have time to breathe, eat, or take a bath myself."

After a short pause, Mama continued in her slow, deliberate, staccato voice, "What do you want from me? There is only one car. I have to go to work. We have to eat. Guess who's going to drive? Me. Please, Abby, leave me alone. I'm sorry about college. You have to face the facts and go to work. Work, save your money for a car, and go to college in a few years."

Then, she clenched her teeth and growled, "Go get Mandy before I scream!"

I was getting ready to plead when Mama's eyebrow flew up, and she gave me that *look*; the discussion was over. I mean o-v-e-r. I went and got Mandy. I picked her up, and she immediately quieted down. I put her on my hip and started singing "Satisfaction," swaying my hips to the beat.

When she fell asleep, I went to bed and decided "tomorrow would be another day." My grandmother told me Scarlett O'Hara said that in *Gone with the Wind*. Scarlet had ways to get what she wanted. Whatever it took, I would find mine. I did not want to end up with a life like Mama's.

* * *

The bedroom door opened with a loud creak and a bang, shattering dream images into tiny mirrored fragments.

Already unnerved, I shrieked at the touch on my shoulder. "Sorry, Mom. Dad wants to know if you're sick."

At first, I didn't recognize who it was. "What time is it?" I asked, a little shaken by the shock back to reality.

"Mom, it's almost five. Dad's hungry, and Krista and Katie are downstairs wondering if you're mad about our talk. Krista said she's sorry for running out like that. You know how she is," Chelsea reached over and touched me again. "Can I help with supper? I hope we aren't having spaghetti. You know I hate it. Mom, are you all right? You don't look so hot."

My head was pounding. All I heard was tunnel garble. I wanted her to go away, but she continued staring at me. Responding seemed to be the only answer. "Chelsea, I'm fine. I'm going to take a quick shower, and then I'll be downstairs. Please reassure everyone I'll feed them. And, no, we're not having spaghetti. Why don't you start setting the table and put some cheese and crackers out for starters. I need a few minutes to freshen up, okay?"

I could tell Chelsea wasn't convinced. She lingered a couple of minutes, assessing the situation, then said, "Sure, Mom. See you downstairs. I love you."

Krista

I knew it was childish to leave the kitchen so abruptly, but sometimes I couldn't stand the way Mom, Katie, and Chelsea whitewashed everything. My muscles and demeanor softened as I entered Mom's bedroom. I loved being back in the comforts of home, surrounded by love and common understanding. Still, these comforts made slipping into our old roles all too easy. I would become the "I'm woman, hear me roar" self that permeated my persona during the teenage years. Katie, who had proven her intelligence in college, fell into her flighty sister role. The familiarity of home, which to our family was wherever my parents lived and never a specific place, held a sense of security for me.

Seeing my daughter playing with Mom brought back beautiful memories, but also a gripping fear I'd never do the job as well as my predecessor. I hoped Janie could find a way to love me anyway. My childhood memories collided with the present as I heard Janie's happy glee. I remembered feeling my life had been full of closed doors and missed opportunities—doors swung shut without even the slightest push. All I ever wanted was consistency outside my family bubble, to walk into the local diner and say, "Just give me the regular, Daisy." Too often in life, there is a great divide between what you want and what life has planned.

Bursting with thoughts, I wandered out of Mom's bedroom to find a computer and check e-mail, desperately looking for a note from my husband. No such luck. I paused at Mom's wall of sayings. She loved to collect snippets of knowledge and inspiration. I read one she must have recently posted:

> *When one door of happiness closes, another opens, but often we look so long at the closed door that we do not see the one which has been opened for us.*

According to Mom's handwriting, this quote was from Helen Keller. I leaned back to soak in the words—a rare reprieve from parenting. It didn't take long before my mind began to wander . . .

* * *

My life sparked with parents living in a college dorm room for married students. There's no doubt some education was going on there. In the midst of pure knowledge, the tiny glimmer in my father's eye produced me, a seven-pound bundle. My mother was not only prepared to be a mom; it was her true destiny. The man she married had what she felt she lacked, all the attributes she wanted her children to have and emulate: intelligence; good looks; charisma; and, above all, self-esteem and confidence. She wanted to undo the terrible wrongs her parents and other authority figures had done to her while also protecting us from things she would never even tell us existed. Unfortunately, children tend to ignore all the pretty words and good intentions and merely emulate what they see.

Both my parents came from a small economically depressed town in North Carolina. My father, the only hope of escape for my mom, had the support of the town behind him. Sick of poverty, he was determined to go to college and make a better life for himself. He had the love of a great woman and the drive of a great man. That was all he needed.

Two months before my arrival, Dad graduated from college and moved to Illinois for a job. Mom, seven months pregnant and not knowing a soul, joined him. She also found herself in line every morning in front of the donut shop across the street. Between those donuts and me, she gained sixty pounds. The day I finally arrived, Mom's immediate feelings were warm and loving while Dad's focus was on time and money. Throughout my life, I'd hear Dad refer to himself as the pragmatist and call Mom a romantic. Sure enough, as my mother imagined my first words, my first steps, and the first time I'd tell her I loved her, my father was already lost in a career, chasing images of dollar signs.

My mother's expectations wore me down early. She wanted the kind of love only a child could give her, a love she desperately wanted and never had. I was independent and quite the queen bee and bore this expectation as a burden. My mother's mantra of don't-touch-that-again translated in my mind to trying it just one more time. Mom swore I was only satisfied with a bright red badge of courage on my hand.

The competition between the family and Dad's career never ceased. He was consumed with climbing the corporate ladder, which had rungs in different states and resembled an upwardly sloped monkey bar. Each rung was no more than three years and no less than three hundred miles away. Life was like being thrown on a roller coaster you initially wanted to be on but couldn't stop to get off. Dad's company and his achievements brought many material things to our family, but also left an emptiness that took years to understand.

Each family meeting to discuss the next move had a general theme, and usually occurred around the dinner table. My parents thought good food deadened the impact of life-altering news. It helped that Mom was a wonderful cook and truly believed food nourished the soul and not just the body.

"Girls," Dad would say, "I have a great opportunity in (insert city name here). You'll love it. We'll have a (insert object to entice us here)." He said this with a painted on smile.

Mom sat next to him and tried to look excited and supportive, but I always saw through the charade and knew she had roped Dad into spearheading the discussion. After all, he was precipitating the move or, at the very least, responsible for it. Thus, he should throw out that huge boulder of information and see where it landed and what ripples it would cause.

I could only imagine the discussion my parents had the night before.

"You want to keep uprooting this family, by dammit, you tell them. I've got enough to deal with just cleaning up the mess afterward," Mom would say, burying her head into her overly dry nonmanicured hands.

"I'm doing this *for* you guys not *to* you. I've got enough pressure on me at work without having to deal with this shit," he would reply as he chose his suit, tie, and shirt for the following day.

"Do you know what the girls are going through here? Do you have any clue?"

"They'll be fine. You're always making this harder than it should be," I'm sure Dad said, his defensiveness evident.

Depending on the move and the particular situation, this conversation would deepen, and the guilt would fly both ways. Then, after the dinner announcement, Mom's attempts at family reconciliation would start with the same speech, "Girls, you know your father only wants what's best for us."

Why was it that parents always tried to break bad news with unbelievable solidarity? I often wanted to check with good old Webster on the meaning of *best*. Mom tried hard to make it all right. But my greatest pressing desire was to yell at her to get to the point already! Unfair as it was, our reaction to her was always more honest and combative than to Dad. It's sad how programmed girls are from the beginning to act properly for men. Growing up, I was allowed to play and be myself until Dad came home and graced us all with his presence. Then, I had to sit at the dinner table and act like the most perfect child. Mom did not just create a fantasy world for us; she made sure she had one for him as well. Hiding her own fears, she would take on the arduous task of pumping us up while also making Dad come out blameless.

For so many years, I loved a man who didn't exist. He was a cinema version comprised of Mom's wishes and reality, mixed with a little childhood idealism. Not that I wouldn't have loved who he was, I just wasn't given the opportunity until adulthood.

Mom ended up being the dartboard for all of our emotions, but particularly mine. "Full of piss and vinegar," Dad always said. I tested every parenting skill she possessed and even the ones she went to classes to hone. My willfulness sent her to classes such as "How to Successfully Handle a Teenager." I was allowed, hell I was encouraged, to let loose my fears and anxieties. She absorbed all of them as they screamed out of my body. This release allowed me to support Dad's every move (no pun intended), resulting in comical exchanges. Don't get me wrong, most of the time he would receive some of my attitude, but in a watered-down version. My anger at him came out in the age old man versus woman battles.

"Krista, get me some iced tea," he would say.

"You have two legs, don't you?" my reply was openly hostile and begged for an ensuing battle of the wills. Dad rarely disappointed.

Mom would immediately jump in and say, "Honey, I'll get it." Often, my sisters would also try to come to the rescue.

"No, dammit, Krista'll get it! I've worked all day, and she can get me some damned tea!"

"I wasn't born to be your slave!" I would shout as I poured the vile liquid.

This went back and forth. In the end, I would get the tea, but not until we were past Dad's breaking point. I should have earned several Academy Awards over the years, and Mom certainly earned her wings watching such idiocy. I would become a two-headed Hydra, beast and angel. Whenever Mom was around, I'd let loose about the cruelty of upheaval and how much I dreaded yet another feather in Dad's cap. Later, for Dad, I would smile and have rousing cheerleader conversations—outside of getting his damned tea, that is.

I imagined a battered wife who kept smiling as her husband bashed her brains in. She would tell all her friends what a bastard he was while absorbing every punch he could land. Oh, I'd stand up to my father on the trivial matters; but then I'd do my best to soothe the situation and be the apple of his eye.

Beyond the dichotomy of my behavior with my parents, I felt a need to become what each new location expected of me. Or, better yet, what the kids in each new place expected. I became a chameleon, constantly sizing up my surroundings and morphing to fit in. After awhile, it was second nature. In the beginning, I didn't see it as a weakness; I saw it as taking control of my life. Labeling me confident as a young girl didn't quite cover it; I often spoke before much thought. My quick tongue gave Mom indigestion, made Dad ready for battle, caused my sisters to clear a path, and, later, taught me one of my most valuable and life-altering lessons. My favorite aunt called me the Mouth of the South, and my favorite uncle swore I'd be the next Gloria Steinem.

My mind always seemed on the edge of exploding due to its constant activity. I couldn't control it, couldn't stop it, and often wished I had an on/off switch for it. There was always so much to do, to think about, to . . .

* * *

"Hey, Krista, I bet Mom needs some help in the kitchen," Chelsea said from behind me. I jolted back and wondered how long I'd been drifting. Leave it to Chelsea to want to bring us all together after even the slightest argument or discomfort. The guilt of missing time with my family coupled with the realization Janie needed a diaper change and to be nursed spurred me downstairs. I followed Janie's squeals of delight and found Dad sitting on the floor, tossing her up in the air. When he was with Janie, I saw something different in him—something I really liked.

Abby

Feeling groggy and unsteady, I tottered into the bathroom. I closed my eyes, splashing cold water on my face. Grabbing a towel, I dabbed the moisture off and applied a thin layer of lotion. I gasped when I looked in the mirror, seeing a tired old woman staring back. I barely recognized her, but enough of her younger self was there to call her by name; it was me. My dark brown hair appeared lackluster, dry, almost parched. The deep mahogany gold tones lacing through it when I was younger, the ones grandmother said looked like a fox's tail, were gone. Wild gray strands sprouted from each temple, giving a disheveled look to my usual coiffure appearance.

Peering closer, I realized there were lines I had never noticed before. The triple C marks around the edge of my eyes mimicked the deep, horizontal eleven-shaped wrinkles between them. Several ruptured veins on each side of my nose alar and cheeks blared red, giving my skin a dull, sallow cast.

Fearing the Candy Man might surface if I stood there longer, I left. *Enough of this,* I thought. I grabbed my glasses off the desk and went downstairs.

When I entered the kitchen, I began grabbing first one size pan and then another. I *had* intended to do spaghetti, but would never own up to it. Within minutes, I had a new game plan. I quickly turned the thawed ground beef, onions, and peppers into my mother's famous meatloaf, not forgetting her secret ingredient, a finely shredded apple.

Krista slipped in behind me, hugged me tightly, and said, "Sorry."

I smiled and returned the hug. "No problem. How about giving me a helping hand?" I pointed to the bag of potatoes on the counter and asked her to peel them.

We got into a quiet rhythm. Before I knew it, Chelsea and Katie were beside us finishing the salad and toasting leftover rolls. The kitchen was small, but we fit in it with the ease of a working unit.

When we heard the familiar "Here I am!" we knew who it was. Bob slid across the kitchen floor in his sock feet, giving his impression of Ed McMahon's famous introduction, "Here's Johnny!"

Giving away to annoyance, Krista crinkled her face, saying, "You think you're so damned funny, don't you?" Then laughing, added, "You'll never grow up!"

Chelsea and Katie were bent double, laughing, too. Bob ignored Krista and took bows.

With one hand firmly planted on her hip, Krista continued her scene evaluation, "You know that's truly more than I can stand."

"Get over yourself, Krista. Loosen up!" Bob finished his production with two pirouettes.

"Stop it, right now, both of you." I said. "Let's eat supper before it gets cold, or I'm gonna throw it in the trash can!"

Everyone quietly moved to the table, sitting in the same chair they had occupied as a family unit. Our new addition, Janie, sat in a high chair close beside her mother. I took my seat, bowed my head for prayer, and wrapped myself in a welcoming cloak of familiar warmth.

Several seconds of rustling went by before Bob began his prayer; and in my impatience, my legs started to bob up and down, a coping mechanism learned as a child when told to sit in the corner and be quiet. It was then, for the first time, I felt Chelsea's legs bobbing in tandem with mine. When I willed mine to cease their frantic movement, Chelsea's relaxed as well. The cloak tightened its grip, and I pressed in further.

"Heavenly Father, bless this food we're about to receive and forgive us of our many sins. Amen," Bob said, slowly and rhythmically. I knew he was calculating the exact timing when the girls would join in, reciting the ending with him. When all the voices joined and blended, it was his I heard above the rest; and I could sense his cloak tightening, too. He loved the cadence, but also enjoyed the fact they could never remember his prayer until he started it.

When the prayer ended, Krista slammed her hand against the table in disbelief, "I swear, the man has been saying the same prayer my whole life, and I still can't remember it until he begins. I must have some kind of mental block. It is beyond annoying!"

We all laughed, each confessing to our lack of recall. This fueled Bob. Now he had our full attention, knowing it was the perfect timing to begin telling jokes. For the next few minutes, I watched as he turned on the charm they loved. Even Janie reacted to the revelry. It was as if they'd never left home. Their faces came alive around the table.

I studied them, admiring each. They had grown up to be wonderful women, sharp, energetic, and competent. Pride swelled; but painfully I still saw some of me in each of them, parts I had hoped to squelch and burn out of their genetic makeup.

"Mom, are you okay?" Chelsea asked. She was the daughter in tune to everyone, noting if anyone was not involved or interacting.

"Yeah. I'm just tired. I think I'll do the dishes and head up to bed. Maybe try to finish the book I'm struggling with."

"I guess that means you aren't up for games?" Chelsea asked playfully.

"I promise I will be tomorrow."

Truth was, I couldn't wait to escape. The whole day had left me out of sorts. I needed to be alone and regroup. Maybe if I could get some rest, I'd be better company.

In my mind, I knew it was a cop-out. My whole family was with me, something I yearned for; and here I was, running off to do the same thing I did every night, be alone. Shrugging my shoulder, I pinned my out-of-sorts feelings on dredging up the past. It always left me glued to my chair, not in excitement, but more in dreaded apprehension. As much as I felt I had grown past its hold, the pit was always there waiting to claim me. I slipped quietly out of the kitchen, seeking the refuge of my bedroom.

Entering my room, I spotted another posting, this one by Betty Ford:

Don't compromise yourself. You are all you've got.

Earlier I showed it to Bob, half-joking about restitution for plagiarism. How many times had he heard me say those exact words? I told him if I had had more notoriety, the credit would definitely have been mine.

I kicked my shoes off and fell across the bed. I tried lying very still, listening to the steady thump of my heart, hoping to nestle into nothingness. Usually this anchored me when I was feeling too cemented to the past, but tonight it didn't work. I tossed and turned, then relinquished.

Sitting up, I spotted Tom Clancy's book *Clear and Present Danger* on the nightstand. I reached for it, hoping it would keep my mind from wandering. But there was no escaping the past. My mind drifted right back to the place I had exited earlier . . .

* * *

Monday came and went, and I did not go to ECU. When I got up that morning and faced "Tony the Tiger" at the breakfast table, I knew life had not changed. Same table, same nicotine stained pine-paneled walls, same ceiling circled in brown from countless leaks, same tear in the Naugahyde on the chrome kitchen chairs.

I went through the day trying not to think about tomorrow. It was too painful. Mandy tried to get me to smile. She had just turned one and was such a ham. Normally, she could make me forget my troubles, but not today.

For several days, I moped around. In the end, I got tired of punishing myself. Mama was right; I had to get a job. I bummed a ride from the old lady down the street. She took me to Little Washington to apply at several of the factories. I didn't want to work in a factory. It frightened me to even think about it. I heard tales the old women told about long hours, bad treatment, filthy conditions, and bosses who prodded and poked. Even knowing this, I felt I had no choice. I could not stay home and do nothing.

After my fourth interview, I was hired to answer phones. In addition, they would train me to do light office work. The last part sounded intriguing; I could dress up and look like a professional. In high school, I took several business classes, typing and accounting. Then I remembered how poorly I had done. Determined to make a good impression on my new employers, I decided to drag out my textbook and study.

Now the most pressing problem, I still needed transportation. This was becoming a thorn in my side. The old lady reminded me about it on the way home. She kept saying, "If you think I'm going to haul your ass all the way to Washington every day, you might as well figure differently. Get you a man. That's what you need. A man!"

I for sure didn't need a man. I needed a car.

The next day my luck changed. I walked to the Ford dealership and asked to speak to a car salesman. Turned out I was talking to the owner. He was very nice to me, said I was a real sweet thing, and he had the perfect car for me. I thought it was beautiful, a 1968 red Ford Torino, and the man assured me he would fix the wrecked parts. Now, all I had to do was convince Mama to cosign the loan.

I boldly rushed into the kitchen, totally out of breath, and announced, "Mama, I've got a job!"

"High time."

"I've got a car, too."

"What fool gave you a car?" She stopped, swiveled around, lifted her left eyebrow, and said, "No siree! I am not signing a loan. I'm up to my eyeballs in debt already. Go away!" Her conversation was finished.

I followed right down behind her, pleading, "Mama, I need your help. I promise I won't let you down. Mrs. Winfield, my new boss, told me she would train me and pay me a dollar an hour for forty hours a week. I'll only take out enough money to pay for gas. I'll put the rest of my paycheck on my car payment. I won't let you down."

Mama stood in one spot for a long time. She did this a lot lately. It was hard watching her stare at the floor or out the window without saying anything. I had learned long ago to be still and wait. One time I said something too quick and Mama had a brush in her hand. You do not want to be near Mama and a brush when she is angry. It's unforgettable.

"All right," she said, sounding defeated, "I'll sign for you. I must be totally crazy, but I'll do it. We'll go to the bank tomorrow. If you fail me, you might as well be sending us all to the poor house, do you understand?"

I couldn't believe my good fortune. I had gone from a college potential to a loser to a professional with a car. I wanted to call and tell someone, but everyone had already left for college. I knew Bob would be impressed. I would have a car to sport around on his first weekend home.

Bob loved my car. I told him he could take it to school sometime. I knew it would be a treat for him since he had to bum rides home from college.

Work wasn't what I expected it to be. I didn't know anything my co-workers were talking about. The woman training me got angry, a lot, "What rock did you climb out from under?" I got mad about the rock comment; it made me feel stupid and dirty. The intelligence part didn't sting like the idea I was unclean. Mama always impressed on us the idea of wearing clean clothes, especially clean underwear. It was how people judged you, and you never knew when you might get into a wreck. But for the life of me, I never understood how wearing clean underwear would make a wreck go any better, but Mama made it into a golden rule.

Every time my boss pointed to something and said, "See this blah-blah-blah. This is a blah-blah-blah" I just wanted to puke. To keep from crying, I kept pleating my skirt with my fingers. All the thingamajigs she was talking about were so foreign.

One day, after work, Mama suggested I go over to the neighbor's house and talk to them about going to night school. I think it was her way to make up for my disappointment. To be honest, they weren't really neighbors; they were Mama's renters. I didn't like Henry; but his wife, Janice, who was a schoolteacher, was nice. The kids were okay, but sort of bratty like most kids. Mama told me to be nice to them. Seems Henry's father owned a big dairy farm up in Maryland. I didn't care, but Mama did.

When I got over there, I cupped my hand on the screen door and hollered, "Anyone home?"

It wasn't formal where I lived. Aunt Lila said we were all home folk. Most of them seemed like home trash to me. They lived paycheck to paycheck and kept their outhouses open in case the water got turned off. She was very proud of the name of our area of town, Doodle Hill. She said it was named after a mound of bugs. People went on and on about it, but I laughed every time I heard it. Doodle bugs were plain old roly-poly pill bugs. Looking around at some of the people roaming the streets, I knew the name was appropriate. Not for a beetle that is, but for their posture that folded up whenever life presented them with a challenge.

Just then, I heard Henry yell for me to come on in. As soon as I got through the doorway, I saw him sitting at the kitchen table with four crushed empty beer cans in front of him. He was chugging down his fifth. Immediately, a red flag appeared in my head. I got the same feeling when those construction workers uptown wolf-whistled or yelled names at me. I never talked to anyone about how it made me feel. I wasn't even sure why I felt that way. Jolene said the guys were just complimenting me. They must have thought they were real cute trying to guess my name. *Wrong.* My name was not Pussy.

"Where's Janice?" I asked.

Henry's eyes looked glassed over, red and fiery. "She ain't here; over at Merle Ann's, I guess. What you want?" Henry said, letting go a disgusting burp.

"Mama told me to come over here and find out about going to night school."

Henry pushed away from the table, slow and deliberate. "So, you wanna go to night school?" He wiped his mouth on the back of his sleeve and started toward me. He was bigger than I remembered.

My heart thumped in my chest. Alarms were so piercing I couldn't think straight. Heat rose from my chin to my forehead. And he kept advancing, drawing closer and closer until I thought I would suffocate.

Not knowing what else to do, I started walking backward, "Look, I'll come back later when Janice is home. Tell her I came by, okay?"

Henry caught my arm, "Not too friendly, are you?" His voice was raspy as he reeled me closer and closer.

"Look, I gotta go. Mama's expecting me. I don't want her to get mad." Twisting to get out of his grasp, I could feel his fingernails digging deeper into my skin.

The more I talked, the more menacing he became. I could smell the sweat and cheap beer on him. My nostrils stung with its putrid, foul odor. Then my stomach began to tumble and roll. I swallowed hard not to vomit on his stained ECU pirate shirt, but figured it was too nasty to make a difference. I lowered my eyes, watching his feet move closer to mine.

"You're some fine piece of ass."

His mud-caked brogans were almost touching my loafers. I whimpered.

"I've been watching you out in the yard in that orange polka dot bathing suit. A man like me could really show you how to use those curves."

His shoes touched mine, and I struggled with all my might; but he kept clinching me with his hairy hands. *Where was Janice? Where were those bratty kids? All they did was whine. Where were they now?* That's when I realized the only one who needed to act was me. He yanked me into his arms and kissed me, over and over again, smashing his full lips over mine until I could barely breathe. He freed one of his hands and began running it up and down my body, with the other he dug in hard. Jerking around only incited him to wrestle me tighter. I hated his slobbery lips and his moans. I hated feeling his manhood against my stomach. For a moment, I thought I would faint. Instead, I started to cry; and when I sniffed, I bit his lip.

"Damn! You little whore! You bit my damn lip."

And with that, he hit me upside my head and pushed me to the floor so hard I couldn't tell what hurt more, my head or my bottom. I sprang up like an innerspring coil and ran home as fast as I could.

I ran past Mama doing the laundry and didn't stop until I slammed the door to my room, and locked it. My breaths were coming in short, panting bursts. Bright circular spots danced and floated, filling the outer perimeter of my vision. My knees weakened. I remembered the school nurse telling a girl to sit down and put her head between her legs to keep from fainting. It was the last thing I remembered until . . .

*   *   *

"Mama, are you listening? Why is it so dark in here? It's already nine o'clock. We've been up for hours. Dad said to let you sleep, but this is ridiculous!"

I didn't want to answer. I wanted to feel nothingness envelope me again. Shadows danced, absorbing and releasing energy, pulsating in tune to each breath tone, drawing me closer and closer to its conical end. Every nerve ending freed as I began my descent. Then the noise happened again.

"Mama?"

I forced my eyes open and stared into a woman's face, the darkness only added to my confusion.

"Hey, you scared me. You okay?" Krista switched on the lights and began opening the drapes.

I struggled to regain composure, "I'm sorry. I think I was in the middle of some horrible nightmare."

Ignoring my state, she ranted, "Mom, I'm so tired of Dad's crap. Who crowned him the king of knowledge? I have had it with him. He never listens to anyone.

Pushing myself up, I swung my legs off the bed, "What in heaven's name are you talking about?"

Krista continued, "Can you believe him? Here I am trying to give him sound financial advice, and he totally blows me off. He would rather listen to an outsider than to me. In fact, if it weren't for me, he wouldn't even have an advisor."

I held my hand up for silence, "Krista, please! Your dad trusts your opinion. If you hadn't talked to him about a financial advisor, he wouldn't have one. You're way too sensitive." My head was reeling.

I knew she needed me to reach out to her and soothe her. What I wanted to do was to scream for her to suck it up. Hadn't we all had to deal with life's challenges? Why couldn't she leave well enough alone? Sometimes, I thought she would swallow me up.

I paused, gauging my response, "Look, I'm going to go freshen up so I won't scare anyone. When I get downstairs, let's treat everyone to coffee and hot chocolate." I patted the bed for her to sit down next to me. When she did, I reached over, putting my arms around her, "Don't fret over Dad and his opinion. He's just bantering with you. You know how much he loves and respects you. And, you know how much I love you."

Krista softened, "Mama, why is everything so hard for me? I feel life is always a battle."

"Battles are started. They usually don't begin over nothing. Downstairs. Coffee. Hot chocolate. No war. No battles. Okay?"

"How about some Depression cookies to go along with the hot chocolate?" Krista responded with a nod and hug.

"Coffee for me, hot chocolate and Depression cookies for the rest of you. Sounds like a winning combination!"

Krista

Hot chocolate. Only Mom would think hot chocolate could save the world and tilt the balance of good and evil. If she could, she would hand out cups of hot chocolate and freshly made cookies to soldiers entering battle and feel certain it would shield them from harm. As an adult, I loved her for it. As a kid, I wanted to scream for some focus. There were times I did not want Dorothy on my side; I wanted the Witch.

But I sat there like always and sipped the hot liquid while throwing back a few saltine crackers with melted marshmallows and peanut butter; she called them Depression cookies. She said these were all her folks could afford when she was little. *Whatever.*

Mom wanted to give us the world, while reminding us not everyone was as fortunate as we were. Instead of simply encouraging us to donate our old clothes or be more resourceful, she would constantly tout examples of the less fortunate from newspaper articles or friends to emphasize her philanthropic nature and our responsibility to mankind. While we were blessed to live in a nice neighborhood and have plenty of conveniences, I yearned for close friends and belonging to a community. I wanted a mother whose priority was to stand up for me, for Christ's sake, not just the poor souls of the world. Instead, she was a travel agent for guilt trips. And Dad didn't understand how any kid with our privileges, when he had none growing up, could have *any* problems.

As I remembered other situations where Mom defended whom she deemed innocent, I could feel old anger rising. It didn't take long for one particular incident to come flooding back. My mind started playing scenes, and I felt myself slowly transporting, as if all the years in between hadn't happened. It was that raw to remember.

PART II

Coming to the Table

Chapter 1

Summer 1985

Krista

Hands on my hips, my voice way too loud, I let loose my tirade, "Mama, you don't understand. She steals my babysitting jobs by pretending to be so sweet. Can't they see through it? She's so freaking fake."

Mom braced herself before responding, "Listen, Jean will be here shortly to discuss the situation, and she's bringing Susie. This wasn't my idea. The two of you brought it on yourselves with that screaming match at the top of the hill."

"I won't speak to her. I won't deal with her crybaby ways. She'll have tears coming before she walks in the door." *Too late*, I realized when I heard the doorbell chime.

Jean walked in cuddling a sobbing Susie and headed for the couch. Both Jean and Susie ignored my presence in the room. "Abby, how are you?" Jean asked. She pulled Susie in tighter and gave her a kiss on the forehead before continuing, "As you can see, Susie is very upset," and then for the first time she turned to me. "From what I understand, Krista, this is all over some babysitting issue."

Oh, the drama of it all, I thought. *What nerve to walk into my house and put on a show!* I was so angry. "I was working in this neighborhood way before she moved into town," I said to Jean and then turned to Mom. "Just because she's a measly six months older doesn't mean she can take my jobs." Not only was Susie older, she had sprouted boobs and was several inches taller than I'd probably ever be. Her features were sharp and defined, and her long blond hair was always perfectly coifed.

"But, Mom, that's just not . . . ," the theatrical sobbing began before Susie could finish her plea. She leaned into her mom showing us all she was too beaten and exhausted to continue.

I smelled blood and went in for the kill, "Are you two going to fall for this? I was here first, I had the jobs first, and I wish she'd just stick to her people. She can't have mine!"

Mom gave me the evil eye. It was the same upturned brow that had terrified her younger brother, Teddy. Now she was using it as a clear warning her patience was running thin.

"I know you want to help her be independent and strong, Abby. But, if you would just give me five minutes with her, I'd have her straightened out. Five minutes!" Jean said to my mother and then shot me a determined look suggesting it wouldn't even take that long.

"You are not my mother!" I yelled and then turned to the woman who was. "Do I have to cry to get some attention around here?"

I hated Susie. Within a few months of us moving to Tennessee, Susie's family had also moved into the neighborhood. New kids usually huddled together like dogs in the pound, even if our only common bond was our newness. It was a survival technique not to be more noticeable than any of the other new kids. Still, secretly I hoped the in crowd would see me shining among the pack and choose me.

It had not helped at all when our mothers fast became friends. I wanted Mom to have friends, but not the kind that turned her against me. Nobody else could have her.

"If she's just gonna cry, I'm out of here," I said and jumped up to leave.

Mom grabbed my arm and whirled me around to face her, "Whoa, young lady, I think you need to apologize to Susie. There are plenty of jobs to go around for both of you."

The sting in my arm was nothing compared to the hurt of Mom giving Jean and Susie what they wanted, "Great, take her side. Perfect! She can have whatever jobs she wants. I don't care!" I yelled. I looked Susie dead on for effect, "Sorry I made you cry AGAIN, Susie. Can I go now?"

Mom nodded. I exited stage right and ran up to my bedroom. I couldn't resist slamming the door. I knew Mom was going to apologize for both of us. I waited for Jean to come upstairs for her treasured five minutes, but she never did. A little while later, I heard the front door close lightly. I waited for Mom to come to me. As the minutes passed, I knew she had other plans.

Silence killed me. Fighting and battle, those I could handle. In my mind, those solved things, but silence was deafening. Mom knew it, and she used it against me. She would not take the bait like my father. She always felt for the underdog—the poor, pathetic loser who would weep in a corner and look up at her with sweet, puppy-dog eyes. My mother did not like me; I could tell. Many days, I did not like myself either, but I was not going to resort to crying to win. I would rather be a loser than be a cheat.

Sitting in my room, I stared at the clock and imagined all the ways Mom would come crawling through the door in apology. It couldn't take long for her to realize Susie was the runner-up only to Katie for an Academy Award. Instead, about an hour later, I heard the familiar cry to come to dinner. Great, now Dad was home, too. I'm sure she wasted no time telling him about his daughter's latest mouthing off.

I slowly edged down the stairs and turned the corner to the kitchen. There they all were. Dad was waiting with his empty glass, Mom was helping Chelsea, and Katie chatted sweetly about her day. My stomach turned, but I dutifully took my assigned seat. The drama unfolded almost immediately.

"Your mother tells me school is over in just a few weeks" Dad said.

"Yeah," I answered. In my mind, I continued to launch a tirade at him, *Some other details that might be important to you, Dad, like my name is Krista, and I'm in the seventh grade. And I am not about to get you any tea, so don't even ask!*

Oblivious to my rantings, Dad forged ahead, "Krista, get me some tea. I've got something to say to everyone."

"Sure, why not," I said under my breath, "Get some tea, apologize to cry baby..."

"Krista, is there something you need to say?" Mom asked. She did not miss much. Even when I thought she did, she was merely overlooking it to avoid a conflict. Dad, of course, had no clue. *Take me on and make my day*, I thought.

"No, ma'am," I acquiesced.

I trudged over to the counter, and Dad began to weave his tale of a fabulous new city. I heard the details, but nothing was grabbing me. He had thrown a couple of juicy worms out, but I eyed them suspiciously and wasn't ready to bite. After all, I had just entered middle school. I was only recently settled in enough here to make friends, and I had built up my babysitting clientele—though I was doing better on that front before the crybaby. Still, thanks to some steady inflow of cash, I didn't have to always suck up to my parents for every little thing I needed.

Although he danced around saying the words, I knew moving was a foregone conclusion. This time it would be a coastal town in North Carolina. *Man, we are so lucky*, I thought. *We would have a beach and all kinds of fun in the sun. We would be absolute fools to want to stay in Tennessee.*

The camping and boating used to entice us here were no longer fun and cool. We would have the beach and beautiful weather. And, to top it off, we would only be three hours from our grandparents. We could now see them *so* much more than once a year, as if that would make it better when they forgot our birthdays and Christmas.

Unfortunately, there was no reason to pour out any of this thinking to Dad. This wasn't a democracy. We were moving, like it or not. Why waste time telling him how much it was going to ruin our lives? Especially after he and Mom had used up so much energy painting such a beautiful picture.

I wanted to run. My life might be inevitable, but I didn't have to watch the same old scene again. "I'm done, can I be excused?" I asked. They nodded.

I ran up to my safe haven for the second time that day. I opened the drawer to my corner desk covered with yellow flowers and pulled out my small pink diary. I reached for a pen and lifted the key out of its secret hiding place.

Dear Diary,

Just what I needed. Guess, what? We are moving again. Who cares! It's not like I have that many friends here anyway. You think Dad would just get a clue! At this rate I'm never going to have any real friends, and God knows I'll never have a boyfriend!

Not that I'd have one anyway. Can't wait to hear the NC kids tell me I'm ugly.

I wish Mom could stop him. If she loved me, if they loved me, they would make it stop.

I closed the cover when I heard a knock on my door. I stuffed the diary into the drawer and prepared for Mom's reasoning. Instead, it was Katie.

"Krista, I don't want to go," Katie said in a whimper. She cuddled at the foot of my bed. Her skin was more golden than mine, and her face much more angular. Except for a couple of inches in height, our three-year age gap was barely noticeable.

I felt resigned and knew truth was all the help I could offer, "I know, but we have to. Didn't you hear him? It's going to be perfect." Truth wrapped in pretty lies.

She trusted me, and I should have said more. But I couldn't be convincing right then, and I knew all she really wanted was a diversion.

I offered her one. "Let's feed Oscar. You can bring me the food," I said.

Katie jumped off the bed and ran to do my bidding. We walked over to the hexagonal bowl housing my Chinese fighting fish and sprinkled in some food. Katie's big brown eyes stared in wonder. I would do my best to make this move easier on her and Chelsea; they would have me on their side. Katie was only in the fourth grade, so this move would not devastate her world like it would mine. At least the fish would die a quick death devoid of lingering agony. We would flush it before the move.

Katie had me in her corner, but I had nobody. Nobody that understood. As I had learned from our previous upheavals, Mom would soon be busy with all the preparations for the move. Plus, she would get caught up in the drama of it all like she always did. And I would survive because I always do.

I heard the phone ringing and jumped on it as all teenagers do. "Hello!"

"Hey, Krista," Courtney said. She was my best friend, and I was going to miss her so much. "Guess what? Mary told Dean you liked him. What are you going to do?"

I am going to move, and Dean will never even remember I existed, I thought. "Nothing. It doesn't matter," I said.

"Doesn't matter? You've got to be kidding! You've been in love with him since the first time he wore those gray parachute pants."

I wanted nothing more than to slide into the conversation and chuckle. But it would've been only a short-term fix and caused me more pain in the end. *Couldn't she tell my world was crumbling?* I had to cut my losses quickly. So what if it was another couple of months before we moved? I had to prepare. They weren't going to see me cry.

"Dean's not all that anyway," I lied. Courtney asked me if I was losing it. "Did I mention we are moving again?" I figured it was better to run quickly over the hot coals.

"Move? Seriously?" Courtney asked.

"Yep, moving to the coast. It'll sure be better than this place."

Abby

I dreaded meeting with Jean and Susie. No one warned me preteen flare-ups were part of the parenting package I signed up for when I left the hospital with my beautiful bundle. The babysitting battles that ensued between Krista and Susie were more than a nuisance; they were threatening my friendship with Jean.

Jean was a lovely woman. She had moved a few doors down from me, and the chemistry between us was immediate. We were both stay-at-home moms and loved many of the same activities and pastimes. As soon as the girls boarded the bus, we would phone one another, laughing and planning our day together. Having someone to pal around with made the move more palatable for me.

But my last phone call from Jean had not been friendly. It bristled with anger, "Abby, Susie and I need to come over, now. I'm not going to have my daughter torn up all the time. Do you understand?"

I bit my tongue. "Jean, I know the girls have issues over the babysitting, but I think we should . . ." I stopped in midsentence, realizing it was useless. The only way to stop the accusations was to have them over and hash it out.

From the minute they arrived until they left, I felt helpless. No solution appeared to work. None. I knew I was being tested from both ends. There were two choices, stand tall for my daughter or raise the white flag.

"Abby, there are plenty of babysitting jobs for both girls, but I do think Susie is older and has more experience. You remember how we used to feel when we left our babies, don't you?" Jean leaned in and pressed one hand to my knee.

I knew the sign and hesitated. My gut screamed for a swift removal of my guests. These tête-à-têtes were becoming a weekly squeal-all, mother-daughter confrontation that defined and strained our relationship. I did not need this kind of friendship and neither did Krista.

"Jean, I'll talk to Krista." Facing Susie, I added, "I know you and Krista can work this out between the two of you." I smiled and reached to touch her shoulder. Susie was a nice kid who knew how to work her mother. Sometimes I wondered if the girls would be better friends if Jean and I were not.

"Abby, I'll talk to you tomorrow morning. How about lunch?"

"Sure, give me a call." I walked behind them to the door. We waved, smiling as if nothing had happened.

Krista ran upstairs, and I was left dreading the next confrontations, Krista today and Jean tomorrow. For a minute, I considered the avoidance route, run like hell and never look back. But I found this strategy didn't work. The next option sounded like a winner. I would cocoon myself in the house, forget friends, and stop playing the role of fixer. I needed someone in my life to fix my problems, not add to them.

I looked upstairs and saw Krista's door closed. Damage control was needed before things got more explosive, but my prayers for a reprieve were answered when Bob called.

"Hey, I've got some good news and some bad news. Which do you want first?" From his tone, I knew he had an objective.

"Tell me the bad news first. I want to end my day on a happy note." Surely Bob's news couldn't be any worse than the dreaded conversations with Krista and Jean.

"Al called me into his office and told me Aircraft Engines in Wilmington, North Carolina, called about my availability. I think he's pulling my chain. Frankly, I don't think there's anything to worry about, but I thought you should know."

I shifted, "Bob, you've got to be kidding. We've only been here twenty-two months. We can't do another move, not now. This year has been hell on us, you know that. If they approach you again, tell them no!"

"Look, I didn't go asking for this, Abby. You know I've got to be receptive to every break I can get! You need to trust me."

"Not at the sacrifice of this family, Bob. When do we come first?"

"You're being totally unfair! I won't even discuss this with you right now," he sputtered.

"Fine! Tell me the good news."

"I received confirmation on the management course I applied to in New York. It will mean a lot to my career. I leave in two weeks. I don't know how I'll manage. My calendar is packed, but I've got my secretary moving around appointments to make it work." He seemed oblivious to the heavy breathing on the other end of the line.

"You mean the four-week course you mentioned? The one I wasn't supposed to worry about?" I reiterated, hoping I had heard wrong.

Ignoring me completely, he continued, "I was glancing at the brochure and saw a break on the second weekend. Spouses are invited to join their husbands that weekend or I guess I can come home. We don't need to decide anything now. Listen, I've got people standing at my door. Love you." Bob hung up, shutting down another day for me.

The receiver was still in my hand when I heard Chelsea scream. *Oh my God*, I thought. *What's wrong now?*

I ran. When I cleared the corner into the den, I slammed into Katie. She was wearing a hideous Halloween mask to torment Chelsea.

"Girls, stop it right now! You almost gave me a heart attack. Katie, put that ugly mask back in your father's closet. Chelsea, stop screaming!" I picked her up and crooned, "It's only a mask, baby."

My day was going from bad to worse, and I hadn't even left the house. Life seemed to be shredding into tiny little pieces. I didn't know how to pick them up and put them back together or find the lost ones. It was hard not to feel trapped and overwhelmed. The thought barely left one brain lobe before guilt washed over me, making me feel worse.

I sat on the floor and pulled Chelsea into my arms. Holding her tight, I let go of the spinning and focused on her golden curls and soft skin. When she turned her face up to me, I studied her petite features—her upturned nose and her large blue-gray eyes that Bob always said were flecked with fairy dust.

Instinctively, she lifted her hand and started rubbing my collarbone. Chelsea's fetish, an obsession with a dual result, was actually soothing for both of us. It gave me time to unwind or, like now, to contemplate my next course of action. Drawing up a plan was easy; executing it was challenging, especially when it kept changing. There wasn't enough paper or patience to keep up with it.

It's not like I didn't try. I'd wake up in the morning determined to take one day at a time. Then several days piled on top of one another, and I was buried. Time? Never enough. Going to the toilet proved an endeavor.

I put Chelsea down in front of the television to watch Big Bird. *If life could only be this easy,* I thought. I rushed into the kitchen to check on supper and finish my to-do list for the following day. Taking a baked chicken out of the oven, I heard Bob enter, greeting Chelsea.

A few minutes later, he lumbered into the kitchen, "Man, I'm beat. Another day like this and I'll gladly stand in the soup line."

I could have been mad and riled about our earlier conversation, but looking at him deflated my anger to a puff of smoke. "Hey, why don't you go upstairs and change. Supper's almost ready. On your way down, tell Krista and Katie it's time to eat."

I did it again. I fell for his *I'm-so-worn-out-and-dejected, please-be-nice-to-me* look. From the moment he walked into the room, his face betrayed his intentions. He thought playing the wounded victim might lessen my reaction to a monthlong course and a possible move. He was dodging the inevitable.

He knew my first reaction would have been negative. That's the point where we would switch places. I'd play the victim, and he'd put on the here's-how-it-is posture. My feathers would ruffle, and I'd clam up, never telling him my true feelings—how lonely I would be or how I trembled to think of handling the children, house, and everything else that could come crashing down on me while he was gone. In the end, we would both end up feeling misunderstood and alone. Why? What kept the edge, the refusal to be honest and open?

"Mama, what's for supper?" Katie asked.

"Uh, baked chicken and rice." Looking at her reminded me how beautiful her large brown eyes were. Impulsively, I reached over, giving her a big hug. Her little body melted into mine, and I held her tight. She was quiet and reserved, often getting lost between Krista's shenanigans and Chelsea's needs.

"Katie, do you know how much I love you?"

"Uh-huh," Katie said, squeezing my hips with her tiny arms.

"Will you help me set the table?"

Katie squeezed one more time and went to do her chore.

Chelsea walked in, sat down, picked up her fork, and slammed it on the table, "I'm starving!"

"Me, too," echoed Bob.

"Where's Krista?" The words were no sooner out of my mouth when she entered the kitchen. "Oh, there you are."

She looked pouty. I realized I should have gone up to her before now. Sometimes, I forgot she was only twelve. She was so independent, but times like this reminded me she needed more reassurance and love. I vowed to carve out time for her later.

Bob turned to Krista, "Get me some tea."

Krista glared and stalled, but walked over to the refrigerator to fulfill his request. I was thankful she didn't mouth off at him, knowing how upset she was.

Without one iota of sensitivity, he blurted, "Have you ever heard of Wilmington, North Carolina?"

Drawing a defiant bead on him, I silently dared him to bring up anything we had not discussed. "Bob, let's talk about this later. We've got plenty of time to plan our family vacation." I hoped to ward off any discussion about a move.

Bob paid no attention to my warning. "I hear Wilmington is great. There are two beaches close by. Wouldn't it be nice to live near the ocean instead of only seeing it once a year? And to make it even better, it's closer to your grandparents. If we happened to move there, wouldn't that be great?"

Silence. Total, stark silence.

I bowed my head. I couldn't face Bob or the children. Burning tears threatened to spill, but I blinked hard. At that moment, I hated Bob and his job. I hated his jolly *guess-what-girls-our-life-is-changing-again* optimism. It was crushing. It was brutal. He was wrong to have presented something so traumatic to our children this way, especially without warning.

Krista broke the silence, "Whatever, Dad. Can I go upstairs? Maybe I'll start packing or something." Without waiting for an answer, she got up and left.

Bob squelched any further discussion, "Abby, I don't want one comment. I know what's best. If this move happens, and it looks like a good possibility, we need to prepare them. They'll come around, they always do," Bob said, shoveling in another mouthful of food.

I didn't answer. I got up and started picking up plates of uneaten food. I kept thinking, *I'm throwing away good food while children in China are starving.* But my anger wasn't about starving children in China. It was about my children. They weren't starving for food, but starving for some consistency in their lives, starving for our attention, and starving for their mother to make things right for them.

An overpowering need to talk to Krista and Katie consumed me, "Bob, watch Chelsea. I've got to check on the older girls."

"If you need any help, I'll be right here," Bob said, knowing I wouldn't ask.

Despair rippled through me like a current. I pushed forward, dragging up the stairs to do damage control. I had no answers for their questions. I only hoped my words would come out reassuring, then listen and take the blame. Blame because I supported Bob, and later, blame because I didn't. It was a no-win pit I was about to hurl myself into.

I stopped at the top of the stairs, feeling dread rising in my throat. When I saw Krista and Katie talking, I felt I'd been offered two reprieves today. Watching them made me thankful they were close. I knew Krista gave Katie a lot of time—time I didn't have. She was like a surrogate mother. Fiercely protective, she often chastised us when she didn't like what we were doing with Katie. In some ways, I relinquished this part of mothering. Wrong as it seemed, it helped take some of the responsibility off me. Maybe that's why Krista appeared older to me. I depended on her. Maybe I depended on her too much.

Tiredness threaded my back into knots, causing my shoulders to pull tight. I sat down on the hope chest at the end of my bed and stared at the blank screen on the television. My gaze never left its repose until I heard Bob coming down the hall talking to Chelsea.

I started to rise and assist, when he shouted, "Abby, I'll get Chelsea ready for bed."

I lingered longer than I should have, but finally got up to finish my earlier task. I peered in on Katie first. She had fallen asleep on top of her bed linen, fully dressed, holding a blue crayon in one hand and a Barbie coloring book in the other. I took the crayon and coloring book and laid it on the nightstand. Next, I picked her up, gently stuffing her under the covers, kissed her forehead, and turned off the lamp.

Heading toward Krista's room, I heard her laughing and realized she was on the phone. Deciding I'd been pardoned again, I turned toward my bedroom and almost ran into Bob exiting Chelsea's room.

He grabbed my arm, "We need to talk." I followed, too tired to protest. What's one battle over another?

The bedroom door closed behind me. I walked over and sat down on the bed before facing him, "So, what do you want to say?"

He stood by the door, quiet. His face was not filled with anger or an agenda. I couldn't tell if he was gauging his words or ready to cross the battle wary barrier we usually drew. Never taking his eyes off mine, he moved toward me. I gulped. This was new. No emotion rose or raged, until he reached for my hand, pulled me into his arms and kissed me.

Pulling away, he whispered, "Trust me."

CHAPTER 2

Krista

Looking through the dingy school bus window, I felt the emptiness of losing another place. I wasn't losing a home; my family knew better than to get attached to a location. It was worse than that. The persona I had created would stay behind; so in essence, I was dying. I never knew what I would have to become on the other side. The North Carolina girl was an illusion I had yet to create, and the real me was a reality I had yet to know.

Late at night, I would often think about the many characters I had created over the years. I could imagine how actors felt. They become each new role only to lose that part at each premiere. Now, through the cruelty of fate, I would lose another me. I was heading off to start a new movie, a new role.

"Krista, did you study for the math test?" Courtney asked.

I looked at her and felt the separation beginning. She was blurring, and I was starting to forget her. I blinked several times to combat the fuzziness. "I looked over the stuff last night. Not worried about it."

"Well, I am. I wish I had your knack for numbers."

"It's the only thing you can trust," I said to her. "Two plus two is always four."

"Yeah, yeah, yeah. But what the heck is x?"

We laughed. Excelling at school was never a challenge. Existing at school was. As we filed out of the yellow bus, I remembered I had not cared when I showed up here, and no one would care when I left. While I was absorbed in questioning my existence, chattering females in the hallway buzzed in my ear as I made my way to my locker. I clicked it shut and was heading to the math test when Dean stopped me.

He put one hand on my shoulder and looked into my eyes and said, "So, you think I'm cute?"

A crowd had gathered, and I felt heat spreading from my neck through the rest of my body. *God take me to the next place now. Why must I suffer one last humiliation before I am a memory?* I wanted to run, to scream at him to please let me down easy; instead, I stared at him feeling like a simple bystander to the scene.

"I only date pretty girls. You know any?" he said and turned to high-five the second most popular guy in school and his best friend. Several girls snickered behind them while several more looked at me with pity.

I simply walked away. Not a word. Neither Mom nor Dad would have believed it. After all, I had a comeback for everything they said. But to defend myself to these kids, I had not one word. I melted into the scenery because I knew how to do that. This was the cherished popular crowd, and I had never gained access.

Later that day, I got off the bus resigned to my life. Suddenly a pack of ten girls yelled, "Surprise!" I was astonished yet thrilled to see my friends. *Why do people seem to appreciate you only when you are on your way out?* It made leaving harder, and I did not want the distraction from my planned withdrawal. We had a party, throughout which I promised everyone I would write about the beach and all my new friends. Several were openly jealous about my upcoming adventure. I did one of my greater acting jobs to date and convinced them I was thrilled as well.

Once the crowd dispersed, I sat with Courtney on the deck of her above ground pool. Her feet dangled off the edge, barely skimming the water. *No ocean would be as comforting as this pool water*, I thought.

"Can't your dad keep his job here?" she asked. She pushed off with her hands and slid into the water.

"It's not that. He's on a fast track. He's going to be the big guy in Wilmington. Some sort of really important position." *Did I believe any of this?* One thing I knew, if I was ever quizzed on exactly what he did, I would have failed my first test.

She swam back over to me and put her arms on the side of the pool, "But, come on, how many places have you already lived? Maybe your dad should think about you guys for a change."

Defensiveness to protect him replaced any anger I was feeling, "My dad does think about us. He wants to give us everything!" Courtney rolled her eyes, so I said what shouldn't have been said, "You're just jealous! Your dad lives with his new wife, and you never even see him or your new sister." *There, that should do. That should wipe away the last string holding me here.*

When I arrived home, I ran up to my room. I didn't want to have a conversation with Mom about my friends. Who cared? I was leaving so no point in dwelling on them. I didn't have the energy. And I didn't care to be in the way of the men packing up our lives into taped-up brown boxes. It was too much. Safe in my room, I pulled out my diary. I had to purge these feelings, get them out before they consumed me.

Dear Diary,

This is the last chapter to our Tennessee saga. They threw me a party. What a slap in the face. I wanted to slink away. It was fun, but it makes it harder. Forget the whole thing. I don't know why I try each time. Everything is harder when you start to like it.

We leave tomorrow. It seems like yesterday this move was only a maybe. Why can't Dad make it all stop? What does he need that he can't find here? Can't he remember being a teenager? Doesn't he know how much this all hurts? No, of course he doesn't understand. He was born in the same house his parents still lived in. He has no clue.

Leaving always happened so quickly. Before I knew it, I was looking around my empty room in a daze. A lone dust bunny rolled over to the corner where I stood. Through my window, I could see the wooden swing set that would become part of the bargain for the new owner. I realized memories lived within these walls even though we no longer would. I crept from my room when Mom called. I slid my hand down the banister trying to soak up every image. There was such emptiness, a void. The moving truck had left this morning, carrying all our treasures and possessions. It was sad seeing everything we owned summed up in a few boxes. I learned to take a mental picture of each item packed because I never knew how it would come out of the box on the other side.

I walked through the living room and down the hall to the door leading to the garage. I stopped. I knew once I walked through the door this chapter of my life was over. The normal panic began to set in; my hand shook when I tried to open it. The vines on the remaining wallpaper grew out to encase me, trying to keep me. My feet felt like lead. It took focused effort to convince my brain to push my arm forward and grab the doorknob. Midreach, Dad opened it. He grabbed my hand and guided me to the car, but it felt more like abduction than aide. He was yanking me from all I knew and throwing me into the abyss, so great of him to be here to help now. He had been gone on and off for weeks between his new job and training, and had only recently come back to collect us.

The car engine started, and it seemed like a roar. We pulled out of the driveway and made one last trek out of the subdivision. I tried to memorize our house and each major landmark in town. I knew I would never be back, but I didn't want these images to wither away. I remember a teacher asking a riddle once, "If a tree fell in the woods and no one was around when it hit the ground, did it really make any noise?" Even if everyone else forgot I lived in Knoxville, I wouldn't.

We had a ten-hour car drive ahead of us, and the car was packed. The Ford LTD's dark burgundy interior mirrored my mood. Katie and I had drawn our battle lines in the back seat and dared each other to cross over. We made a very comfortable pallet on the floorboard for Chelsea. She was little, but she kept complaining about the hump and wanted to sit on the seat with the big girls. Kids. The tension in the car was thick and the silence deafening.

About an hour into the trip, Mom reached into her purse and started digging around. All of us in the backseat perked up like wolves smelling the oncoming lambs.

A chorus of female voices purred in unison, "Mama, do you have anything to eat in there?" When she didn't respond, I leaned forward and slung my arm over the seat into adult territory.

"Is that gum?" I asked when I saw the glint of aluminum wrapping.

"Yes, Krista. Sit back."

All hell broke loose. "Mama, Mama, I want gum. Please," Chelsea said with the innocence of a five-year-old.

"Me, too. Me, too," Katie chimed in.

Mom whipped her head around, and insanity blazed in her eyes. I jumped backward and pinned my back to the seat. She had an intensity of displeasure I knew too well. Usually, it was a result of my own actions. Still, I wasn't prepared when silver bullets started whizzing past. I threw my hands up to shield my face. Three shots and all hit their targets. Katie and Chelsea were whining, and my father's hands tightened on the steering wheel. He was trying to make himself invisible, but Mom was too quick and flung a piece of gum at him, too. Then, as if the whole incident never occurred, Mom gently returned her purse to her side and laid her head on the headrest.

Dad never said a word. He knew better. The silence took root again. I looked out the window. We were passing through the Appalachian Mountains. Looking at the scenery kept my attention, and it was a welcomed relief. I imagined God dragging a stick through the earth to create this highway through the mountains. To my right were sand ramps for trucks unable to stop. I knew because I asked Dad; Mom was still sleeping.

The first few hours of these trips were always as anxiety ridden as the last few hours. I never really wanted to go back where we started or continue to somewhere new. These were the moments I secretly wished the car could be our home, and all the things outside would vanish.

My anger and frustration were slowly overtaken by boredom. Every parent's worst fear on a long car drive was three bored kids. I leaned forward and whispered to Dad to turn on the radio.

He ignored me at first, but I persisted, "Come on, Dad. I am so bored."

"Can't you just go to sleep like Katie and Chelsea?" he asked, never looking my way. *Traitors*, I thought when I looked at their dreamy little faces.

"I want some music!" I whined; and to ensure I'd get it, I added, "Fine, I guess I'll just keep talking to you for entertainment."

Those words obviously horrified him because he jolted forward and turned on the radio. The music filled the car. He searched quickly through the stations without even pausing for opinions, and soon the voice of Johnny Cash filled the air.

'Oh, Dad, not him! Not country music, *please*! Anything but that!" Dad flinched. He glanced at Mom with despair, but she was still sleeping.

When the commercials started, Dad fiddled with the radio. He was clearly afraid I would take advantage of the momentary silence and speak to him again. Too late. In an effort to make sure there would be no more burning rings of fire or drunk men on bar stools filling the void, I started asking him about Wilmington. He gestured toward the sleeping figure beside him and then put his index finger up to his lips. The conversation was over.

This car trip held a special treat for us. Not only were we heading to a new town, but we were also going to see our relatives. Dad was going to drop us off for a week, and later we would make the three-hour trip to join him at our new home.

I finally closed my eyes and decided to sleep. When I awoke, it was getting dark, and we were pulling into our grandparent's town. The town where my parents had met. We made the familiar trek through downtown, prompting a comfortable feeling that actually calmed the sense of dread at visiting kin. We pulled up to the street separating our two grandparents' houses. The battle began when the tires stopped rolling.

"We'll stay the night at Mom and Dad's," Dad said. He opened his door, and the light poured in.

"There's no room there, Bob. Let's just go to my parent's house. They'll still be awake and expecting us."

I knew the rest of the conversation by heart. I would never marry a man with his parents this close to mine. Talk about problems, this situation screamed for a divorce every time we were here. We girls would stand in the middle of the street and literally be pulled on each side by a parent. As if we weren't going through enough, now we would have to endure the in-law battles. Thankfully, Dad would be leaving to go back to work soon. We would just stay at Mom's parents' house and be ignored, but at least the battle would be over.

The week flew by. Usually it didn't; but since we weren't eager to go to the next destination, time played tricks on us. Mom gathered our stuff and corralled us back to the car. I tensed at the thought but numbly helped Katie gather her backpack of boredom-reducing goodies before heading out.

Mom pulled over at a trucker's restaurant on a back road halfway to Wilmington. We were starving and applauded her decision. We loved greasy food, and I definitely learned early on Mama's Kitchen and Country Cooking restaurants were the best.

We settled into the booth and waited for menus. The waitress brought over a basket of biscuits to hold us over until the main food could be served. She flipped open her order book and asked, "Good day, ma'am. How're ya'll doing?"

"We'll need just a minute," Mom said without looking up.

As kids do, we jumped into the biscuits headfirst. I was the first to attempt to bite into one and almost lost a tooth. Mom noticed and grabbed one herself. After holding it for what seemed a second, she summoned our waitress back to the table.

"Excuse me, Miss. Could you come here a second?" I saw my mom's eyes as she said it, and I feared the worse. I wanted to warn the waitress, "Run, run!"

Mom took the server's hand, turned it palm side up, and smacked the biscuit into it. "I wouldn't feed this to my damn dog!" The waitress stepped back, startled, and then began apologizing. Mom waved her off. "Please get me some coffee and milk for the girls."

The woman slinked away, and I excused myself to go the bathroom. After I was out of Mom's sight, I walked over to the register where our waitress was ringing up a *polite* customer's order, and I apologized. "Mom didn't mean it. She's not upset at you, she's just mad at my dad."

Abby

Within a couple of weeks, we knew Bob had gotten the Wilmington job. Six weeks later, he headed to his new post. Door open, door shut, alone again. His "trust me" didn't seem so comforting now.

The summer heat raged, adding to our mounting anxiety over the move. Krista and I were in one conflict after another. If it wasn't the continuing saga over the babysitting jobs, it was her crisis over leaving her friends. One day she was ready to go; and the next, she was having a meltdown.

The day before the movers arrived, Bob flew in to help, tanned and overly nice. The first thing he offered was to treat everyone out to dinner. I was surprised he didn't comment on how gaunt we looked. For weeks, I'd been working to clean out the refrigerator. The cupboards were down to crackers and cereal. When food was mentioned, the girls almost broke out in song and dance. Dad became a hero. I got labeled Stalag13's POW commander in chief.

We woke to rain. By the time the truck arrived, I was in a tether, "I will not have my furniture leave this house! My God, Bob, it's raining cats and dogs out there. No way! You go tell the driver if the rain doesn't stop, the furniture doesn't get loaded. Do you hear me?"

"Abby, we have a moving van in front of our house. I have taken off work because you insisted I be here today, and by damn it, we are loading it! These men are professionals. Let them do their job."

I fumed. Looking out the picture window, I could barely make out the houses across the street. The rain came down in sheets. Runoffs from the gutters hit the sidewalk, bouncing up in foot high splashes. At one point, Krista asked if she could use her boogey board to surf down the driveway where the water flowed so rapidly the drains were overflowing, creating white caps at the curb.

I saw the truck parked in front of the house, the driver sitting in the cab, and the three loaders standing in the cargo bay. They weren't talking or carrying on like I had observed on other moves. Clearly, this crew didn't run together. Earlier, I overheard one guy laughing about how the driver spotted two of the men standing on a downtown street corner, asked them if they needed work, and threw them an Allied Van Lines shirt. One didn't even speak English.

As soon as I heard this, and without consulting Bob, I panicked and called our moving representative. She tried to reassure me this was customary. I told her this was my life's possessions, everything I owned. She said the company had insurance. I told her I didn't care about their frigging insurance. I wanted my furniture to end up on the other end of this move in one piece. Her response was a dial tone.

I would have gone to Bob; but when I looked back out the window, I saw he had joined the losers in the truck. Now he was cavorting with the enemy. I threw up my hands, deciding I didn't care. Let them load the damn truck in the rain. I hoped everything would get moldy and the company's damn insurance provider would have to pay. Maybe I'd even get some new furniture out of the deal!

I walked into the kitchen and picked up my Diet Coke off the counter. One swig and the hot liquid almost came back up. How many times in my life would I have to go days without a refrigerator, and McDonalds for breakfast, lunch, dinner, and snacks? Tossing the flat pop down the drain, I looked at the new wallpaper on the walls.

Three months ago, I bit the bullet and squeezed some extra funds out of savings to redo the kitchen, selecting a look-alike wallpaper I had seen in a national magazine. The matching border print displayed oak baskets filled to overflowing with colorful spring flowers. I even found matching placemats and towels. When I finished the project, I felt so decorator smart.

But that was then, and this was now. Nothing I had here would go into the new kitchen. In fact, I detested the wallpaper there, large red poppies with black foliage on a stark white background, way too contemporary for me. Bob bristled when he heard me tell the real estate agent how ugly it was. Later he remarked if it was stuck on the wall and not torn, we would live with it. I don't know where he gets off saying *we* all the time. *We* are going to go into that house on the first day, find a loose edge of wallpaper and pull as hard as *we* can.

Mass destruction in comic proportions filled my mind. Detonating the new house might be the answer; maybe I'd get a new one, one I liked. I envisioned Road Runner with explosives in his hand, standing in that ugly kitchen, poised, ready to ignite . . .

The front door slammed, bursting my daydream. Then I heard footsteps pounding the stairs. I leaned outside the kitchen door, shouting, "Bob, is that you?"

When I didn't get a response, I decided I had better check it out. It was probably one of those jerks from the truck with the runs, coming in to use my bathroom. I stopped. Come to think of it, it wasn't my bathroom anymore. If he crapped all over the place, so be it. A devious smile spread across my face; but my curiosity was peaked, wondering who the intruder was.

I waited like an assailant. When no one came down, I went up to spy. Krista seemed to be the only one upstairs. Her door was partially open, and I could see she was sitting on the floor by an open box. Her mahogany hair had fallen around her face. She looked young and vulnerable. I watched her reach into the box and pull out her diary, hugging it to her chest. I wanted to go in and share my day and hear about the surprise party at Courtney's, but I heard her crying. My heart sank. I knew the move was going to be the hardest on her. She would be a teenager in a few weeks, and I knew girls at this age might not be welcoming or accepting.

Something told me to leave her to her privacy. To be honest, I didn't know what to say to make things better. I felt like a child, too. I didn't want this move. I could have cared less about the money or any of those goal-objective, bullshit discussions Bob always saddled us with. No one ever asked me what I wanted any more than we asked the girls. I slipped quietly down the stairs not to disturb her. When I got to the bottom, I heard Gary and Margaret talking to Bob.

Margaret was one of my dearest friends, and she was battling cancer. Even now, she radiated past the tiredness that was becoming commonplace. Her crystal blue eyes were mischievous, always leaving the observer questioning her true intent. But there was nothing pretentious about her; she wrote the book on genuineness.

"So, how was the trip to Crotonville?" Gary asked Bob.

Gary was a wiry man, military stiff, and military thinking. He was proud to announce he had sported a jarhead crew cut for thirty-seven years. On first inspection, one might be leery of him, but it didn't take long to see through to his lion-sized heart. Chelsea was the one who pointed out the lion in *The Wizard of Oz,* saying it was Mr. G., and the analogy stuck.

"You wouldn't believe it! What a setup! Jack came to our last meeting. Intense doesn't even begin to describe the man. But I can't complain, I got four weeks of great food to go with the BS'ing and brainwashing," Bob laughed.

"Yeah, I remember the routine. Not a bad getaway, is it?" Gary reached over, giving Bob a male-bonding slap on the shoulder, signifying their camaraderie.

Pasting on my warmest smile, I rounded the corner and exclaimed, "Well, if it isn't my two favorite people in the whole world." I hugged Gary and gingerly reached for Margaret.

There was no doubt she was trying to look happy, but I knew she was as sad as I was. She had been in remission until three months ago; but now the cancer was back, spreading like brush fire. Her bones were deteriorating, too. Two weeks before, Gary had given her a big hug and broken three ribs. We were all treating her like precious glass.

"Well, what do you think?" Margaret asked, whirling around.

"Think?"

"My new wig? Gary helped me pick it out today. Wouldn't you know it, I got in the shower yesterday, and the last of the locks went down the drain." Her eyes glistened. Gary reached over and put his arm around her.

I couldn't help myself, tears started flowing. *How in God's name was I going to leave her or face the fact I might never see her again?* Bob put one arm around me and the other around Margaret. We formed a tight huddle, crying for what felt like an eternity.

Chelsea invaded the scene and announced, "I've gotta go poop." She refused to wipe herself, hollering every time she had to go to the toilet.

Crying collapsed into laughter as we disengaged, scrambling for Kleenex. While I attended to Chelsea, Margaret, Gary, and Bob moved boxes around for makeshift chairs.

"Mama, why is everyone crying?" she inquired.

"Well, we're sad to be leaving Mr. G and M&M. We love them very much."

"I'm sad, too, Mama," Chelsea said, puckering her bottom lip and whimpering, sensitive to everyone's feelings.

After Gary and Margaret left, we spent the next few hours watching it continue to rain, swearing out loud, and seeing each piece of furniture and every box head toward the truck. As soon as the loading crew left, Bob and I piled the children into the car and headed out onto I-40 toward our new life.

Two hours into the trip, I reached for my purse to get a piece of gum. As I peeled the aluminum wrapper off, it crunched. Every ear in the backseat zeroed in on the sound. From behind me, beginning in a low hum, elevating to a piercing falsetto, was the word *gum*.

I ignored them, wanting to wallow in self-pity and remorse. I did not want to deal with children right now. Couldn't they see I was upset?

As they screamed, begged, and taunted, all I could think was, *You want some gum? Well, you're gonna get gum!* I shot each child a death glare. Then, in one fluid motion, I pulled pieces of gum out of the packet and threw them over my shoulder. I didn't care where they landed or who they hit. I wanted them to shut up and leave me alone. When I realized there was one lone piece of gum left, I flung it at Bob. Who needed gum anyway?

After the soothing tirade, I calmly laid my purse down beside me and leaned my head against the headrest. I wanted to absorb the pain and loss. I kept seeing shutter-shots of Gary and Margaret leaving our house; Jean and Susie crying when we went by their house to say good-bye; and friends and church members calling and sending letters, telling us how we would be missed.

At some point, I must have fallen asleep because the next thing I knew my head whiplashed forward. That's when I heard, "Come on, Dad. I'm so bored."

I tried repositioning my head against the seat rest, acting as if I hadn't almost suffered a chronic neck injury. I was wide awake and starving; but I continued to keep my eyes closed, listening to Krista and her dad. When I realized he was managing, I settled back into a fitful sleep.

The next thing I knew, Bob was announcing our arrival in Williamston. Pulling in front of Mom and Dad's home, I realized how much I had missed them. I was actually looking forward to spending some quality time with them. Bob ended up staying overnight before heading back to Wilmington to work, and we settled in for a couple of weeks of reconnecting.

Bob called the night before we were to join him. His jubilant, high-pitched voice was more than I could take. My reaction was the same, dull and uncaring. I wanted him to join the suffer crew. He ignored my silence, rattling on and on ad nauseum until I finally interrupted, "What in the world's got you going?"

"Oh nothing, just had a really interesting day," Bob announced, too cheerfully.

"Interesting, how?"

'Well, you remember Maureen, our real estate agent?" he asked.

"Yeah, I remember her."

"We went to lunch today. She wanted to bring me up to speed on our new neighbors and some interesting issues going on in the community. I met her husband, too. Great chap! I really enjoyed meeting him as well."

My temper raged, "Lunch? Bring you up to speed? Great chap?"

"I really appreciated them taking the time to fill me in. You know, giving me a heads up. By dinner, I had more information than I could process."

The idiot actually kept talking. "You met Maureen and her husband for lunch and dinner?" *Why on Earth was he making me go through the twenty questions game?*

"Well, no, he met us at *O'Charlie's* for dinner, and we had a few beers. Hey, I'm sorry; my secretary just motioned for me. I'm late for a meeting. Kiss the girls for me. Can't wait to see you all tomorrow." The phone went dead before I could say another word.

Steam billowed from every portal of my body. A few beers? What the hell was going on? He was acting single, out partying; and I'm stuck in Williamston, hauling Mama and Grandmother around, listening to them bicker about everyone and everything. And in between their squabbles, I'm dealing with three kids professing boredom every thirty seconds. My ears were ringing; my nerves shot. And Bob's out getting wined and dined!

I was blind furious all night. I tossed and turned. When the clock went off, I was wide awake and ready to go. I told the girls once to get ready. Usually, they didn't listen, but today they jumped to attention without one word! Their extrasensory radar channeled to alarm.

Mama and Dad left early for work, leaving me no obligatory courtesies. No sweet daughter-face needed. I firmly planted my mad-as-hell-won't-take-no-crap-from-anyone face on, before heading out.

An hour into the drive, I realized how childish I was acting and started looking for a place to eat. The girls had to be starving, and I needed a coffee fix. Within minutes, I spotted *Hazel's Country Vittles*. I pulled off Highway 17 and announced, "Who's hungry?"

Internally, I raised the flag, calling a truce. There was no sense in making the situation completely unbearable for the girls. We were quickly seated; and while I helped Chelsea with her selection, the waitress put a basket of buttered biscuits on the table. I told her we'd need a couple of minutes, so she winked at the girls and walked away.

Hearing Krista and Katie grumbling, I looked over to see what was up. Krista laid her biscuit down, one bite taken out of it. Katie did the same. The flag went down. Very slowly, I reached over and picked up one off the table. Katie gasped. It was so hard you could have broken a tooth on it. My temper rolled out of its protective covering, plastering itself squarely on my face. Krista whispered, "Uh, oh," when I motioned for the waitress.

"Yes, ma'am? Y'all ready to order?" she said, in her sweet Southern accent. Her innocent blue eyes and perky little blonde ponytail did not change what I had to do.

"Honey, hand me your hand," I said. When she did, I turned it palm side up, smashing the biscuit into it. "I wouldn't feed this to my damn dog!" She chilled under my death glower, stepping back as if I had a loaded magnum in my hand. But I was not finished, "Get me some coffee, milk for the girls; scrambled eggs, bacon, and toast to go around." She hurried away, her face red, her eyes brimming with tears.

I didn't care. All my good intentions to take off my mad-as-hell-don't-take-no-crap-from-anyone face ended. Now, it was glued on.

Later, I reflected on the incident and had to give it to that young waitress. She came back every five minutes to refill our glasses and ask if there was anything we needed. She was courageous, if nothing else. I hope she doesn't remember me. No, I hope the incident changed her life forever; and she'll become a judge or lawyer one day. I only hope I don't end up in a courtroom with her, ever. Women never forget an injustice, especially when it's been fired on them, unaware and calculated.

Chapter 3

Krista

After the restaurant incident, we quietly continued our trip. It was ridiculously hot, and the car was the last place we wanted to be. When we pulled into Wilmington, all the dread and fear gave way to pure curiosity. Katie and I pressed our faces against the window, trying to catch the first glimpse of our new neighborhood; the warmth of the incoming sun's rays starting to defrost all of us.

"Mama, what does our new house look like?" Katie asked. There was even a twinge of excitement in her voice.

"Well, you'll see soon. But, I can tell you this, each of you will have your own room," Mom responded.

This would be a first, and I could tell Katie was thrilled. I had gained independence in the last move, even though my room was always used as the guest room for some relative I didn't even want to see. When they invaded, I would have to bunk in with the other girls. Now, with each of us having our own room, why would Mom kick me out of mine if she could kick out Katie? I had earned a step up in rank. *And clearly fairness was the mantra of the world.*

Chelsea had no stake in Mom's revelation and was playing in the backseat with her favorite toy, a tattered stuffed worm that glowed in the dark. Thanks to her blond ringlets and chubby face, Chelsea embodied a cherub. Most who saw her were drawn to her. I was thankful Mom remembered to pull Chelsea's GloWorm out of a box at the last second. Angelic or not, Chelsea would have quickly become tuned in to her devilish side without her dear worm.

We pulled into our subdivision and onto a street rowed with houses. Although I saw many different colors of siding and brick, each house we passed was a traditional two-story with a two-car garage and nice landscaping. No surprises. The whole backseat perked up as we passed each structure. I wanted to control the smile creeping onto my face but couldn't. I was still angry, but curiosity was winning. With each house we passed, I grew more nervous, excited, and disappointed.

Finally, we took a right and pulled into a driveway. I noticed two things right away, Dad's car and a basketball hoop. *Dad definitely owes me a couple of games of HORSE. Actually, at this point, he should just let me win.* The dark red brick and yellowish cream siding distinguished our house from the rest on the block. Also, our house was different because we accessed our garage from the back. All the other houses had garage doors facing the street.

We piled out of the car, so excited to start exploring. New house, new subdivision, and new city. The concern of finding new friends and fitting in had yet to consume our thoughts. It was time for a new beginning. The recent ending we had suffered was miles behind us. The moving truck, as Mom told us in response to our many questions, was due the next day; so we would be staying in a hotel one last night.

Carefully studying every nuance as I marched up the garage stairs, I was taken aback by the overwhelming scent of salty air. We were ten miles from the nearest beach, yet the smell permeated everything. I could feel my skin and clothes become heavier and my hair coarser.

The previous owners' choices in decorating struck us when we entered the house: burnt orange carpeting, wood paneling in the living room, and several almost-humorous paint choices in the bedrooms. Mom would only tolerate this for a little while before she'd go absolutely crazy. I knew from past experience we would come home from school each day to a newly decorated room. Within weeks, it would have her stamp all over it, and not until then would it feel like it was ours. Right now, we were just renting someone else's place.

Katie and I separated from the others and headed upstairs to see our new rooms. Being the oldest had definite rewards. I was gifted the largest room after Mom and Dad. Naturally, the largest, with a private bathroom, would go to them. But this trip, Katie was in all her glory because she got to pick from the two remaining bedrooms. She didn't take long to decide. As soon as we hit the top of the stairs, she took an immediate left, plopped down in the middle of the room Indian-style, and announced her territory. I was waiting for her to thrust our family flag into the carpet and stake her claim.

"This is my room!" she proudly yelled.

I realized what an important moment this was for her and fought the urge to continue to my own room. Instead, we explored hers for a while. We soon moved on to my room just outside my parents'. Almost a teenager, I really didn't want this kind of proximity, but wasn't willing to give up my much-deserved space just to get away from them. After about an hour of exploring, Mom and Dad suggested we head into town for some lunch and sightseeing.

Chelsea was too new to know what was coming, but Katie and I had been looking forward to this moment ever since we found out about the move. Every move, without fail, my parents tried to make it an adventure. We would ride around and wander through the town, often stopping for new clothes or other bribery items. They wanted us to stop being angry with them. In Knoxville, since beautiful lakes and campgrounds surrounded us, they had bought a boat and camper. We had some great times with those, and hated seeing the inevitable For Sale signs posted on them when we found ourselves uprooted again. It was like testing out a toy in a department store and loving it but not being allowed to take it home.

Heading down Market Street toward Cape Fear River, we stopped at a traffic light. I looked out the window to see three black guys hanging out at the corner. The crazy music blaring out of their boom box and their clothes intrigued me. In Tennessee, we might have had three black people in our whole school. Plus, I had never seen such clothes. Their pants were hanging down so low I could see their underwear, each wore a thick gold chain (one had a huge gold dollar sign hanging from it), and one had his hat on sideways. I found myself staring.

Noticing me, they started to make obscene gestures and yelling, "What ya' looking at, little girl? Like what ya' see? Come on and get you some." Then, one of them grabbed his crotch in what I could only describe as a Michael Jackson move.

"Don't look them in the eyes, Krista!" Dad yelled, sounding panicked. "Never look them in the eyes. Do you hear me? You just stay away from them at your new school."

"But, Dad."

"Dammit, you heard what I said!"

Startled and confused, I looked at Mom and tried to keep my eyes focused inside the car. She gave me a comforting glance and gently laid her hand on Dad's shoulder. He tensed in response. I felt clueless and disturbed by the whole scene. Now, on top of being worried about a new school, I was terrified by the kinds of people I would encounter.

We finally arrived in downtown Wilmington. Across the water was a huge battleship. We walked around the shops and Mom bought each of us a new outfit and a new bathing suit. Fitting, as they promised to take us to the beach soon. Exhausted from the bribery and family bonding, we had a nice seafood dinner and then headed to the hotel. *Everything was starting to look up*, I thought and rested my head on the foreign pillow for what would be the last time, for at least awhile.

The next day, we went to our house first thing to join Dad. Dad had headed over earlier to greet the moving truck while Mom sat with us in the hotel for their free breakfast. Couldn't miss that! As soon as we got close to the house, the bright orange moving truck glowed from our driveway. Several guys in polyester carted boxes and furniture in and out. I rushed up to my bedroom. I saw all the boxes marked

Krista's Room. Excited to see the contents again, I ripped opened the first box and tore away all the protective tissue paper, making sure everything survived. I always feared my stuff wouldn't make it; or, thanks to some major screw up, I would find myself in foreign territory with none of the familiarity I had collected over the years. Or worse, I might end up with some stupid boy's stuff while he'd be fingering my treasures in Texas or some other random place. I felt a huge sense of relief with each item I unwrapped. Finally, I found my glass unicorn collection. To my dismay, everything was *not* as it should be.

"Mom," I yelled after my discovery, "Hurry!"

I could hear Mom tearing up the stairs, and I saw the look of panic on her face as she shot into my room.

"Look," I said through a cloud of tears. I held out the pieces of two prized unicorns. Well, technically, I now had one piece for a horse collection (since one unicorn's horn had been amputated) and another for a three-legged unicorn collection.

Mom gently took the pieces and stroked my hair. Neither of us spoke. We both knew each move had its casualties. After a brief period of mourning, Mom took my hand and led me downstairs. Our dog, Sugar, a mix of so many breeds that you had to love her originality if nothing else, bounded behind us and sniffed each corner. For the rest of the afternoon, I helped Mom unload boxes in the kitchen. Our bond strengthened with each new broken glass or chipped plate we found. We didn't say much, but just being close to her was lifting my spirits. Chelsea ran in and out, showing us all the toys that had made it. The sweet thing thought it was Christmas. Katie was upstairs relishing having her own space. And Dad was hiding somewhere, afraid one of us would pounce on him at any moment. There was a sense of peace; the important pieces of our lives had made it, maybe a little worn and slightly broken, but overall intact.

The week flew. I woke up each morning to find another room transformed. It was like a bunch of fairies had put the house together while we slept. With each change, the structure became more like home. The place was feeling more and more like it deserved hanging *The Simmons* wooden plaque outside.

Sunday morning rolled around, and it was time to begin the battle of the churches. At every new location, we would sample the various churches before deciding where we best fit or where *we* felt most comfortable. Mom and Dad decided, that is, us kids never had a choice. Usually, several people from Dad's work would invite us to their churches. This Sunday morning was no exception. After settling on a dress and combing my hair, I headed downstairs feeling confident. I looked forward to Sunday mornings when I could read the comics and settle into the safety of our cocoon. Skipping around the corner, I noticed breakfast was ready. Dad heard me approach, lowered his paper, peered over the edge, and furrowed his brow. I immediately felt self-conscious and tugged at my skirt.

I slipped into my chair, wanting to make myself small enough to hide. But I knew better. Dad never let anything go.

"Krista, are you ready for church?" he asked.

"Come on, Bob," Mom said, trying to halt his assault. "Girls, hurry up now, we need to get going."

"Abby, I swear to God she is not going to church looking like that."

Dagger to my heart. I stood up and pushed my chair back as hard as I could. Then, with much dramatic flair, I stomped upstairs and slammed the door to the bathroom. In the background, I could hear Mom pleading, "Bob, can't you just leave her alone?"

I stared at myself in the bathroom mirror. All I could see was the little girl who had changed into the forever difficult and messy teenager Dad just admonished. I said out loud to this image, "If he thinks I don't look good now, he's got another thing coming!" With that, I found Mom's teasing comb and methodically began to create something I knew would make him go ballistic. After putting the finishing touches on my creation, I ventured downstairs.

Dad stared at my work of art and then startled me with the boom of his voice, "What the . . . Abby get in here."

Mom glanced at me and looked straight at Dad. "Bob, we don't have time for this." Then, she leaned down to me and said, "Why do you have to make this more difficult?"

Strike two, another dagger to my heart. I wanted wanted to punish Dad for making me feel ugly but never intended to catch Mom in the crossfire. I was such an idiot sometimes. Whether I tried or not, I always seemed to find the perfect way to ruin everyone's day. As the family headed to the car, I wanted an escape. A sudden bout of the flu, anything to head in another direction. I longed for a place where I could just be me, not some compiled version of what others thought I should or needed to be. Unfortunately, I didn't know how to find that person. Two parts of me were fighting for control. One wanted love and acceptance and was willing to do *anything* to get it, and the other was hell-bent to find herself regardless of the consequences. It was a constant struggle between acceptance and discovery.

Deep in my own thoughts, I sauntered behind the others to the car. Katie and Chelsea were chattering away while Mom and Dad were discussing some grown-up crap. Isolated and alone, I felt like this happy family had picked up a homeless kid from the side of the road. Me. There was a distance. Hard to describe and horrible to feel.

Abby

As we neared Wilmington, there was a beautiful stretch of scenery that reminded me of central Georgia and the many trips we took there to visit relatives. Fruit and vegetable stands speckled the roadside, advertising their succulents. Motorist slowed down to read the daily specials. Seagulls flew low, squawking, searching for morsels left behind by tourists. Dilapidated, abandoned Texaco stations and small Ma and Pa shops lay crumbled and rotting, barely echoing their once vibrant existence, their hallmark as community gathering centers. What was and now became a shredded tapestry, bound on each corner with an invitation to resurrect its former beauty.

Further along, thickets of tall strapping pines pointed upward spearing blue cloudless skies. In between, open meadows overflowed with colorful wild flowers. Their stalky limbs swayed in the gentle breeze. Slowing at a four way stop, I glanced down a narrow clay drive. Live Oaks danced intermittently down the stretch that led to an abandoned plantation home. Each giant dripped with Spanish moss. Its branches spread outward, still offering gifts of shade and beauty.

My eyes traveled from one side of the road to the other, wanting to absorb every sound, scent and scene, to imprint this new imagery in my mind in order to begin the healing process. People who never moved might not understand this type of loss. With each upheaval, we experienced most of the traditional stages in death—denial, isolation, anger, bargaining, depression, and finally, acceptance. My snag was getting past the depression. The loss felt like a purgatory with no finality, no heaven with its golden gates or hell with its burning fire. It was a place of no resolution and little comfort.

That's how I felt traveling toward Wilmington. I knew my thanksgivings—a successful, loving husband, three beautiful daughters, and good health. I desperately wanted to look forward to a new home, a new adventure, and meeting new people, but I was stuck in the pit. This time, I had dug myself in deep. I couldn't find the familiar footings carved into its sidewall or see light peeking over its rotund opening.

Katie's voice pierced my introspection, drawing my attention to her, "Mama, what does our new house look like?"

"Well, it's been two months since Dad and I were here. It's hard for me to remember what it looked like. I remember there's a swing set outside with an attached clubhouse. That'll be nice, won't it?"

"Uh-huh," Katie sighed.

I glanced in the rearview mirror at Krista. She was being quieter than normal. I was thankful. Krista's persistent routine of jarring questions and accusations would not have helped. Thankfully, Chelsea was content playing with her GloWorm. If we had forgotten it, we would all be in the pit, clamoring for peace. The pit was not big enough. Right now, the burgeoning weight of my self-pity wouldn't allow anyone else inside.

I looked up just in time to see the sign for College Road. We had finally made it to our destination, and our new home. Despite myself, I felt a tinge of excitement. "Girls, we're almost there!"

"What's that?" Chelsea thrust her finger by my right ear, pointing to the left.

"Oh that's the college. We'll take a ride through there this weekend. We'll add it to our family day of exploring. Wouldn't it be nice to think one of you might go to school there one day?"

My long-term enthusiasm poured out unaware, but Krista quickly stamped it out, "Sure, Mom, that'll be me. I'll breeze through junior high and high school in the next year and apply there. That way I'll be certain this is my last move!" Defiant, she crossed her arms over her chest and huffed in disgust.

Ignoring her, I pointed ahead, "If you continue down this road, you'll eventually end up in Carolina Beach. Dad and I drove down there on our visit, and it was nice. We found a donut shop on the boardwalk to die for! That's a must when we get out this weekend!"

The traffic subsided, and I made a quick turn to the left, "Here we go; we're entering our new subdivision. What do you think?" Everyone peered out the window in anticipation.

We pulled in front of the house, and I spotted Bob's car in the driveway. I couldn't believe he'd actually made it on time. What a relief! I didn't have keys to the house and assumed I'd have to sit in the car and wait for him. I put on my prettiest smile and waved. When I rolled down the window, I was greeted with a firm, moist kiss.

"I see you made it in one piece. I've really missed you girls. Come on, let's look at your new home!"

The car doors sprung open; and before I could say a word, all three girls were out, squealing and running toward the house.

"Well, I guess you got your answer," I laughed.

Bob opened my door and put a hand out, "May I escort you to your new home?"

"I think that would be very nice, Mr. Simmons."

Bob slipped his arm around my waist and pulled me in close, "I've really missed you." There was a faint glimmer to his dark eyes that made my heart flip.

I didn't feel the need to say anything. Bob's words were perfect. I could feel myself being hoisted up out of the hole, little by little. My step lifted lighter, and I was actually looking forward to examining the house. Everything felt right. My energy soared, along with my optimism.

"Have you been inside yet?"

"Yeah, I came by before the closing to do a walk-through. By the way, I did notice a few things were missing."

"Missing?"

"Yeah. The previous owners took the towel bars in the upstairs bathroom, and overhead fans out of two bedrooms."

"They did what?"

"Listen, Abby, these things happen. I was upset, too, but I'll fix it this weekend." Bob should have stopped, but he didn't, "Maureen said they'd come over later and take a look at it with me. I don't know what I would have done without them the last few days."

"I beg your pardon?"

"Abby, I told you how helpful they've been. And, they want to meet the girls. Do you have a problem with this?"

"Yes, as a matter of fact, I do. This is our first night here. I thought we'd do something special, like take our children out to dinner, spend time answering their questions, getting them adjusted. I don't need to deal with Maureen and Tyler."

"It's Turner."

"Whatever." The word rolled out in flat multiple syllables. Then I whirled on my heels and thundered toward the house, hoping every towel bar had been jerked out of the wall. I didn't care; I didn't like the house anyway.

My walk-through was quick and abrupt. I tried to block out the burnt orange carpeting, the hideous wallpaper, the gaping holes in the wall where towel bars should have been, and the wires dangling from the ceilings. I barely noticed the beautiful oak mantle or the winding staircase. On my inspection trek around the outside, I glanced at the sliding glass door in the den. What I saw made me loosen my reserve and stubborn hostility. I realized in the process of being cynical, I almost missed the chatter of happy children who had staked out their territory and were running from room to room, checking out their new surroundings. All they saw was a home.

We went back inside, catching the girls in full throttle. Over giggles, Bob asked, "Who's up for seafood and a drive downtown?"

Wanting desperately to recover, I ran toward the back door, yelling, "Last one in the car is a rotten egg."

To the girls' delight, Bob selected an all-you-can-eat seafood buffet. After too much shrimp, french fries, and ketchup, we rolled out to the car and headed downtown. The drive was beautiful. Cresting over a hill on Market Street, the pink-purple tones of the sky greeted us with open arms. The Cape Fear River glistened in the distance, rippling, merging skyscape and landscape into a picturesque scene. Salt air permeated the car, filling us with anticipation. Bob rolled down the windows; and we pointed, ooing, and ahing at all the new sights.

Bob stopped at a red light. I immediately spotted a sweet little kitchen shop and was deep in planning a shopping trip when I heard, "What ya lookin' at white girl?" At first, it didn't register what was going on until I heard Bob yell at Krista.

The light changed, and Bob accelerated. I looked back and saw a young black boy standing on the street corner. My glance went directly to Bob. His lips were pursed, his forehead furrowed. Instinctively, I put my hand on his shoulder to waylay any further comment. Then I swung around to Krista, giving her a smile and shrug, hoping to disarm any alarm her father's response may have caused. I had forgotten Wilmington was more ethnic than the midwestern, predominately white towns they were used to living in. I made a note to myself to have a discussion before school started. I certainly didn't want them to be fearful of other people from lack of understanding.

The next few days went by in a blur. The moving truck arrived with bad news. It had wrecked traveling over the mountains, and the furniture had incurred damage.

"Damage?" My face pinched into a deep scowl. The term bad news was taking on a whole new meaning.

A midget man in a wrinkled blue shirt spit out a wad of tobacco and explained, "Yeah, we had to get a truck to go out to the crash site, unload your stuff, and put it on this here rig. Hey, lady, we were real careful." Every syllable threaded around his tongue and was released in total ambivalence.

That's it. I went into immediate shutdown. The only thing getting me through the next few hours was experience. When you see a box, unload it and put it in the nearest cabinet, closet, or shelf. Ira Levin would have been proud of me, the robotic, complying female. But I saw myself, at the moment, as Porky the Pig jumping out of a snare drum shouting, "That's all folks!" The only difference would be the voice—mine, deadpan and joyless.

By Sunday, we were half in, half out of the house but eager to try a new church. Bob said we had been invited to one nearby. I asked him what denomination, and he said he couldn't remember, but the gal at work told him she was a Christian. I learned not to comment any further. Bob felt obliged to attend a couple of churches his co-workers invited him to. Who was I to complain? I didn't know anyone to ask about churches.

"Girls! Breakfast! We don't want to be late for church."

Katie and Chelsea made their grand entrance. Katie had put on her yellow gingham dress Mama had given her and placed a pretty butterfly barrette in her hair. At my pleading, she had helped Chelsea get dressed while I tried locating my clothes in boxes. Chelsea looked precious. Her red jumper and white blouse made her blue eyes pop.

The girls slipped into their places at the table. I poured cereal in bowls and put hot cinnamon toast on a paper towel in front of them. Bob almost stabbed me with his empty coffee cup when I passed him. I grabbed it out of his hand and headed toward the coffeepot. It was nice one of us was able to sit and enjoy coffee.

From behind me, I half-heard Bob's snide comment asking Krista if she was ready. I twisted around to see if she wanted Alpha-bits or Capt'n Crunch but stopped in midsentence to stare. She had put on her new lavender and cream ensemble we purchased before we left Tennessee. It looked darling on her, but what I saw in front of me did not look darling. Krista had parted her hair down the middle, slicked it flat, and shoved it behind her ears. It hung in limp, greasy strings. One stray piece at the crown stood up like Alfalfa's. Her navy blue glasses were cocked sideways on her face, one side pushed up high enough to fan her hair comically up and away from her eyebrow. The other side so low her eye was peeping over the frame. Krista's lips were puckered, and she had both hands on her hip poised for war.

I didn't know what to say; and by the time words surfaced, I had blown it. Krista marched out of the room. I didn't want our day to start this way. I wanted to run after her, but I was stunned. Krista was such a pretty girl; but for some crazy reason, she could not figure out how to put it together—clothes, shoes, hair, none of it. She had no interest in how she looked. I dare say the only thing she saw in the mirror was the new zit on her face, her latest unhappiness.

"Bob, can't you just leave her alone?"

Bob never lowered the newspaper to respond. He did manage, however, to point to the oven. Smoke seeped in a steady stream between the door and frame. The last piece, my piece, of cinnamon toast lay on the pan charred beyond recognition. No coffee. No breakfast.

Fitfully scrubbing the blackened remains of toast from the pan, I heard Bob burst out, "What the . . . Abby, get over here, now!" I jumped afoot, thinking fire had spread beyond the oven. Whirling around, I stood in utter amazement. Krista had returned.

She stood in front of the table more defiant than I had ever seen her. Her hair teased and tossed in a multitude of directions. To demonstrate her anger, she had smeared on a bright red lipstick.

I didn't bother looking at Bob. I didn't need to. I picked up my pocketbook and Bible and walked right out the back door. I didn't even stop when I realized I had forgotten to put on my panty hose or take off my bedroom slippers. Let Bob dare say something to me about my looks. I was past target practice. I was poised, ready to aim and fire.

Chapter 4

Krista

The gray clouds began to clear, and curiosity won over melancholy. We turned left out of the subdivision and traveled down College Road for a few minutes before pulling into a small white church. We filed into the quiet and pristine sanctuary with the other families, picking a pew toward the back. We didn't want to stand out as the new people. With school starting soon, there would be plenty of time to be out of place. Before we could get settled, several members of the church swarmed us and greeted Dad. When he introduced us, I knew he considered skipping me, but he didn't. It was amazing how quickly he fit into a crowd. Watching the number of people who came up to him, another newcomer would have thought he was a deacon of long-standing. They seemed drawn to him like he was a celebrity.

In the midst of Dad's social coming out, the minister stepped forward and motioned for the piano player to begin the first hymn. A silence settled, and everyone dutifully returned to his or her place. The hymn began. People suddenly filed down the aisle with various musical instruments. It had the same transitional effect of moving from yoga to kickboxing without missing a beat. The preacher produced a shiny set of cymbals out of nowhere and began clanging them to the sound of the music. I was startled, and my parents exchanged a look of confusion.

As the singing reached fever pitch, a lady a few rows up fell to the floor and began throwing her body around and speaking in some foreign language. Then another person did this and another until I was scared. I turned to see Chelsea clinging to Dad while Katie, already in love with the music, slid out into the center aisle and danced with every fiber of her being. A few of the people close to us turned to watch her and smiled. Clearly they thought the spirit had moved her. Katie always knew how to work a crowd; she was like Dad that way. I was surprised he didn't jump out there and join her.

"Mama," I said in the faintest of whispers. I had to tug on her sleeve several times to break the spell she was under.

"Krista, what is it?"

"I don't like this. These people are weird."

"Please just sit and behave."

Then she turned to Dad and arched her brow. I couldn't hear what she said to him, but he got a familiar look. The one you see on a kid when asked why they did something. He tossed any blame, and he didn't care who it landed on as long as it cleared him.

An hour and a half into the festivities, I felt a light touch on my shoulder during a prayer. I ignored the sensation for a second, afraid Satan himself would grab me if I dared open my eyes. Or worse, someone would ask me to share with the group. Having sweated out several prayers since we'd arrived heightened the fear. Finally, I peeked through an eye to see Mom motioning for me from the aisle. I was relieved, realizing it was time for our getaway. Dad had Katie in a vise grip as she leaned toward the pew with arms outstretched. She did not want to leave. Thank God they did not begin another song right then, or she would have been overcome with the strength to stay. We crept toward the back door and ran for the car. I looked back, still terrified those people would notice our absence and form a search party.

"Bob, I think we should get the girls some lunch," Mom said as Katie begged for more dancing. "Katie, settle down!"

"But, Mama, I want to dance. That was just the best! Can we go again next Sunday?"

I didn't listen for Mom's response. I knew it would be a cold day in hell before we went back there.

Of course, this kind of fun was just the beginning. The next week was my birthday, and I knew we'd celebrate as a family. Since I didn't have friends here, there wasn't a party. Katie's birthday was only eight days after mine, so we celebrated together. Even with all the boxes surrounding us, Mom made sure we both felt cherished and celebratory. She planned a nice afternoon at the beach with a picnic lunch, and Dad joined the celebration for dinner and presents. It was a nice day of escape.

On the Monday morning following my birthday, we planned to visit my school. Even though Monday came the same time every week, its wrath still hit me unexpectedly. The weekend was usually such a euphoric high; and then, boom, it was Monday, and the real week began. There was no way to avoid it. Occasionally, I would attempt a faked sickness to avoid Mondays at school, only to discover that Tuesday then became Monday—what a vicious cycle.

I peeped through my eyelids and let the light shine in and warm my body. I could hear the muffled sound of Sugar snoring. Just as I was sitting up and swinging my legs over the edge of the bed, I noticed a faint tapping on my door.

"Yes," I said. Mom poked her head around my door, "Honey, you need to be ready in twenty minutes. You know we're going to sign up at school today."

"Oh, Mom, can't you just go to Chelsea's school today, or Katie's?" I pleaded.

Mom put a quick end to my pleading by walking toward my dresser. I jumped off the bed and laid a hand on hers. "Okay, okay," I said. "I'll be downstairs soon." Thankfully, that stopped her from further investigation.

"I want you to wear your aqua blouse and khaki pants," she said with a quizzical look. "And, please wash your hair *before* you put it in a ponytail."

After the door shut behind her, I pulled on the drawer. The clothes were stuffed in like trash bags on garbage pickup day. I knew if Mom had seen it, I'd be in for two wonderful experiences in one day—seeing my new school and a crazed Mom who would remind me she was not my personal servant. Thank God the outfit she suggested had not yet been put away and was wrinkle free. It was still piled in the corner along with all the other clean clothes Mom brought to my room a few days ago. I figured as long as they were in a neat pile they were easier to spot.

A quick shower and ponytail clip later, we hopped into the car and headed to Williston Junior High. I attempted to plaster a smile on my face but gave up and just stared out the window. Chelsea and Katie were along for the ride as always. *Couldn't Mom occasionally find someone to keep them so we could do something alone?* I didn't need any witnesses to the ensuing fiasco.

The longer we were in the car, the more I noticed my surroundings. The houses seemed to get smaller and closer together as we drove. To further escalate my feelings of claustrophobia, the street also gradually narrowed. When we stopped at a traffic light, I could have reached out and touched the car next to us. I even closed my eyes a couple of times as Mom seemed to barely miss a mailbox or two.

Twenty minutes later, we pulled up to a brick building with a huge courtyard. We ambled up the stone walk and through two heavy doors. The school was older than my previous ones. There was nothing interesting about the building. It was a big square with no defining features except the darkness provided by burgundy brick and dark green paint. Every window had a machine of some sort hanging out of it. I made a mental note to ask Mom about those later.

Nearing the entrance, I stopped to read some writing on a couple of cement benches outside. It looked strange, but I was intrigued.

The bright paint colors spelling "Brian wuz here" and "Tracy loves Brendan" was the only decoration I had noticed at the school so far. There were no flowers or pretty pots. Something about this place was very uninviting.

When we entered the school, a glassed-in room was immediately to our left. One lady sat at the desk. Right outside the room stood a couple of benches. Mom sat Chelsea and Katie down and opened their knapsacks.

"Girls, sit here while I get Krista enrolled. We'll be right back. Don't you dare move," she said and ushered me through the doors.

"Excuse me, I need to register my daughter," Mom announced.

The lady looked over and pushed a stack of paperwork at her without uttering one word. While Mom sat down to fill out my life history, I looked around and noticed the bulletin board. Several photos were up depicting scenes from the previous school year's activities. I had never seen so many shades of people wearing such little clothing. *I guess there isn't a dress code*, I thought. In my last school, your shorts or skirt had to touch the ground when you were in a kneeling position. I looked at my pasty-white legs and shuddered.

"The principal will see you both now," the lone lady said, motioning us toward the office.

"Hello, I'm Mrs. Laughton. Welcome to our school."

"Hi," Mom said and reached to shake her hand, "I'm Abby Simmons, and this is my daughter, Krista. She *is* ready to start the eighth grade." She shuffled through her purse, withdrew some papers, and handed them to the principal, "I have her seventh grade report card and immunization records here."

"Why don't you two have a seat and let me look these over," she said as she took the information from Mom. After a quick glance, she started rolling off the classes she thought would fit. It was amazing how she knew everything I needed from a few sheets of paper.

"I see you're an A student Krista. We usually do some testing, but I think you are ready to start algebra. However, we are short on some electives for you. A lot of the classes have filled up already, so let me see. Umm, we do have a choir team, the Troubadours, but you have to try out for it. What do you think about that?" Mrs. Laughton looked straight at me for a response.

Wow, I was going to get to speak for myself. Novel idea. "I can handle anything you throw at me. Algebra is no problem, and I want your advanced English class. No problem with the singing either. I sing in church."

"Are you sure you can handle all of this? What do you think, Mrs. Simmons?" she asked, switching her focus.

"I know what I can . . ." I interjected.

Mom shot me a glance. I crossed my arms firmly against my chest. I knew I could handle anything this school or any other threw my way; I didn't need these two adults deciding my life for me. Plus, I hated Mom's duplicity. At home, she encouraged my opinions and thoughts. But God forbid I stand strong in front of other adults. You'd think she'd be proud I had such faith in my intelligence. Instead, she was just embarrassed that Mrs. Laughton saw my independence and strength.

"I think Krista will be fine. Is there anything else you need?" Mom asked.

"No," my new principal said and then turned to me, "Krista, we'll look forward to seeing you in a couple of weeks."

"Umm, Principal, I mean Mrs. Laughton, can I ask a question?" I asked hesitantly, knowing my father's response would have been, "You just did."

"Sure," she said with a smile.

"Do you have a dress code?" It had been weighing on my mind since I saw the pictures. At my previous school, you had to wear all shorts and skirt at knew length. Often the principal would make boys and girls get down on their knees. If your skirt or shorts touched the ground, you passed. Otherwise, home you went.

"We discourage profanity on T-shirts or wearing too-tight clothing. Now, if you have any other questions before school starts, please feel free to call."

After some quick pleasantries, we went out to rescue Chelsea and Katie and headed to the car. I was in a fog. Just across from the school were some of the most run-down houses I had ever seen. And I knew we had passed two other schools on the way to this one.

"Mama? Why is my school so far from home?" I asked.

"Krista, they have busing in North Carolina," Mom said after a bit of hesitation.

"What does that mean? Katie's school is much closer," I said. I could feel her eyes rolling even though I couldn't see them.

"Well, it's sort of complicated."

"Complicated because you don't want to tell me or complicated because you don't think I can understand?" I spit out in anger. I had no patience for this stuff. *Give it to me straight,* I wanted to yell. I had a gut feeling it had something to do with the black people Dad warned me about. Plus, I had seen nothing but black people on the streets around the school. And as we were pulling away, I noticed a Quickie Mart with bars on the windows. It was all strange and unsettling, and I just wanted some honesty. "You know, Mom, I'm almost thirteen. I am not a little girl anymore."

"No, you're an *ugly* girl," Katie yelled from the backseat.

"Oh, yeah, well you are stupid and fat! I'd rather be ugly than a fat moron," I snapped back.

"All right, girls! Enough of that! We are going to Katie's school now, and I don't want to hear another word until we get there," Mom shouted.

"She started it!" I yelled.

"Yeah, well, I am finishing it," Mom snapped.

Abby

Mama said there'll be days like this,
there'll be days like this Mama said.

The musical group, The Shirelles, weren't kidding, but Mama should have been more emphatic about how bad some days could be.

No, I didn't go to church without panty hose or in my bedroom slippers. Once we got into the car, I feigned forgetting something and ran back inside to finish dressing. Bob was too busy to notice. His focus was holding the car door, making sure the girls didn't hit his new toy, an older Nissan 300ZX. God forbid if one of us smudged his *baby*.

Going to a new church was not high on my priority list. There were only so many churches anyone should attend in one lifetime. From a child, I dreamed of going to a little white church situated in a beautiful meadow, similar to the ones depicted in Norman Rockwell's paintings. I imagined walking through its quaint double doors and straight to the family pew where loved ones gathered. People would file by and wave, reach over and touch your shoulder, or present a hug, ask about your children or your aging grandmother who was unable to attend anymore.

When we pulled up in front of the church, it was a rectangular metal building with a flattop roof. Concrete lined a football length of yardage on all four sides. Not quaint. No meadow.

We arrived early, sitting toward the back of the sanctuary. I insisted on this location, a vantage point to sit and watch. Groups huddled, laughing, greeting, and hug-wrapping. The familiarity was sincere, extended to people you share joy and grief with. I smiled and then withdrew when the tenderness, the nearness, overwhelmed me. Casting my eyes downward until my lashes fluttered against my cheek,

I pressed my back further and further into the pew, hoping to blend away.

I would remain in this posture until the opening music began or someone inquired, "I don't believe we've met." It was getting harder and harder to paste on a genuine smile.

"Bob? Bob Simmons?"

Out of the corner of my eye, I saw Bob put his hand on the back of the pew and twist around, "Matt Gusterson, how're you doing? I haven't seen you in years!" Bob extended his hand to the man behind us.

I looked away after tossing a quick smile. Damn, he knew someone everywhere we went. I tuned them out. He never introduced me anyway. Most people knew me as Bob's wife, the girls' mother, someone's sister, or someone's daughter. Often I would interject my name to hear it, to remind myself I had a separate identity from my family.

The music started, and I raised my eyes toward the pulpit. On either side of the podium, screens were being lowered. I could hear the choir entering the sanctuary from the rear of the church. Words appeared on the screens in front, and the congregation stood to sing. We followed suit. It was a dear old hymn, "Standing on the Promises." My spirits lifted; and I began to sing, blending my voice with those around me, sensing a warm connection beginning to form.

Katie stood close, swaying to the music. Krista was still turning pages of the hymnbook looking for the song, and Chelsea was in Bob's arms watching the incoming procession. I began to sing with intense jubilation. I could feel my body matching tempo with Katie's. Once the song was over and the choir seated, a tall broomstick-thin, dark haired man rose from behind the podium and walked forward. He grasped both sides of the stand and stared deep into the eyes of the congregation, moving his head in a sweeping fashion not to miss a single soul.

"Are you filled with God's spirit? I say, children of God, are you filled with His spirit?" His delivery was slow and methodical, getting more excited and animated with each word.

I could feel a stir in the congregation, a low hum building into a grand crescendo of "Yes, Lord, we have the spirit!"

Directly in front of us, an elderly black woman stood up, her hands lifted as high as she could reach, "Yes, Jesus, I am filled with your spirit." The first thing I spied was the tail of her skirt folded squarely into her butt crack, exposing a brightly colored slip and thigh-high hose rolled down below her knee.

My head swung in Bob's direction; he had seen it, too. I shook my head when I saw the mischievous grin on his face. I knew the look; it meant trouble. To taunt me, he acted as if he was going to reach over and pull her skirt down. My left eyebrow flew up daring him. He pulled his hand back, muffling a laugh.

People began to rise, raise their hands, and mutter praise expressions—an echo of different words, different pitches, sliding over and under each other, but never blending. Some jumped up, others sank to their knees. I couldn't turn my head fast enough. To my left, a young man and woman walked by, reminding me of catatonic zombies. Their arms outstretched as if on a predestined mission. Following close behind was a young girl who looked to be about eight. She held her head at an odd angle, causing her Adam's apple to protrude. As she passed our pew, I could hear her uttering unfamiliar blubberings, too.

Krista tugged on my sleeve, but I was glued to the theatrics in front of me. *What in the world was going on?*

Turning toward Bob, my defined look of admonishment quickly switched to an evil eye stare down.

Shrugging, he spit out each word on an eight count note, "Like I knew." His eyebrows appeared glued to his hairline, giving an expression of complete surprise.

Right then, our attention was drawn back to the preacher. He had grabbed two tambourines and was moonwalking across the pulpit, crying, "Praise God! Hallelujah! Praise our Almighty God! Bring us power. Bring us Your eternal Almighty power!" He abruptly stopped; faced the congregation; threw off his suit coat; and started doing jumping jacks, swinging the tams over his head. People left their pews, moving out into the aisles to dance and gyrate. My mouth dropped; and without looking at Bob, I grabbed his hand, clutching it tightly. Krista grabbed my other hand, and Chelsea buried her head into Bob's shoulder and started to whimper.

An emergency siren flared! *Oh my God, where was Katie?* Panicking, I scanned the commotion, searching. I should have known. Katie was smack-dab in the middle of the aisle dancing, having the time of her life. The tension sprung, leaving me feeling like an accident victim—hard hit and dazed. I didn't know what Bob had gotten us into, but I was getting my family out of this fiasco.

I waited for the right moment to escape. I nudged Bob, who was still holding Chelsea, out into the aisle. He grabbed Katie in the middle of a pirouette, and I tapped Krista to make an exit. We ran for our lives. We never looked back, fearing Bob's friends would notice and come searching for us.

Safely inside the car, I steadied my breathing and checked on the girls. Each sat stoic and quiet. To restore mental order, I grabbed my purse and took out a pen and paper to begin a to-do list.

First thing on the list was to plan Krista and Katie's birthday party. Sadness snagged my hand midline when I realized there would be no party this year. Shaking it off, I decided a special event with the five of us would be better. Second thing, talk to Bob about party plans. Third thing, forget Bob, do party yourself.

It wasn't long before we forgot Sunday's fiasco. School was fast approaching, which meant buying clothes and supplies. We were almost unpacked and were making an attempt to adjust to our new home and community. Several of the neighbors had walked over with muffins, cookies, casseroles, and offers to help. They acted genuine. I would do my perfunctory smile and gracious thanks. It was predictable. I wanted to say, "Go away! Don't waste your time on us. We'll be gone soon!" But, I didn't. It wasn't their fault our lives stayed torn up.

Walking into the bedroom, I halted, shaking my head at the scene before me. There was Bob stretched out on the bed, a large binder lying across his legs, late night news muted on the television, and his latest Cussler book opened up beside him. His reading glasses were perched on the end of his nose, his head propped against the headboard, and his mouth relaxed, slightly open. I studied him, watching his chest rise and fall. Slipping in beside him, I snuggled close, feeling his body heat, taking in the faint aroma of English Leather. Missing him filled my senses. I craved him, wanting him to turn toward me and take me in his arms.

I lightly stroked his chest with the end of my nails, hoping to caress him awake. Speaking softly, I opened up with a topic dear to his heart, "Bob, how's your new job going?"

"What?" His voice cracked, barely audible.

"Your job, how's it going?"

He yawned and straightened. I slid over, propping up to see him better. Half-tone, he responded, "Same old, same old. It's hard to be the new manager, work with a new team, and try to meet the company's objectives as quickly as they think you should. I feel pulled, knowing what I'm doing is not always what the company expects me to do." He yawned again.

Moving my fingers in gentle strokes down his arms, I hoped the talk would end and venture to more productive avenues. "Will this moving cycle ever stop? Do you think we'll be here longer than two years?" I sincerely wanted to know what he thought, even though I knew he didn't have the answers.

"Abby, I don't know. Your guess is as good as mine. I do what I do best, manage people. The company sends me where they know I will add value. I don't know their time frame. I don't think they do either."

I edged in closer, dropping my shoulder to a provocative curve, "Will there ever be a time when we can just say no to a move?"

"Abby, as long as I feel responsible for this family, three college educations and three weddings, I don't think I will be upsetting the cart. There's too much pressure on me."

I put my arm around his chest and squeezed tight. I hadn't realized the pressure and responsibility he felt for this family.

Pulling up to face him, I whispered, "Bob? It's been a long time since we made love."

I puckered up, closed my eyes, and went in for a kiss. He jerked up, pushing me aside and blared, "Damn, did you hear that, Abby? Damn Democrats! What the hell do they think they're doing? I can tell you this, I don't like the way they're trying to make Reagan out to be the bad guy. Have they forgotten what he's done? Before he took office, there were seven million Americans unemployed. We went through the worst years of inflation in over sixty years . . ."

I think I heard the rest; but when I woke up the next morning, I couldn't remember. I was tired. Fatigue was the norm lately. I couldn't understand why. Poor Bob! At least on top of everything else he was going through, he didn't have to worry about living with a sex goddess. All I'd managed to do was frustrate myself and pass out in the middle of his passionate, political review.

I struggled out of bed and walked into Krista's room to get her up, offering to help her select her clothes, but she rushed me out. I sensed she didn't want me to see the horrific mess spilling from her dresser drawers. Too late, but I went along with the ruse. Today was registration at their new schools, and they were dreading it. No need to start the day by setting up a war zone.

After dragging everyone up and out, we headed to Williston Junior High. I dreaded Krista's response. The school was located downtown and had been the blacks' school before busing. Everything about it was sad as if its purpose had been taken away.

Inside its giant lobby, we stopped at a glassed-in office to register Krista. She had been pretty quiet up until now, but I could feel the tension rising. We had already had several conversations about what she wanted, and she had made me promise not to interfere with class selections. Krista was smart, but the moves had done her a marked disservice. She was cynical. With each school she faced new rules, classes, objectives, and challenges. So far, she had met each, but this was different. She also faced the biggest hurdle in her life, a new breed of kids, teenagers. If you added raging hormones and a more diverse ethnic group to the mix, I feared a stark awakening was in store for her.

After our initial meeting, Krista had the schedule she wanted. The principal impressed me by directing most of the conversation to her. She seemed genuinely interested in what Krista thought and surprised at Krista's responses to her academic needs. I was in awe at how she handled herself with such confidence and commitment. Right before we left, Mrs. Laughton asked if I thought Krista's schedule was going to be too much for her. Before I could even answer, Krista busted in with a smart remark.

Embarrassment slapped me in the face. Adhering to our former agreement, I let Krista talk, trusting her judgment, and now she wouldn't even allow me parental confirmation. She knew I was her ally, fostering an open forum of free expression, encouraging independence and strength. Maybe her intention wasn't to set me up; but what she unknowingly did was feed my insecurities, making me feel like a fool.

How many times would I have to look at disappointed adults who wanted to send me to parenting school to handle my child? Anger hit, and I shot Krista a shut-up glance. Would I ever win with this kid?

In the middle of internally arguing with myself on how to handle Krista's brashness, she asked, "Mom, why is my school so far from the house?"

"North Carolina has busing."

"What is that?"

I stalled. How in the heck do you explain to a preteen the idiocy of busing? How do you begin to describe the hard feelings surrounding busing without feeling a part of its lunacy?

Dammit to hell, where was Bob? Why did I always have to face these challenges alone? I didn't have the answers to these complex questions.

Right then, we passed a bricked building on the corner of Castle and Glen Street. Bars ran the length of the windows and doors. A dozen black men were propped up against its walls, talking, gesturing, or jiving to music blaring from boom boxes held up on their shoulders. A police officer walked back and forth, eyeing each patron. Obscene, suggestive language decorated the storefront in multicolored spray-painted words.

Seeing this added to my anxiety. How could I tell my children the world wasn't like the fairy tales I read them? How could I talk about something I didn't understand myself?

Indira Gandhi once said, "You cannot shake hands with a clenched fist."

Right now, I felt the clenched fist was parenthood. I didn't know how to merge the image of what I wanted for my daughters with what the world offered them. I felt the last vestiges of my own youth shatter.

Chapter 5

Krista

Walking down the driveway, I waved to Mom as she pulled away with Chelsea and Katie. She was escorting them to school their first day. Mom didn't notice me, but Katie turned to wave out the car's back window. She managed a smile, but I knew it wasn't real. I envied her ability to try. I'm sure Mom wished I would not only envy it, but emulate it. At least every once in a while.

I was officially a teenager now and well on my way to being an adult, but I wasn't seeing the perks. Still, in five years I would be going off to college. I grinned when I realized what that meant. For the first time, I, not my father, would be picking a location. The thought both excited and terrified me.

Even though the bus stop was only five houses down from ours, I walked with calculated slowness, kicking hard the first stone daring to cross my path. For a moment I considered diving into a nearby group of trees and taking a few more breaths before the oncoming scrutiny would begin.

Rounding the trees, I noticed three other girls waiting for the bus. A deep and learned sense of fear grabbed me; and for a moment, I lost my breath. The sound of the bus knocked the wind back into me, and I hurried along. Panting for many reasons, I secured my place in the back of the line. The bus door creaked open, and I hesitantly placed my foot on the first step. My head was down, but I still managed to notice the bus driver's eyes. I could have sworn she gave me a look of reassurance as if she knew what I was about to endure. I climbed in with exaggerated steps and turned the corner to find a seat. *God, where's Mom? Why did I have to do this alone?*

The scariest thing was the fulfilling destiny of seat selection. It might seem childish and unreasonable, but it was monumental. I had a lot of experience in these matters. If I sat next to the class nerd, and he or she befriended me right away, then I was stuck. It would be hard to save myself. If I was lucky enough to sit next to Ms. Popular, and managed *not* to embarrass myself, then I might find a way to fit in. I realized there were several empty seats, and panic set in. My thoughts were running amuck. *God, couldn't you make this easy for once? How many times do I have to do something as difficult as finding the perfect bus seat? Please make it obvious where to plop my butt.* I was worried enough about my first day, so I took the easy way out. I took a leap of faith and sat by myself in a completely empty seat. *Hah, God, take that!* I handled it. Let someone pick me for a change.

Two stops later, my new life began to form. I noticed a shy, overweight girl with glasses coming down the aisle. I took fate into my own hands for the second time that day. I pretended to drop something and bent over to pick it up. She moved on. *Hah, again!* Then as I was leaning back into a sitting position, a blond sat down.

"Hi, I'm Cindy," she announced. Although she was a bit overweight, she had a very pleasant face. Her eyes were a lighter blue than mine, and her hair was like wavy caramel. Her skin was a few shades darker than mine, but by no means a golden tan. She had on the newest Guess jeans, ones I had only seen in a catalog, and I could tell she had spent hours on her hair and makeup.

"Hi," I said sizing her up. "My name is Krista."

"Are you new?" she asked. "I don't remember seeing you last year." It never took very long.

"Well, then I'm new or used to be invisible, right?" I said. Humor, I'd found, was an easy way to see people's reactions without putting yourself out there too much.

"So, I guess you *are* new. I should show you around. I mean, we wouldn't want you falling into the wrong hands or anything," she said with great confidence. I liked her already, and she had just read my mind.

I stuck with Cindy most of the day. Luckily, we had most of the same classes and the same homeroom teacher. She introduced me to several other girls and told me the best places to buy clothes. I dare say it was an easy first-day transition. In addition to meeting Cindy, I found out the new busing policy meant there were many unfamiliar faces around school making my newness less noticeable.

I apologized to God for doubting him on the bus earlier. And I considered letting Dad off the hook for forcing me to move here.

Usually, at the end of my first day of school, I walked to the bus with my head hung low and tears threatening. But not today. Cindy and I chatted the whole way home, and I hated for it to end. Even the North Carolina heat, combined with bus leather and forty-five other students, could not dampen my mood.

Stepping off the bus, I waved good-bye to Cindy and skipped home. Mom's car was in the driveway. I raced in for the smile, hug, and Little Miss Debbie cake I knew would be waiting for me.

"Mama, I'm home!" I announced.

"Chelsea's sick, so Mom's upstairs with her," Katie said, sounding sheepish.

What, no homecoming parade? No concerned Mom who had been waiting all day to make sure I had survived. I considered asking Katie how her day had gone; but feeling cheated from such a treat myself, I decided to find Mom.

"Mom, guess what?' I said. When there was no answer, I started looking everywhere, "Where are you?"

"Krista, I'll be with you in a minute," she said sternly, poking her head out from Chelsea's room.

Okay, yeah, whatever, I thought and then walked into my room and slammed the door.

Knowing Chelsea's emergency would have Mom occupied, I pulled open the closet doors and pushed the clothes aside. On the initial inspection of my new room, I noticed my closet had a shelf about waist-high to the side. Instead of putting clothes there, I made it a secret desk. Then to really set the mood, I taped (an infraction in our house) posters of Madonna and Duran Duran on the inside wall. I stashed my diary in a canister on top. This was my secret and personal retreat. In a moment of true brilliance, I managed to pilfer a lawn chair to keep in there, knowing Mom couldn't remember every single piece of furniture she was supposed to have. After this many moves, I'm surprised she even remembered her name.

I pulled out my "great mood" pen. It was the kind you could push down a plastic piece and get red, blue, black, or green ink. I chose red. I opened my diary and made the inaugural entry for this location.

Dear Diary,

I had an unbelievably great first day. I met this girl named Cindy and her friend Kerry. And both are actually popular. Finally, after all these years, I think I'm going to be important.

Of course, Mom doesn't know this because she's so busy with Chelsea. Note: I'm happy.

Mom tapped on my bedroom door as I was shutting the gateway to my secret hiding place.

"Come in," I said. I kept my voice steady because I didn't want to give Mom the satisfaction of my happiness after her earlier dismissal.

"Krista, I'm sorry. I'm having some issues with Chelsea."

"No big deal," I was determined not to give into her easily. Nope, now she would have to beg me for information.

"Well, tell me about your day," she said and sat at the foot of my bed.

I noticed the exhaustion in her face, and I cracked, "Mom, I had the best day. I met this girl named Cindy, and her friend Kerry, and we ate lunch together and rode on the bus together, we even . . ."

"So, it went well then," she said. She seemed more relieved than happy.

"Yeah, just fine," I replied. She had taken the wind out of my sails. "So, what's for dinner?"

"I'm not sure. Do you have any homework?" When I nodded my head *no*, she continued. "Well, come on down and help me. Your father will be home soon."

I was never sure how seven o'clock was soon, or how anyone was supposed to wait that late to eat dinner, but I followed her. Katie and Chelsea were playing in the living room, which connected to the eating area. Mom and I managed to put together a pasta dish. Since I was not interested in helping much more in the kitchen, I maneuvered into the den to play with Chelsea, and Katie went to help Mom. We were all startled when we heard the garage door open.

"Girls, why don't you get the table set? I need to be with Chelsea for a bit," Mom said. There was always panic when she knew Dad was home—like she was in the military and about to get her mattress inspected.

Except for Mom, we were all seated when he came in. As usual, he disappeared for fifteen minutes. This was the dinner ritual. Here we were starving because we were waiting for him, and he pulled a Houdini. Since his plate was always the first served, we sat staring at each other. When he finally graced us with his presence, he said his tired old prayer and then eating was allowed. Thank God for small favors.

"So, how was your day, Bob?" Mom asked.

I could hear the pin drop, and I fantasized I had the ability to stop time. One question and the crap just started pouring out of him like liquid soap added to a dishwasher. There was always some idiot who had not done things properly or, worse yet, some fool who dared admit he was a Democrat. He'd never take a breath to ask about any of our days. Mom would try to refocus him, but he refused to be sidetracked.

"Bob, the kids all had big days today, not just Chelsea with her incident," Mom said. I realized I still didn't know what had happened with Chelsea.

"Abby, I was in meetings from one o'clock on. And, I had this damned new voicemail system to get used to. I told Sylvia to just write my messages down for me. Damn Democrats came up with that phone system, I just know it," Dad replied, oblivious.

Mom was either too tired or too hopeless to try again. The company-talk continued until I threw up the white flag.

"Can I be excused? I forgot I do have homework," I said. That homework crap would get them every time.

"Me, too," Katie chimed in.

"Okay, girls, you're excused," Mom said. It was like parole after an expected life sentence.

Katie followed me upstairs, as did the dog. Still uneasy about the way she had been acting earlier, I stopped to talk to her.

"Hey, how was your first day?" I asked.

"Okay, I guess. Are the girls at your school tanned and skinny?" she asked.

"You mean like me," I joked as I gestured toward the white pants I was wearing, or rather the pale white legs hanging out of my shorts. Joking aside, I put my arm around her, squeezed her a little, and said, "It'll get better. Hey, you want Sugar to sleep in your room tonight?" I wasn't quite sure what was bothering her, but maybe her first day just hadn't gone as well as mine, and I knew that particular misery all too well.

When her face lit up, I turned and happily went to my room. Not only had I smoothly started school, I had done my major good deed for the day. The sun was surely smiling on me. Maybe this was going to be the move where I finally gained acceptance and found my niche. Yes, something told me right then, North Carolina was going to change my life.

Abby

I didn't dare mention my concerns to Bob. Racial issues were a sensitive subject with him. His family had been involved in riots during the Civil Rights Movement of the Sixties. According to my father-in-law, Martin County was handling things just fine. That is, until the infestation of Northerners fractured tentative relationships between blacks and whites by injecting poisonous thoughts into old wounds and rupturing them wide open. A coalition of local men formed to take care of the intruders. There were nights of fighting, car bashings, foul language, and injuries. Peace attempts were wrapped in various forms of dishonesty and political agendas. The division cratered into fault lines of diplomacy. As one issue melded, other canyons widened, leaving most people, innocent and hooligans alike, caught on the edge or falling through the cracks, unprepared and unaware.

My family talked very little about this part of our country's dark history. I was thankful I had been spared. Some of the rhetoric and debates I'd heard made me feel uneasy and fearful. I never understood how one man could stand down another based on nothing but the color of skin, variation in speech, or differences in mannerisms. Wasn't that what made us special and unique? Focusing on differences rather than similarities didn't make sense to me. But I knew both whites and blacks who firmly held their ground and their opinion. It was a subject I withdrew from completely; but going forward, I would have to face my daughter's questions. There would be no way around it.

Several days later, the five of us celebrated Krista and Katie's birthday. It wasn't what I wanted for them, especially Krista. Thirteen was a big year. It should have been special. I could see her disappointment when she saw there were no candles on the cake. But when you share a birthday party with a younger sibling, where do you put the candles? Do you line them up like soldiers on each side? Well, I tried, and it looked like an excuse for being uncreative. I finally settled for looking thoughtless.

School started, and the routine began. For days, I had been physically struggling to keep up with my work. When I had time to think about it, I surmised it must be the flu. But whatever it was, it continued to drain me.

"Abby, you're going to be late getting the kids to school. You need to get up." Bob halted, an impatient look crossing his face, "Hey, are you sick or something?"

For a minute, I couldn't figure out what was going on. When I finally managed to open my eyes, an overwhelming rush of nausea hit me. I struggled to sit up and struggled harder to get my legs over the side of the bed. Bob stared, hands on his hip, during the whole ordeal.

I finally ended his turmoil, "Bob, please go get the girls up. I need a few minutes."

"Abby, I hope you're not sick. I have an important meeting this morning with Arnold that cannot be canceled."

"No, I'm just tired. Tell them I'll be downstairs in a few minutes."

Relief washed over me when Bob left the room. I didn't want him to see me this way. He freaked out every time I got sick, afraid I'd ask something of him.

I managed to make it to the bathroom, splash water on my face and brush my teeth. I avoided the mirror. Yesterday that dreadful old lady glared at me, again. Her eyes were hollow and rimmed in deep purple circles, her complexion sallow and blotchy. I moved away quickly not to embarrass her with some rude comment or jarring remark. I thought it best to leave her be and hope she went away.

Bob reentered the bedroom, walked briskly by me and finished getting ready for work. I tried hard not to look at him, not to show him my metamorphosis. I struggled to get my clothes on and run my fingers through my coarse, unruly hair.

"Abby, have a good day. I'll be late tonight. The big wigs are in from California. It's amazing how they think they can run out here every few months and whip everyone into shape. We'll march to their tune until they leave, and then we'll figure out what needs to be done." Bob planted a kiss on the top of my head and left.

Everyone rushed around, and we made it out the door in time. I started to ask Krista if she wanted me to drop her off at the bus stop, but figured she would rather walk and not look like a baby. I had to keep reminding myself she was a teenager.

Katie and Chelsea were quiet on the way to school. I didn't try to make conversation; my stomach still much too high in my throat. Instead, I concentrated on the road ahead and my destination.

When we got to Katie's school, she jumped out of the car, "I want to go in alone."

Her comment took me off-guard. "Are you sure, honey? You know I'd rather go in with you."

"No. I want to go in alone." And with that, she left. I stared at her for several minutes, stunned. Her words fell like a gavel, but I refused to ponder the implications.

"Chelsea, are you ready to go to your school?"

"Yes, but, Mommy, will you walk me to my class? I don't want to go in by myself." Chelsea pouted, then smiled.

"Absolutely! I can't wait to see your room, meet your teacher and your classmates."

Mrs. Wendell was standing at the door, greeting each child with a warm smile and a friendly pat on the back. "Chelsea Simmons, I'm so glad to see you this morning. Why don't you show your mommy around?"

Chelsea nodded, pulling my arm excitedly. She could barely contain herself as she whizzed from one center to the other. Her excitement mounted by the second. When she found the aquarium set up with two box turtles, she squealed.

We stopped and looked around when we heard a "Hi!" The soft voice did not prepare me for the mass of dark curls springing up and flowing about her small head. She was swaying to and fro, her deep violet eyes sparkling, wide, like her smile. "Ya wanna play?" Each word bounced off each sway.

Chelsea looked up at me. I could tell she was torn between showing me her room and getting acquainted with her new friend. I squatted down and said, "Chelsea would you mind if Mommy looks at your room later? I need to go home and put the dog outside to potty."

Chelsea threw her little arms around my neck and kissed me. Then she faced her new friend and grabbed her hand. I watched as they skipped over to the reading center acting like long time friends. How wonderful to be five!

When I arrived at home, my intentions to go to bed were soon forgotten when I spied six unpacked boxes. That's where I was when I got a call from the school about Chelsea.

"Mrs. Simmons, this is Frances Meede, the secretary at Pine Valley School. Mrs. Wendell asked me to call you about Chelsea. We think she has had a seizure."

As soon as she said Pine Valley School, my heart started to race. I knew exactly what had happened. "I'm on my way."

Before school started, Bob and I discussed and decided not to tell the school about Chelsea's epilepsy. The medication had her seizures under control, and the neurologist was hopeful we could begin a six-month withdrawal program beginning the first of the year. We did not want her activities curtailed or have the teachers and staff treat her differently from the other children. Now I questioned if we had done the right thing.

Arriving at the school, I was met by Mrs. Wendell. "Mrs. Simmons, I don't know what happened. I was on the playground, but didn't see anything. From what I gathered, Chelsea was under the monkey bars when one of the children accidentally fell on top of her. Going on the description given to me, I assume she had a seizure." She was clearly distressed.

I followed her back to the nurse's office where I spied Chelsea asleep on a cot at the back of the room. Sitting down beside her, I pushed damp ringlets of curls away from her brow.

Resolving to let her rest for a few minutes, I looked up at Mrs. Wendell. She alternated between kneading her hands and popping her knuckles. "Don't be upset; accidents happen. In a little while, I'll take her home. She's going to be fine."

From behind us, I heard pounding footsteps and labored breathing. Mrs. Wendell whispered, "It's Mr. Stalls."

When I saw her jump, I looked up. The doorframe was filled from top to bottom and side to side with his bulk. He held his hands waist level, clutched, twiddling his thumbs. His lanky brown hair fell into a point on his forehead, spearing his brooding close-set eyes, "Mrs. Simmons, is Chelsea all right?" His voice was deep and raspy.

"Yes."

His stance made me feel he had more to say. And he did, "I pulled your daughter's records. There was no indication she had a seizure disorder. Am I to assume this was her first?"

Guilt washed over me. Frankly, I did not know what to say. It wouldn't have surprised me to see the police walk in and handcuff me. "No, this was not her first seizure. It was a hard decision to leave it off her paperwork. I realize now how foolish it was." I lowered my eyes from his accusing ones.

"You have consciously put your child and our staff in jeopardy. I'm surprised you would withhold such serious information. I cannot imagine what your reasons were," he chided.

Scold and *suffer* are stern terms until you are in the middle of them, then they are hell. "I don't know what to say. The reasons don't justify what has happened here. All I can say is I made a bad judgment call. It won't happen again."

He didn't say a word. With one last sharp scowl he left, hammering the floor with each step, pausing often to hack up phlegm.

I got Chelsea home, tucked her in bed and nestled GloWorm close. Satisfied she would sleep a couple of hours, I reflected on the day. I had blundered. The impact of what had happened hit full force like a fist landing squarely in the gut. I stretched myself out on the floor, face down, imagining a guillotine above me. Its blade, metal sharp and shiny, hovering, ready to complete its gruesome task. I stretched my right hand up imagining the cord, wanting to yank and sever the guilt consuming me.

Overwhelmed and near tears, I had a burning desire to hear Bob's voice. I needed his assurance, some comforting words. Quietly I got up and edged down the hallway to my bedroom.

I dialed his work number and crossed my fingers he would answer. I heard his breath before his greeting, and began rambling, tripping over each word, "Bob, this is me. I mean, it's me, Abby. We were at school, that is, Chelsea was at school. She had a seizure. The principal acted like a Nazi general. Her teacher was upset. I just got home. I think we made a big mistake."

"Woah, Abby! Please slow down. What are you talking about?"

I gulped, blowing blocks of air between each sentence, thrusting my words out loudly, exaggerated, "They called me from school.

Some kid on the monkey bars fell on Chelsea. Mrs. Wendell didn't see anything until she heard the kids screaming. She was so upset. Then, he came into the room and glared. No, I mean he was scary. Oh, I don't know what I'm trying to say."

I closed my eyes and consciously tried to concentrate, to harness my incongruous thoughts. With deliberate effort, I continued, "He sounded mad because we hadn't told the school about Chelsea's seizures."

Bob's voice strained, "Abby, please answer my questions. Is Chelsea okay?"

"Yes."

"Do you have her at home?"

"Yes."

"Continue to stay focused, okay?"

"I'm focused."

"Here's what I want you to do. Make sure Chelsea does not have any obvious injuries, got it?"

"I did that already."

"Great! Now, don't forget to keep an eye on her for a couple of hours. Check her eye dilatation. Are you still with me, Abby?"

I heard a crack. The avenging guillotine dropped. When the dust settled, it was not my head rolling on the ground, but the unreasonable grip of guilt. Calming myself, I fanned the words to regain control, "Bob, please don't treat me like a fool. I was scared. The situation is under control. Don't worry!"

"Well, if you're sure. You didn't sound like yourself. Look, I'm sorry about today. I want to talk more about it, but I've got to go. I'm already late for a meeting. We'll finish this tonight."

I sat down in the middle of the bedroom floor and cried. I was scared and didn't know which way to turn. Trying to handle a crisis alone was daunting. Did I have Superwoman's *S* printed on my chest?

If I thought leaping off a tall building in a single bound would lift my spirits, I'd be in midair.

When the sobbing stopped, bile filled my esophagus. I cupped my hand and ran to the bathroom. The last of my energy and reserve lay puddled in and around the toilet as I wretched and dry heaved. Lying on the cold tile floor, it suddenly hit me. Was I pregnant? A chill ran up my back and My teeth began to chatter. *I do not need this right now, God.*

I don't know how long I lay there; but when I looked up, Katie was standing in the doorway. "Are you okay, Mommy?"

It was all I could do to roll up and struggle to put a normal face on. "I'm fine. I must have eaten something that didn't agree with me. Would you remind me later to call Mrs. Adkinson and thank her for bringing you home today? By the way, Chelsea's asleep in her room. She's not feeling well. Go get yourself a snack. I'll be down in a few minutes."

"Mama, I'm home!" Krista announced from downstairs.

Before I could answer, I heard Katie yell to her from the upstairs hallway, "Chelsea's asleep. Mama's in the bathroom."

A reprieve! Right or wrong, I could not deal with another episode today. I didn't want to answer questions, and I couldn't take another discussion about how horrible Krista's life was. Then I heard her call for me again. I didn't want her to find me cleaning up vomit. But fate had another plan. A wave of nausea slammed me toward the toilet, leaving me spent and uncaring.

Chapter 6

Krista

Several weeks into school and I still couldn't believe my good fortune. Cindy and I were bonding with ease, and I was actually fitting in. This move, my transition was smoother than my father's. One for the record books. Even if the popularity wasn't enough to make me happy, which it was, there was also a new sense of power humming through my veins. Other kids were putting a tremendous amount of effort into being my friend. In most locations, I had to lurk among the shadows for a long time to learn the ins and outs of the new school and the people who basically ran it. Once I determined what people wanted, I transformed and started begging them to accept me. This place was different. Cindy was the pot of gold at the end of the rainbow. Running into her the first day of school was one of the best strokes of luck in my life thus far.

Although well into fall, I was still able to wear shorts and short-sleeved shirts. Thanks to the self-tanning product, QT, I had even found a tan (albeit orange, but a tan nonetheless).

On this beautiful day, Cindy and I were heading to the cafeteria. Williston had a great setup for lunchtime. Students went into a centrally located cafeteria, got food and then took it out to a huge, outside courtyard to eat. The only occasional problem was police cars parked in the open area to maintain the threat of peace. The courtyard had been the scene of many racial clashes. Sixty percent of the student body was black, and I still hadn't fully adapted to it.

Many of my white counterparts weren't accepting it well either. But aside from the sometimes looming presence of policemen walking around and the scattered signs with Nigger or Honkey written on them, lunch was a very social time. Cindy and I always had a group of girls clamoring around us.

"Krista! Cindy!" Mary Beth yelled. She was a tiny blond with dark brown eyes and always trying to buddy up to us. "Did you see Felicia this morning with that fluorescent green coat? She actually seemed proud of it. So sad!"

"Was that green? My eyes were adjusting, so I really couldn't tell. Oh wait, there she goes now," I said with a giggle. Then loud enough for Felicia to hear, I said, "Put on your sunglasses girls. Don't let the green glow blind you." The sweetest sound of laughter and acceptance flooded around me. The girls loved my quick wit and humor. I loved the attention and new feeling of superiority.

"Oh, look. Poor Felicia is crying. Are you crying Felicia?" Cindy asked, her voice laced with mock concern.

A small part of me noticed Felicia run away in tears, but mostly I reveled in my newfound niche. Every classroom I walked into, every trip to my locker, there were girls and even some guys waiting to hang out. Of course, I was usually with Cindy; we were a force. We knew together we were better than apart. Kerry, Cindy's best friend before I showed up, tried to make it a threesome. But Cindy and I were the real pair. And Kerry knew it.

Cindy and I went to my house from school. Katie greeted us. Cindy usually came home with me. Her mother was often unavailable, and mine wouldn't allow me to hang out at a house with no parental supervision. So here we were again. Katie, who wanted so badly to be part of everything I did these days, begged for even a scrap of our attention. She hadn't brought any girls home yet, and I kept hoping she'd find some friends of her own. But, usually, we'd take pity on her and let her hang out for a while. Today, we were in a great mood. I had a crush on the next door neighbor, Jeff; and Cindy and I had a lot of heart drawing and squealing to do.

Katie snuck into the room, trying desperately to be part of the action, "What ya' doing?"

"Krista's got a crush on JEEFFFF," Cindy said. She was very considerate to make Katie feel part of us—another reason to like and trust Cindy.

"Stop it, Cindy," I said with a fake scold to my voice. "Plus, it's not like Katie understands anything about boys. She's only ten."

"I do, I do. I'm not so little, you know," Katie argued.

"Okay, okay," I acquiesced, "you can come and hang out for a little while."

We sat around on my bed. I started drawing Krista loves Jeff surrounded by hearts on my book covers. Cindy had done similar hearts for her crush on Steve. Katie tried to make us giggle by having Ken jump Barbie's bones. In the middle of Barbie and Ken's naked kissing, Mom knocked on the door and eased her way in.

"Hey, girls. Just wanted to check on you guys," Mom said and sat down on the end of the bed.

Thank God Cindy adored Mom or this would have been a major indiscretion. I think Cindy was jealous because her own mother was so busy. Her mother didn't have an actual job, but she was some kind of major socialite. Cindy's only sibling was a sophomore at the University of Wilmington and a beauty queen. Her Mom and sister stayed on Cindy about the right clothes to wear and dropping the last fifteen pounds that hugged and haunted her. I was comforted by Cindy needing me, especially since I definitely needed her.

"Krista and Jeff sitting in a tree, k-i-s-s-i-n-g, first comes love, then comes marriage, then comes . . . ," Katie started.

"Katie, come on!" I said, mainly because I was embarrassed for Mom to be hearing about Jeff. He was our neighbor after all. "Mom, can't you take her somewhere?"

"Katie, Chelsea and I are going to the grocery store, do you wanna go?" Mom asked. I mouthed a thank you and smiled.

About an hour later, after Cindy had gone home, the phone rang. It was my grandmother.

"Krista, is your mother home?"

"No, she took Chelsea and Katie to the grocery store, but she should be back soon."

"What about your dad?" she asked. Obviously the woman had no clue how this house was run. It was only about five o'clock.

"Still at work. Can I take a message?" I asked.

"Well," she said, hesitating, "your grandfather is sick, and I really need to speak to your mother."

"I'll have her call you," I said. I was being unfairly short with her, but these were the same people who missed every single event in my life. I didn't know what more she wanted from me. I was only thirteen. I couldn't jump in the car and go get my mother. But then again, she probably had no clue how old I was. I envied my friends' relationships with their grandparents. Most were excited to visit them in the summer and looked forward to their holiday visits. Usually, they had some cute name for them. With mine, it was the completely formal and unattached grandmother and grandfather.

"Do that, please," Grandmother said, and then the line went dead. No I love you or how has your life been.

Hanging up the phone, I heard the garage door. I went out to help Mom with the groceries.

"Hey, Mom," I said and helped her carry in the milk, "Grandmother called. She said something about your dad being sick."

"What do you mean?"

Her urgent tone caught me off-guard. "I don't know. You're supposed to call her."

"Okay," she said, "you girls start bringing in the groceries, and I'll go call my mother. Chelsea, you come with me."

Going in and out of the house with the groceries, we heard bits and pieces of the conversation. We also saw the tears Mom was fighting to hold back. At one point, Katie and I just stopped and stood. We didn't know what to do. Mom never really cried, so something was wrong. Mom turned and motioned for us to please shut the kitchen door. Dad always joked that we were letting the flies out, but Mom was not too concerned about bugs right now. I asked Katie to go play in the living room with Chelsea. Mom needed me; I could tell. And thankfully, Katie listened.

I stood waiting. Mom put the phone down and laid her head into her hands for a mere second before she picked the phone up again.

"I need to speak to Bob," I heard her tell someone, most likely Dad's secretary. It was clear the woman upset Mom because she immediately followed with, "I don't care where he is. Find him!"

Now, I was getting scared. Mom was emphatic we should never call Dad unless it was an emergency. Even the news of straight A's had to wait until he got home.

After a brief silence, she told Sylvia to make sure Bob called as soon as he could, and then she turned to me.

"Your grandfather had a heart attack," she said but then had to stop.

I tried to give the concerned reaction she wanted, but I didn't know the man. I wanted to do something, anything to make Mom feel better. But I didn't know how.

"I'll take the girls upstairs and play Barbies," I said. It was all I knew how to do. The only thing I was sure I could do for her was to play mommy for a while.

"Thanks, Krista. Please don't pick up the phone if you hear it ring. I'm hoping it will be your dad," she said.

Me, too, I thought. He'd know what to do. All dads knew what to do.

Abby

> *When the moon is in the Seventh House*
> *and Jupiter aligns with Mars.*
> *Then peace will guide the planets*
> *and love will steer the stars.*

Sitting on the bathroom floor, I draped my arms across the toilet seat and shut my eyes. Quiet filled the room, but Aquarius played loud and clear in my mind. It was a mechanism I had used since a child, filling my head with songs when life weighed me down. Music helped diffuse dark thoughts, wafting them on magical notes to faraway recesses.

Chastising myself for indulging in childish fantasies, I did a reality sweep to ground myself. I had to check on Chelsea, make sure the older girls did their homework and finish the work I began earlier in the garage. I struggled to get off the floor, only to face the mirror. There was no doubt, mirrors were becoming my enemy. Mascara ran down my face like murky quagmire, settling in the corners of my mouth, giving a false impression of a smile. The manifestation caused me to pause and smile back, hoping to reassure the pitiful woman who looked back at me that life wasn't so bad.

I picked up a washcloth and removed the blackened tracks of despair. Then the song came back, needling at the outskirts of my consciousness until I couldn't help myself. The hum returned, and the words settled in place.

> *Love would steer the stars. Harmony and understanding, sympathy and trust abounding, no more falsehoods or derision, golden living dreams of visions, mystic crystal revelation, and the mind's true liberations!*

Hopelessness had never prevented the lyrical words of music from lifting my spirit. Liberations! The thought redeposited a smile on my face.

Finishing up, I went to check on Chelsea. As I neared her bed, she slowly opened her eyes and stretched, "Mommy, I'm hungry."

A relief balm spread, "Good, let's go get a snack. Hey, do you feel like going to the grocery store with me? I bet we could find a bag of M & M's."

Chelsea jumped out of bed, beating me to the door, "Come on, Mommy!"

"Hey, wait a minute!" Trying to catch up with her, I realized it was impossible and diverted my attention, "I'm going to see if Krista and Katie want to go."

Shrills of laughter and giggles spilled down the hallway. When Katie saw me, she used my presence to taunt her sister's emerging adolescence, teasing her over a new boy interest. Coming to Krista's rescue, I enticed Katie to go to the grocery store with us.

I could see her relief. Thirteen-year-old girls had a lot to discuss. I remembered. They loved their privacy, secrecy, and scheming. It was part of the maturation process—putting a voice to feelings to see if they fit or not. As a mother, I tried to respect this need.

On the drive to the store, I reflected on the last couple of hours. My life was a constant readjustment, a balancing act. Mothering was not a place for self-indulgences, especially when they threatened to disable you. As a woman, I could use severe draconian measures on myself, berating and bemoaning life; but ultimately, my psyche prevailed. It was the wonder of being female and living with the daily highs and lows of hormone addiction, something Bob would say made his life miserable, unpredictable, and edgy.

When Katie flipped on the car radio, my mind popped back to my passengers. "Sugar Pie Honey Bunch" hit the airwaves; and we chimed right in, singing loudly and demonstratively.

Grocery shopping put me in another frame of mind. I made it a policy that no matter how tumultuous life was, I would not hurry at this task. I walked down each aisle whether I needed anything or not. It was something I granted myself, a chance to slow down. Katie and Chelsea headed to the candy aisle. "Stay together and meet me at the checkout in ten minutes," I reminded.

"Yes, Mama!" rattled behind two figures racing to see who could reach the goodies first.

On our way home, I leaned over and asked Katie, "How's school going?"

"Fine."

"Do you like your teachers?"

"They're okay."

"Have you met any girlfriends?"

"Sort of."

Chelsea reached up and tapped my shoulder, "Mama, I have lots of friends. My bestest friend is Amber, but I like Rosemary, Kathryn, Jessie, Mandy, and Evie. We play together all the time. Sometimes . . ." She became more animated as she talked, her eyes enlarging with each new name and thought.

"That's great, sweetie, but give Katie a chance to talk." I turned my attention back to Katie, "Honey, you really aren't telling me much. I've only met your homeroom teacher. How about the rest? Do you like them?"

She cast her eyes down and shrugged, "They're okay."

"Mama, she said that before. She said, 'They're okay,' just like that," Chelsea interrupted.

"Chelsea, please give Katie a turn. Katie?"

Katie's voice filled with agitation, "Mama, I said they're okay. I don't know what else you want me to say."

Her tone startled me. Always the quiet one, she usually opened up when I gave her my undivided attention. I kept glancing sideways, hoping to detect what was really wrong, but she remained stone-faced.

Before I had time to further investigate her strange behavior, we arrived home. The girls raced to their clubhouse. The previous owners built it over and around an old swing set; and the girls loved it, painting it white with large scrolling pink letters on the front—COOL CATS.

"Girls, I'm going to start supper. Don't stay out too long," I shouted.

Krista rushed out of the house to help me with the groceries. Holding the door open for me to ease by, she cavalierly mentioned Mother had called. A cold settled over my heart like hoarfrost.

I dropped the bags on the counter. My hands were shaking as I dialed Mama's number. A female voice announced, "I'm sorry. You have dialed a number that is no longer in service."

"Shit!" I pressed the hang-up button and slowly redialed. The phone rang over and over again. I was about to hang up when I heard a weak voice say, "Hello."

"Mama, is that you?"

"Oh, Abby, your daddy's had a heart attack," Mama cried out, choking on the last two words.

All my reserves vanished. Tears streamed down my face. Out of the corner of my eye, I saw three faces staring at me. I motioned for Krista to take the younger girls out of the room. She put her arms around Katie and Chelsea and led them away. I was thankful she had a mothering spirit. Sometimes to the point of having both the younger girls screaming, but today, she was a blessing.

"Mama, I'm going to call Bob. I'll leave tonight and be home in a few hours."

"No, your dad is stabilized now. He's sleeping; so, I ran home to wash up and pack a small bag. I'm headed back to the hospital to stay with him tonight. Please don't drive home in the dark. I don't need the extra worry on me right now. Come early in the morning. Oh, wait a minute, here's the hospital number in case you need to call."

I reached for a pad of paper and scratched down the information. "Mama, everything's going to be fine. I'll see you in the morning. I love you. Tell Daddy I love him, too."

My mind went blank. I flopped into the nearest chair, dropped my gaze, and rested it on a black smudge on the linoleum floor. Numbness permeated and spread, coating my emotions in an anesthetized fog. At some point, I lifted my head, picked up the phone, and dialed Bob.

His assistant answered, "Bob Simmons office, may I help you?"

"Sylvia, I need to speak to Bob." My voice sounded foreign, drained, and lifeless.

"I'm sorry, but Mr. Simmons is not available."

Hell flared as I ticked off my response, "I don't care. Find him!"

A classical minuet filled the receiver. By the time she clicked back to my line, my senses were sharp.

"Mrs. Simmons, I'm not sure where Mr. Simmons is right now. I'll have him return your call. I presume you can be reached at home?"

Ice threaded my words. "Yes. I'm at home. Have him call me as soon as possible. Do you understand?"

"Yes, Mrs. Simmons."

I turned to see Krista standing in the doorway, watching me. It was the only thing that stopped me from slamming the phone against the wall. Drawing in an overwhelming rush of fear and anger, I paced my words, not wanting to alarm her, "Sweetie, your grandfather has had a heart attack." The simple act of saying the words out loud was crushing. I continued to look at her. I could tell she was struggling, not knowing what to say. I understood. Most of her life she only saw her grandparents during yearly visits. It wasn't enough time to form a real bond. I guess I had hoped she would show some kind of emotion, but she did what she knew how to do. She came over and hugged me, promising to take care of her sisters.

Krista left, leaving me to face the agonizing reality that I might lose a parent. Sobs burst from my chest, and I sank to the floor. *Please, Daddy, don't die on me,* I prayed. *Bob, why aren't you here? I can't handle all this alone.*

Chapter 7

Krista

Entertaining the girls was the best gift I could give my concerned Mom, so I took them to my room and suggested a dancing competition. Katie grabbed my new *Like a Virgin* album and squealed. I was smart enough to know the material wasn't appropriate for Katie and Chelsea. I wasn't even sure Mom would think it was appropriate for me if she sat down and listened to it. I put on Madonna's "Holiday" instead. Katie and Chelsea danced in a circle with their hands swaying above them. Halfway through, Mom came in and lowered herself onto my bed. She cracked a faint smile at the girls' attempts to strut their stuff. It was unsettling to see her so visibly shaken. She was the rock of the house, and we all counted on her for everything.

When the song ended, I turned it off. We gathered around her on the bed. She took a breath and then said, "Girls, I'm going to visit my dad tomorrow. He's . . ."

"But, Mama, who's gonna pick me up from school?" Chelsea asked.

"I don't want you guys to worry. I'll drop you off at school and be back before you get home."

"You mean you'll drop off Katie and Chelsea at school," I said. Even as the words came rushing out of my mouth, I wanted to slap myself. *Why did I always say something nasty at such inappropriate times?* "Sorry, Mama, the bus is fine." I hoped the apology would erase my initial stupidity.

"Your dad will be backup in case I can't be home in time."

"Mama, I have to bring cupcakes for snack tomorrow," Chelsea said.

Mom pinched the top part of her nose between her fingers, "Oh, honey, why didn't you tell me earlier?"

"I dun'know," Chelsea answered.

This was my chance to further redeem myself, "Mom, I'll help you bake some tonight, and I will come straight home and watch the girls in case you can't be here on time." Mom seemed impressed. Unfortunately, my mouth kept going, "Lord knows Dad can't ever be home early." *Great, that was encouraging. Congratulations Ms. Always-Know-What-to-Say-at-the-Right-Time.*

"Girls, I have to go. My dad is very sick and needs me. Do you understand?" she asked.

We nodded, and a sudden stillness enveloped the room. It didn't take long for my thirteen-year-old brain to figure out that at two hours each way to Williamston, Mom was going to have a long day ahead of her. I prayed Dad could comfort her once he got home. I was at a loss on how to help her.

"Mama, is there anything I can do to help?" I volunteered.

"No," she said, distracted. She got up, "Get ready for dinner, girls. I think I just heard your dad pull in."

Mom left, and I rallied the troops. Sometimes I lived for this big sister stuff. My chest puffed when the girls wanted my advice and leadership. Not that they always appreciated my expertise.

"Okay, let's be good at dinner tonight. Mom needs a break," I said with a regal tone. Both girls nodded as we headed down the hall to clean up for dinner. We marched downstairs in a single line; even the dog joined in.

I helped Chelsea with her plate while Mom finished loading the table with food. There was an eerie silence, and I noticed Dad forgot to say our prayer. A rare faux pas.

"I'll say the prayer tonight," I declared. Knowing I'd screw up Dad's version, I simply thanked God for the food and then added, "God, please look out for our grandfather, because Mama's really worried."

When I opened my eyes, I saw tears glistening on her face. She tried to brush them aside. *Great, I had done it again.*

I helped with the dishes after dinner and then ushered the girls upstairs to get ready for bed. Since our move here, I could not sleep unless my bedroom door was open and music set to play for an hour. As I started to hit the sleep button on my clock radio, I heard Mom and Dad in their bedroom. I could make out bits and pieces of the conversation, and the tension was evident.

Mom's voice came through first, "Bob, I don't know what I'll be facing when I get down there. I need to know you'll be home for the girls."

"I told you I'd find a way to make it work, but you need to call and let me know. What time do the girls get home, anyway?" he asked.

"How dare you not know!" Mom yelled.

"I always have to remind people around here that I work twelve hours a day!" Dad returned. I turned on the music and put a pillow over my head.

The next thing I remembered was Mom waking me for school. Everything sped by that morning; and before I knew it, Mom and the girls were heading out the door. Though I was worried about Mom, I also had my own day to face. As I neared the bus stop, I found myself focusing less on her day and more on my own.

Cindy wasn't there when I walked up. She usually met me at my stop so we could make sure we sat together and have a chance to chat before school. *She must be late.* Then, she didn't even get on at her stop. Funny, she'd usually call if she wasn't going to be at school.

The first half of the day was uneventful. At lunch, Helen sat with me. I knew she felt more comfortable sitting with me without Cindy around.

"I assume you heard about Cindy?" Helen said and dug into her cafeteria special. I was momentarily distracted by the indescribable mass of food on her plate. Thank God Mom had managed to pack lunch this morning even with all the craziness.

"Hear what?" I asked without hesitation. I was halfway through my peanut butter and jelly sandwich and eyeing the Twinkie to my left.

"I overheard a couple of teachers talking, and I think Cindy's Dad is really bad off." Helen had typical Spanish features—true black hair and dark brown eyes the color of chocolate. Her whole family barely spoke English. Matter of fact, Helen had already taught me a few choice Spanish words. My favorite was *crap* (*caca*).

For the rest of lunch, I tried to act normal and not overreact to the news. I was worried about Cindy, but I was also hurt she hadn't called to tell me.

It wasn't fair to Helen. I was caught up in my own thoughts and didn't pay much attention to the conversation she was carrying all by herself. Helen was really sweet and fun to be around, but she was a borderline. Borderlines were people who could talk to anybody in school but didn't really fall into the popular, nerd, or any other category. I knew her plight well because I had been classified that way in previous towns.

If news ever got to Cindy or Kerry about me hanging out with Helen, I'd simply say I took pity on her. But I genuinely liked her. *Maybe I could move her up*, I thought. Yes, I would have to mention it to Cindy. Kerry managed to be in the cherished circle these days; and Helen could run circles around her, especially in personality.

I finished stuffing some things into my locker and headed to fifth period. I was chatting with a girl in my class when Kerry ran up. Somehow, I had managed to avoid her most of the day.

"Did you hear about Cindy's dad?" she asked.

We sat down and got our notebooks out for class. "Of course I did."

She eyed me suspiciously. The teacher had begun his lecture, so she whispered, "Oh, so she did get a hold of you last night. When we talked, she said your phone had been busy all night."

So, that was it. Kerry ended up getting to talk to her because our phone had been tied up all night with worried calls back and forth to my grandparents.

Without any memory of the promise I made to Mom to rush home to be with my sisters, I hurried home to make sure I could speak with Cindy right away. I was shocked to find Mom already there.

"How did it go?" I asked, but one eye was on the phone. *Man, I have to call Cindy and find out what's going on.*

"We'll talk about it tonight at dinner. The weekend is coming up, and we might have to go to Williamston," she said.

"But, Mom!" I started. My argument was sidetracked by the phone ringing. Katie pounced on it like a cat on an injured mouse.

"Krista, it's Cindy," she yelled right into the phone. Poor Cindy probably had no hearing left.

"Mom, I'll get this in your room, if that's okay?" I asked as I started running upstairs.

I picked up the receiver and yelled down, "I've got it!" I waited for the tell-all click, and then I asked Cindy how she was doing.

"I tried calling you last night," Cindy said. She sounded frustrated.

"My Mom's Dad had a heart attack, and she was on the phone quite a bit yesterday. Sorry," I replied.

"Did you hear about my dad?" she asked. "I'm still at the hospital, but I'll be back in school tomorrow. Actually, I wanted to see if I could hang out with you guys this weekend. I don't like being here at the hospital, and I don't think they'll let him out anytime soon."

I knew we'd either be in Williamston or Mom would be gone, and Dad would cringe at the thought of another girl in the house. Still, I left it open-ended, "I'll have to check," I answered. *Wow, I thought, Cindy didn't seem very concerned about her dad.* She always sounded put off by all his medical problems. I was starting to believe she didn't like him very much.

"I guess I could hang with Kerry this weekend then. She's offered several times," Cindy said.

"I'm sure she did," I muttered.

We talked for a few more minutes before Cindy was summoned back to her dad. I sat on Mom's bed and pulled my knees tight to my chest. *When did Kerry start weaseling her way in?* Kerry made me nervous, but I was smarter. I hatched a plan to befriend her. I didn't like the idea of her having any access to Cindy without me. I wanted to make sure my standing was secure. I picked up the phone and gave her a call.

"Is Kerry there? It's Krista," I said when her mom answered.

"Sure, hold on."

"Hey, Krista, what ya' doing?" Kerry said when she picked up.

"I thought I'd see what was going on over there." We talked about superficial topics for a while before Kerry startled me with a confession.

"You know, Krista, I hope we can hang out more," she said. "You and I have a lot more in common than Cindy and I do. Plus, Cindy's always busy with her dad and all her mom's events."

Problem solved, I thought, *she's not a threat.*

Abby

I pressed my hands to the glass partition on the side of Dad's hospital door and peered inside. Mama was asleep in the chair; her mouth gaped wide open. Drool ran down her chin, puddling on the shoulder of her gray flannel sweater. Dad slept, too. His head laid in an awkward position, half-on, half-off the pillow. The light above the bed, dimmed for sleeping, highlighted the lined ridges of his face and cast a dark shadow on his lower body. His shoulders were bare, unnaturally white. One exposed arm lay lifeless across his chest, white bandaging crisscrossed the hand, securing a tube running from it to an intravenous bag. Monitors flanked him on both sides scrolling, beeping and recording his vital signs.

I stood there watching, feeling like an intruder, an outsider. My parent's world seemed surreal and contained. I hesitated, realizing once I entered the room my life would change forever. How? I wasn't sure, but I knew change was on the other side of that partition. The realization was startling and unnerving. I didn't want to be here, didn't want to go through the gory details of Dad's heart attack, or how everyone felt, or what the doctors had said, or how long I would have to stay and help.

Before I could respond or flee, a hand touched my shoulder, "I knew you'd be here." His words and familiar warmth fell over me like a protective cloak.

I didn't turn. Instead, I pressed into the figure behind me. Even though I hadn't voiced it or even begun to hope, I knew he would come. He wrapped his arms around my shoulders and squeezed. Hurt, fear and confusion began to fade.

"Shhh. You're okay." Soft velvet words flowed like heavy cream, coating my cracked armor. He pulled me in tighter. "Don't cry. I'm with you now."

A nod was all I could offer.

"I got a call this morning about your dad and came right over. I've been waiting for you."

I don't know how long we stood there transferring comforting energy from one body to the next. Time froze. I stared at the glass in front of me, marveling at how the fluorescent lighting cast a superimposed, overlaid reflection of all of us, two scenes cast into one. Mom, Dad and I appeared to be centered, surrounded by Drake's strong arms, safe in his keeping and protection. At some point, a nurse edged by us, and her nudge broke the trance. A soft moan escaped my lips; I didn't want this to end, the safety or the comfort. I leaned down and kissed his arms, drawing in an earthy aroma; and with one final press, I let go and turned to meet his gaze.

"Drake," I said, letting his name linger on my lips, "Thank you so much for being here. I don't know what I would have done if you hadn't come. You have an uncanny intuition to know when I need you."

Drake stayed silent, never taking his eyes off mine, nor I his. *Had I forgotten how dark his eyes were? How they looked like two pieces of smoldering coal?* I stared intently, wanting to memorize familiar features, only slightly changed by time. I instinctively reached up to touch the gray hair draping over his ear, and he took my hand in his. His thick fingers cradled mine in the bosom of his palm, "What are friends for?" His trembling hand belied the tone of his voice and the words he spoke.

I didn't answer right away. He had been told often enough we were only friends; so he had said the right thing. But it didn't feel right. Friends? How could he only be my friend when every part of him was so memorable? I wondered, *if I had made other choices, would we be standing here like this? Calling one another a friend?* I doused the words with cold water before they continued to form. This was not the time.

"You are the best friend I've ever had," I bounced over the word *friend* and lingered on the *ever had.*

Again, silence.

Webster describes silence as stillness, muteness, and quiet, but it is more. Silence is defined by the moment, maybe the second. It can enter as a thief in the night or pour out in the brilliance of an early morning sun, peeking around a hillside. It can be the face of a lover or a choice. At this moment, it was all of these.

"Wanna cup of coffee?" Drake finally brought us back, "I'll try to catch you up on your dad's condition." He shifted away, breaking the connection. Someone walked by us, and I watched Drake throw up a hand and nod to the passerby, never letting my eyes leave his face. I didn't want to move, body or mind, from this place or point in time.

Sensing my disarming vulnerability, Drake started to walk down the corridor toward the hospital canteen. Without thinking, I followed him. As we rounded the corner, I picked up speed to catch up with him and ran smack into my brother.

"Abby?"

I threw my arms around his neck and hugged him hard, "Teddy, oh my God, have you seen Mom and Dad yet?"

"No, I just got here. I'm on my way there now." Teddy looked hard at Drake. Word daggers shot from his mouth, "What're you doing here?"

"Teddy!" I snapped.

"It's okay, Abby." Drake paused, nodded, and touched my arm before continuing, "Ted, I came over to check on your dad. My sister's one of the nurses on call. When she realized her patient was your father, she called me."

Shifting from one foot to the other, I bridged the conversation to defuse the mounting tension, "Teddy, Drake and I are going to grab a cup of coffee. Would you like to come with us? Drake's going to fill me in on Dad's condition. Mom and Dad are sleeping right now."

I looked from Ted to Drake, hoping for a sense of resolve, but Ted stood braced for war. He was a large man, not fat, but broad. His shoulders were immense from years working as a logger. The guys on his crew called him Paul Bunyan. Not only was he large and bearded, he stood two-and-a-half axe handles high. Few men would have anchored their eyes on his for fear of being trampled. Drake eyed him without one ounce of reservation.

Flinging my body between them, I squared off at Teddy, "This is not going to happen on my watch, do you understand?"

Shoving his chest forward to show he had the last word, he drove his finger into my shoulder, anger flaring his nostrils, "Abby, you know my limit. Don't push me! I'm going down to Dad's room. That's where you need to be. Are you coming?"

I looked from Drake to Teddy. Why did he always have to make things uncomfortable with his bull-of-the-woods demeanor? He didn't know anything about Drake. In fact, I was sure this wasn't about him at all. This was about Bob.

Ted thought Bob was the best thing that had ever happened to me, and he was going to look after his best interest. Bob was Teddy's idol. He saw him as a successful, rising corporate star with the potential to make lots of money. And money and position spelled "mucho buckos" to him. I figured Drake was a threat. If I did something foolish, Ted knew his chance of getting mentioned in Bob's will was naught. Some brother!

"Abby, look, I need to head out of here and go to work. I'll check on your dad later this afternoon." Drake leaned over and kissed my cheek.

Under Teddy's blazing glare, I hugged Drake more intimately than I should have, saying, "Thanks for coming. What would I do without you?"

Watching him walk away, I had to will myself not to run after him. I pinned my feet in place, my knees shaking from the exertion, until I could no longer see him. Then I remembered Teddy. Glancing down the corridor, I realized it was empty.

By the time I caught up with him, he was already in Dad's room. I heard his loud, booming voice before I opened the door, "What's all this scare about? I'm missing work, in case you've forgotten."

I got to the door in time to see Mom open her eyes, bound out of the chair, and grab Teddy around the waist. Her small, petite frame was immediately dwarfed by his large girth. Teddy pulled her in tight, rocking back and forth to soothe and allay her fears. His anger gone, he looked like an overgrown boy, crooning soft comforting words.

Through it, I stood outside not wanting to enter or to be acknowledged. It took all the courage I could summon to force myself to step inside the room. Mama and Teddy both pivoted when they heard the door close. At first, Mama didn't seem to recognize me; and then she pushed out of Teddy's arms, straight into mine, "Oh, Abby, I've been so afraid. I thought your dad was going to die," she cried.

"Why don't we leave Teddy here with Dad and go get you something to eat?" I offered, trying to give us both an escape plan.

Her face filled with fear, "I can't leave your dad, not now."

"Mama," I pulled her face around to look at me, "you need to get out of here for a few minutes. Ted will have us paged if anything changes. We can be back here in two minutes. Okay?"

She walked over to dad, stroking his arm, looking from him to the machines and back as if the sounds emanating from them would give her an answer. When none arrived, she picked up her pocketbook, lowered her head, and proceeded out of the room, but not before issuing an order, "Teddy, you are to page me if your dad wakes up. Do you hear me?"

"I promise, Mama. Now, go on with Abby."

I took her arm and led her out of the room. Her shaky fingers dug in lightly for support, making me acutely aware of how frail she was. The realization made me wince. I had been so caught up in my own life I had failed to notice the changes. Her hair was whiter, balding slightly at the crown, and the dowager's hump on her back seemed more prominent. She had always been a small woman, maybe five feet tall; but now, she looked smaller than Chelsea.

Concerned, I leaned in, asking "When's the last time you ate?"

It only took one sentence to open the floodgate, "Abby, it was awful. Your dad was watching the football game on TV when he pitched forward. I didn't know what to do. I couldn't move. I don't know how long I stood there before running to the phone to call the ambulance." Pulling me toward her, she whispered, "Abby, I didn't want to touch him. I thought he was dead."

My heart broke hearing her tell what happened. At times like this, I wondered where the handbook on comforting words was. My tongue sat paralyzed, unwilling to sound the smallest tremble of understanding. The stall felt too long. When I finally spoke, I imagined each word being forced through a sieve, "You did all you could. I'm sorry you had to go through this alone."

Her eyes looked wild. Grabbing me she exclaimed, "He could've died, Abby. I could've killed him!" Sobs wracked her body, reminding me of a Middle-Eastern death wail. I wanted to push her away, not knowing what to give or what she needed. Suffocating the screams in my head, I tried to refocus; but remorse, like a secret shame, seeped into my marrow. I watched people file by. Back and forth they marched, never looking, never acknowledging our agony. How could they not sense this type of illness and quarantine us from ourselves?

A code blue blared, and we both pulled away from one another and looked around. I scrambled for tissue, dabbing Mama's eyes, then mine. I hugged her, rubbing her back, as an end, not as a soothing touch. The awkwardness, for both of us, was evident.

A voice behind us broke the strain, "Mrs. Bower?"

We faced a young nurse, holding a patient's chart. She looked young and pert, her face full of youthful enthusiasm. The newness, her freshness, was a sharp contrast to our spent emotions. And it was agitating.

I was the first to respond, "Is something wrong?"

"Dr. Androsi needs Mrs. Bower to return to her husband's room. He needs to talk to her." She smiled, tipped her head, and bounced away, her ponytail flinging from one side to the other.

To observers, we probably looked like two shell-shocked accident victims not knowing which direction to go or what to do if we got there. In slow motion, we tracked a course and proceeded. Teddy met us in the corridor, took Mom's arm, and escorted her in.

There was no way I could reenter that room. Back-stepping, I said, "Teddy, I'm going to go call Bob. When I get back, you can fill me in on what the doctor said."

I exited, took the elevator to the lobby floor, and rounded the first corridor to the pay phone area. Disjointed, burdening emotions jabbed forcibly with each step. Shallow relief fanned me as I sat down in one of the phone booths. I leaned my head against the cold marble wall and closed my eyes. *What in the world is wrong with me? My dad is lying in intensive care. My mother is wracked with fear and guilt. My brother is his usual ass-self. Drake walks back into my life disrupting who I am, tilting my consciousness, raising up warning signs along with hidden desires. This is madness! I cannot do this. Maybe a convent would offer me the peace I desperately needed.*

Just then, the beast showed itself. It wasn't any of this that was bothering me. It was me, only me. I was breaking up over everything. All the decisions, responsibility, and accountability—the whole lot of it stifled me. I was tired of it riding on my shoulders. There were too many lives at stake, too many people wanting pieces of me. What if I fell under the weight? What if I failed from too many expectations? Who would collapse? Who would I drag into the pit with me? Could I take on their failings along with my own?

Questions rambled on until an inner voice invaded, challenging me, "Straighten up! Look at you! You're no better or worse than the rest of us. Do you think you can just give up and run? Life is all about shortchange and shortcomings. Are you willing to miss the bigger picture to be a turncoat?"

My mind halted and stabilized. Lifting my shoulders, I picked up the phone and dialed Bob's work number. "Sylvia, is Bob there?"

"I'm sorry. Mr. Simmons is out of the office. Can I take a message?"

Yesterday I would have screamed and ranted. Now, resignation filled my answer, "That's okay. I'm at the hospital with my dad and wanted to update him. I don't know if I can get home in time for the children. Bob told me to call and let me know if I needed him to be home for them."

"Mrs. Simmons, I'm sorry, but your husband won't be available this afternoon. He has the vice president in from Fairfield and will be conducting a shop tour."

My hand clinched in response, but I said, "Just let him know I called." I hung up quickly, not wanting to reply or to resort to my former self.

I don't know how long I sat drenched in quiet. Then intuitively, I glanced at my watch, realizing I needed to say good-bye to my parents and head home. Taking one last glance around the newly decorated lobby, at the framed prints by local artists, and the people scurrying to and from the elevator, it all seemed like a big soap opera, a giant joke. I shook my head, saying out loud to anyone who cared enough to listen, "If you're bored, I have a life to loan ya! Step right up! It's here for the asking!"

Realizing I had just made a spectacle of myself, I lowered my head; and that damn song threaded through my memory, peeking out of its protective place:

> *Do you know where you're going to? Do you like the things that life is showing you? Where are you going? Do you know? Do you get what you're hoping for when you look behind you? There's no open door. What are you looking for? Do you know?*

When I got back to Dad's room, Teddy was shaking the doctor's hand, thanking him for all he had done. I slipped in, standing on the opposite side of Dad's bed. He was sleeping, his color less pasty. I reached for his hand and squeezed.

In a muted, hollow voice, he responded, "Abby? Is that you?"

"Yes, Daddy." Reaching over, I kissed his forehead. "You scared us. The doctor just left. Teddy can fill us in on what he said."

"I had a heart attack!" Dad cried. Fear and pain tangled its clutches into his face, causing him to grimace and moan. He tried to hoist himself up but fell back onto the pillow, weak and exhausted.

Mom started crying again, wringing her hands in desperation. Ted moved to her side, placing his hand on her shoulder. I sat down on the edge of Dad's bed to offer hope, "Daddy, I know you're scared. I would be scared, too. Yes, you had a heart attack, but you did not die. What we have to do now is understand what this means, and how we can help you manage your health to live a long, productive life." It was a textbook reply, thanks to the pamphlet on the side table, but it was given in genuine love and concern.

But he didn't buy into it, "It means I'm going to die, just like my daddy did."

Teddy dove into action, "Dad you're going to be fine. The doctor said they're going to run more tests to see if you're a candidate for open-heart surgery. If they do the procedure, you'll be good as new."

"I'll be dead sooner!"

Crumbling into a fetal position inside the womb of a sterile green plastic chair, Mama sobbed louder. Tears fell, sliding off the slick surface onto the floor. I stood and watched as they puddled, growing in mass, wondering how long it would take for them to fill the room and float us away to safety.

Ted put his hand on his hip. His eyes slit to narrow openings, which meant his patience was at its end, "Dad, your father died in 1965. They didn't have all the medical advances in heart surgery they do today. They didn't even know your father had a heart condition. For Christ's sake, he never even went to the frigging doctor. You're lucky. Now, people survive heart attacks and live active lives for years."

We continued to talk, answering the questions we had answers for. I got out a pen and paper and wrote down all the ones we wanted to discuss later with the doctor. Finally, I pulled Teddy aside, "Are you staying the night?"

"I don't know. I left it open at work, and I told Sam I'd call and let her know what was going on. So, yeah I can stay. I assume you need to leave?"

"Teddy, please. Let's don't do this. I want to stay. I'd rather stay. But it's the girls. Bob can't get home to be with them. I don't know anyone to call; we just moved there. I'll come back first thing in the morning and relieve you, okay?"

"Do what you need to do. By the way, did Bob get that new assistant he was telling me about?"

"Assistant?"

"Yeah, I called the other night, and he said he was ditching the broad he had. He was looking for a young, sweet thing to help him out," Teddy laughed, throwing his head back.

"Go to hell!"

Teddy continued to laugh and rub his bearded face. I wanted to stick my tongue out and play at his childish games, but decided I didn't want to succumb to his shenanigans. Edging down toward Dad, I promised, "I'll be back in the morning. Please get some rest. I love you, Daddy."

Dad reached over and hugged me, his arms shaking from the effort, "Give Bob our love and kiss those girls for me."

Mom looked grief stricken, shrieking, "Abby, you can't leave us, not now!"

Moving over to where she sat, I knelt down and grabbed her hands, "Mama, listen, I don't want to go. This is killing me, but the girls are going to come home from school to an empty house if I don't leave."

"Bob can take care of them."

"No, Mama, he can't. I'll be here in the morning, and I'll make arrangements to stay this weekend."

As I pulled the hospital door closed, a stabbing pain sent me lunging forward. I grabbed my lower abdomen and moved toward an empty chair beside the nurse's station. Lowering myself into the chair, the pain got so intense I had to immediately stand back up. Fearfully, I called out to a nurse walking by.

"Yes, ma'am?" she said, "Can I help you?"

"I seem to be in quite a bit of pain . . ."

"Abby, it's Teddy. Can you hear me?"

I opened my eyes but couldn't make out where I was. Figures in white circled me, looming tall and shadowy. Light crowned them, casting a soft glow around their nondescript faces. *This must be the portal to heaven,* I thought. Peace enveloped me for the journey. *I'm going home.*

A strong male voice drew me back, "Abby?"

Jolted from an aura of blessedness, I asked, "Where am I?" And as soon as my vision cleared, "What happened?"

"That's what we want to know. Abby, are you on any drugs or medications?" Teddy shouted.

What in the hell was he talking about? Drugs? Medication? Was he being his usual asinine self? And why in heaven's name did he have to shout? I pushed him away, "I've got to go home!"

"Ma'am, I don't think you'll be going anywhere, at least for a while," the young nurse responded, "We're moving you to the emergency room to have a doctor examine you."

Before I could resist, a gurney pulled up, two straggly orderlies piled me on and whisked me away. "Teddy, please get hold of Bob for me," I yelled, watching the rectangular boxed ceiling lights whiz by, hoping Teddy had heard my pleas.

The next two hours were grueling. Tests and questions. More tests and more questions. No one seemed to know what was wrong. Every time I looked at my watch, I panicked. I had to get home. When the nurse walked in, I told her my dilemma. She shrugged her shoulders, saying I would have to discuss it with the doctor. By now, I was done; I'd have to escape. Still woozy and weak, I struggled to get my jeans on.

When I bent over to slip the second foot inside, I almost toppled. It was in one of the twirls to regain my balance that my gaze met Drake's.

Modesty didn't flinch. "Wipe that smile off your face and get over here and help me!"

"I was rather enjoying myself. Haven't had that view in years!" He reached over and pulled up my jeans, buttoning and zipping them adeptly.

I shook my head and laughed, "You wish!"

"Okay, okay, let's not get ugly. What can I do?"

"Drake, I've got to get home. I don't know anyone to call. We're so new in Wilmington," I expelled in one breath.

Drake pushed a defiant look my way, "Call Bob."

"I know you don't understand. Bob is a very busy man. He's in a meeting or something all day and can't leave." I tried to sound genuine, but my voice gave away my true sentiments.

"Only you could come up with an excuse for every time that man disappoints you."

"Same way I did with you? Listen Drake, I don't need this. Just get out of my way," I reached up and touched his shoulder wanting to push him away, out of my life for good, "I can manage life without your help, trust me." I spewed the words, locking eyes with him, hoping my stance would sting.

He grabbed my hand, glowering, "You don't seem to be doing such a good job of it!" His jaw muscles twitched as he edged closer. Obstinacy held me in place until he softened, "I'll see what I can do."

I finished dressing, sat down, and leaned back against the bed's iron headboard. I looked at my watch several times wondering where Drake was, trying to recalculate my travel time with each glance. *Thank God the pain was gone.* To ease my impatience and frustration, I closed my eyes.

Drake finally parted the curtain, announcing, "Okay, kiddo, let's roll."

"Are you sure? Don't I have to sign something? They won't be sending the state troopers after me, will they?"

"No. Make sure you have all your things. We'll talk as we head to the car," he stated, matter-of-factly.

I grabbed my pocketbook and jacket. Drake pulled the curtain back, took my arm, and guided me toward the exit door. I felt like a fugitive, an escapee from prison. Keeping up with Drake was comical. I had to run three paces for his two. Then I saw her. Drake's fifty-seven Chevy.

"Oh, my God, you still have Old Glory!" Stepping up my pace, I got to her before him, running my hand over her high-polished fender, "Sweet! Is this our getaway car?"

"No, I'm just going to take you to yours. By the way, you had an ovarian cyst. Looks like it burst. That's why there's no more pain. You'll need to see your doctor when you get home. I don't want to be responsible for getting you out of here only to have you die from some kind of infection or something." Opening the door, he motioned, "Get in. Where's your car?"

"I'm parked in the visitor's parking area across from the entrance, a red Ford LTD."

Getting in the car, I was greeted by the familiar aroma of pine. Looking at the rearview mirror, right where it had always been, I spotted the pine tree shaped air freshener that had been a joke between us for years.

"I don't believe it," I laughed. "Where do you get that nasty smelling air contaminate?"

"For your information, when something works you don't change it. Laugh all you want, but the pine scent stays. Anyway, there was a time when you thought it was sexy," Drake teased.

"Sexy? You gotta be crazy! I jumped in this car every chance I could to see you, not to smell your retching air freshener." I sat back, inhaling, "We were a crazy pair, weren't we?"

When he stopped and pushed the stick shift into park, I finally opened my eyes, feeling cheated. I wanted more time to go back to those carefree days when laughter was all we were searching for, and the only thing we needed.

"Here we go, princess."

I reluctantly opened the door; but once open, I snapped back to today, to reality. Leaping out of the car, I said, "Thank you, again. I seem to be saying that a lot today, don't I?"

I waited, hoping he would flash me one last beautiful smile, but he didn't. He stared ahead, replying, "No problem. All in a day's work."

I eased into my car feeling drained and lifeless. Sighing, I put the key in the ignition, turned to check traffic behind me, and pulled out of the parking lot. I glanced at the clock on the dashboard and realized, barring no disaster, I would make it home.

The drive home flew by. I turned the radio up to a deafening level; and before I knew it, I was pulling into the school to pick up the girls. Chelsea greeted me with a shout of glee, "Mommy, I beat Allison in a three-legged race today!"

"You did? Wow! I'm impressed!" I reached around, grabbed her, and gave her a big hug. "Do you know how much I've missed you today?"

"Yep."

"Well, let's hurry. Katie's waiting on us," I added.

Chelsea chatted all the way to Katie's school. I nodded at every comment but heard very little. In automatic pilot mode, I pulled into the pickup line, but Katie was nowhere to be seen.

"I wonder where she is," I stated, out loud.

"I dunno. She's mad. She's always mad," Chelsea said, looking out the window. Just then, I spotted Katie with Anna, the little girl next door. Her sullen look and drawn shoulders alarmed me. I knew she wasn't acting herself, but figured she was still missing her old friends. I had hoped getting to know Anna would help her settle in. Katie looked our way, and I threw my hand up to wave hello. I was startled when she stared and turned away with no acknowledgement. It felt odd. I knew she had seen me, but today was not the day to gauge my perspective on sensitivity.

"Hey!" I smiled, adding, "Did you have a good day?"

"Yeah, I guess."

Not to be put off, and to stop the blanket of silence from spreading over us, I consulted Chelsea, "What do you think we could do for your grandmother and grandfather? They're so sad right now."

Excitement and pride spread across her face. "I know. We could draw a big picture and sign our names."

"Katie, what do you think?"

"Sure, why not."

"Would you help Chelsea gather up drawing supplies when we get home?" I tried to counteract her sullenness with some brevity.

"No," fell out of her mouth.

The blanket dropped, wrapping around and sealing us in a tomb of icy silence. Questions aligned themselves in front of my face, but I refused to explore their answers. I drove home praying God would give me a break and fix everything that was falling apart. Silence continued its journey with us, its legs crossed, waiting.

I pulled into the driveway and spied Krista walking home from the bus stop. I waved and headed toward the house. Inside, I dropped my purse on the sideboard and went straight to the kitchen. The last thing I wanted to do was prepare a meal; but once in action, it became a soothing rote routine. I opened the refrigerator door, hoping something would jump out and say, "Fix me." Instead, I stared at a dozen small Tupperware bowls filled with giblets of meats and vegetables and decided I had the beginnings of a pot of homemade soup.

I tossed in this and threw in that, which meant my hips were swaying, and I was singing. When Bob called to check in, my mood had lightened.

"Abby, it's Bob. Hey, I'm sorry about the girls. I've been tied up with the senior vice president of operations. Don't bother fixing me anything for supper. I'm having dinner with the group in from California. By the way, how's your dad?"

I let a few awkward seconds go by to level out my disappointment, "Dad's fine. I wish you'd called sooner about dinner."

"Sorry," Bob said. "I'll catch up with you more tonight. Love you."

I slammed the receiver down and stomped my foot but was immediately sorry. I didn't need another tirade today.

"Mama, my head hurts," Krista whined.

I whirled around ready to pounce on her for creating another excuse not to help in the kitchen or do her homework, but stopped. She looked pale as a ghost. I knew she wasn't feigning, "Oh, honey, I'm so sorry." I pulled her in tight and hugged her. "Go upstairs and lay down. I'll be right up with some aspirin."

I watched as she left. Silence got up and left, too, yawning in boredom.

Chapter 8

Krista

Kerry, Cindy, and I were spending a lot of time together. We had become the Three Musketeers, powerful and thick as thieves. For the first time, I had a real social life and some money to support it thanks to an ever-increasing flow of babysitting opportunities. It's amazing how much people trust a teenager that looks wholesome, makes straight As, and has a mother who tells them she's just around the corner if needed.

In the last few weeks, I had saved roughly one hundred dollars. But I still needed a bit more for the cool Esprit outfit I wanted so badly. Although my parents didn't make me account for how I spent my money, they were practical about how they spent theirs. They bought the basics. If I wanted a specific outfit, it was up to me to buy it. Mom was quick to tell me a name brand meant nothing. *Was this woman ever a teenager?* Then Dad would mutter under his breath about women and their insane need for shoes and clothes. It was humorous to watch him straightening four girls worth of shoes daily while yelling about frivolity. He'd not only make sure each shoe was parallel to each other, but he'd start at the top of the garage stairs and go down by age. Dad's shoes at the top, then Mom's, next came my shoes, etc. Most of the time he was muttering how he didn't have indoor plumbing until he was sixteen. His favorite saying was, "I didn't have a pot to piss in or a backdoor to throw it out of."

This coming weekend Mom had promised to take Kerry, Cindy, and me to the mall. Everything would depend on the Williamston situation. I laid a pretty good guilt trip on her about feeling forgotten in the midst of her parents' melodrama. I had needs. It wasn't like I had a driver's license to take my friends places, and Cindy and Kerry had working Moms.

When I asked to go to the mall, Mom reminded us to hand in our Christmas lists. Even at thirteen, this was an exciting time. In the past, I would sit down with Katie and Chelsea and go through the Sears Wish Book and JC Penney catalogue and circle items for Santa's list. This year, Cindy, Kerry, and I strategically planned our wardrobes to compliment one another. It was the first year when clothes didn't seem the equivalent to getting pieces of coal in my stocking. I was also lucky to have younger sisters still focused on toys. I could get the clothes I wanted and still enjoy playing on Christmas morning.

The girls and I sat around making our lists while Mom prepared dinner. Dad, in his favorite brown Lazy Boy recliner, watched the Dallas Cowboys kick some Redskin ass. His words, not mine.

"Dad, what does it mean again when the ref puts his hands on his hips?" I asked. I loved football, but the best part was sharing moments with Dad. Plus, who wanted to bond with Mom in the kitchen? That was work. I wanted to grow up to be a man—sit around, watch television, and have someone wait on me hand and foot. Want a drink? Wham, bam! There's your drink. Want a sandwich? Wham, bam! Turkey, mayo on rye appeared. What a life!

"It means offsides, Krista. A player from the defensive line jumped before the snap," he said before yelling at the referee to get his calls right. With every bad call Dad said the ref made, I could ask more questions and get more of his attention.

"That's a five yard penalty, right, Dad?" I said, impressed with myself.

"Krista, do you know the best way to learn the game?" he asked without taking his eyes off the television set.

"By asking questions?"

"No, by watching quietly and figuring things out for yourself."

Darn, I should have gotten that one right. Plus, this was the man who would tell me to go to the dictionary and look up a word if I ever asked him what it meant. He said it developed character. *Forget character, just give me the definition!* My internal tirade was interrupted when Mom called us to dinner and the migration toward sustenance began.

Once seated, Dad began one of his famous stories, "A friend of mine said his wife had to take their baby to the pediatrician the other day. When she walked in the office, the doctor couldn't believe how bad the kid smelled." By now, we knew a joke was forthcoming. He loved passing off jokes by acting like he knew the person. He thought he was so slick. He continued, "The doctor noticed the kid had a lot of crap in his diaper, and he asked the mother why she hadn't changed him. The mother said, 'Well, the package said each diaper was good for five to twelve pounds.'"

This started a ripple of giggles. Mom looked tired, sick and tired. Sometimes we did not completely understand his jokes. But usually he did have kid humor since his favorite jokes were about craps and farts. And his laugh was infectious.

"Okay, okay, anybody want dessert?" Mom asked.

"I'm not cheap, but I can be had," Dad responded along with our three raised hands. He then reached out to grab Mom's waist, but she skirted around him.

"Bob, do you think that's appropriate to say in front of the girls?" Mom asked as she rounded the corner into the kitchen to bring out some brownies for everyone.

Appropriate? I thought to myself, befuddled. We all knew the man was cheap, so I didn't understand where there was an issue. A few years ago for Christmas, I asked for a Barbie doll. Instead, Dad got me the no-name doll that came with a thousand outfits in the same box. I told him she'd be the maid to the real Barbies. He told me I was an ingrate. *Can someone please explain that to me, along with the cheap comment? The generic doll came with an apron and broom for Christ's sake, so why wouldn't she be the maid?*

Dad excused himself from the table. When he came to the doorway heading out of the kitchen, he stopped, put his hands on his waist, fanned out his elbows, and puffed out his chest, "I've got to stop working out so much," he said, "or I'm not going to fit through these doors anymore." He then turned sideways in this exaggerated stance and exited the room. We giggled again. He was a silly man. When he wasn't overshadowed by work and stress, he had no shame and resorted to anything to hear our laughter. Chelsea, in particular, loved his efforts. Her favorite was when he'd pretend to walk into doorframes and sign posts.

After the giggles subsided, I went in to help Mom with dishes.

"He thinks he's so damned funny. I clean, cook, and run around all day, and he just puts on a show," she muttered.

"What is it, Mom?" I asked.

Sometimes Mom scared me. Some days, she was not altogether there. "Nothing, let's clean up this mess and start getting ready for bed."

"Krista," she said, catching me making my getaway from dishpan hands, "don't forget that you are going to another dentist this week. And, I think you have a dermatologist appointment next Tuesday."

"Great! Perfect!" I said.

"Where are your glasses, by the way?" she asked.

Damn, she noticed. Again I wondered if this woman had ever been a teenager. "They are upstairs," I told her, bracing myself for the lecture.

"Krista, you need to wear them, not use them as decoration for your bedroom. A bedroom that I noticed is one crumb away from demolition, by the way," she said as I tried to inch closer and closer to the freedom of exiting.

"Mom, you don't understand. Kids make fun of kids with glasses," I said.

"Who cares what other kids think? It's no wonder you have so many headaches. Your father and I did not spend good money on glasses just to have them sit around unused, do you understand?" she said.

Great, Dad pissed her off, and now I'm stuck with her. "Okay, Mom, I get it. I get it. Can I go now?"

"Fine," she muttered.

I stomped up the stairs trying to inflict the sound of pain with each foot fall onto the step landings. I wanted her to feel every step.

Just days before, I had visited the dentist for the first time since we moved here. Mom, after having little dental care as a child, was determined each of us would leave her home with perfect teeth. I hated the dentist. I imagined they went to a special school to develop effective means of torture. With all the medical advances, how could they not find a way to do X-rays without stretching your mouth beyond all reason and comfort?

"Good morning, Ms. Simmons. How are you today?" the hygienist asked as she assembled her tools of pain.

Did she want to make sure she was ruining a good day before she started? What a waste of torture if I was already having a crappy morning! I decided to hedge my bets and tell her my day was fine, knowing it was the most annoying word in the teenage language.

After a long cleaning session, the dentist came in with more professional torture devices. I noticed he was doing less scraping and poking this time and more jaw manipulation. He kept asking me to bite down in all kinds of different ways while he pulled my lips up over my head. I could see the reflection of my teeth in the clear visor he wore to protect his face. I often wondered why they used this or the green cuplike things over their mouths. I needed more protection from them than they did from me.

He called Mom into the room. At this point, I knew it wasn't about a cavity. "Mrs. Simmons, I'm going to refer Krista to an orthodontist. I think she needs braces."

The word *braces* echoed in my head. He might as well have told Mom it was time to send me to the nunnery. That would have been the only other way to completely ruin my high school years.

On the way home, I threw my anger and disappointment like darts—and poor Mom was the dartboard. I couldn't believe my horrible luck.

"Mama, I can't wear those ugly metal contraptions!" I screamed.

"Honey, you may have to. We'll go see Dr. Murray next week. Let's wait and see what he says before we get too worked up."

"But, I already have to wear glasses." Mom didn't know I always took off my glasses on the way to school and borrowed friends' notes to avoid wearing them anywhere on the school grounds. "And, we've already talked about seeing a dermatologist about my zits."

Katie and Chelsea were giggling in the backseat. They had enjoyed coming up with all kinds of new names like brace face and metal monster.

"Perfect! My life is over!"

Arriving home, I ran upstairs and slammed the bathroom door shut. I stared in the mirror and manipulated my mouth in different ways to see what gross things the dentist must have seen. I took off my glasses, and my blurred vision reminded me I needed them. The pink pussy eruptions all over my face were evident, too. Okay, so I needed the dermatologist and optometrist. But I couldn't see what was supposed to be wrong with my teeth. To feel some control, I leaned in close to the mirror and began popping zit after zit. Maybe if I stayed in here for a few hours, I could get rid of every one of them.

Twenty minutes later, I emerged from the bathroom with a puffy, swollen face. I bumped right into Mom in the hallway.

"Krista, if you keep picking at your face like that, you are going to have permanent scars," she said.

I pushed past her and went straight to my room. I ignored her pleas to come and eat dinner and sank deeper into self-pity. My head was killing me, again. To top it all off, I had a huge sleepover the next evening with Cindy and Kerry. We usually spent Friday night at someone's house. This week would be at Kerry's, assuming Cindy's Mom would let her leave the hospital.

I heard the telephone ring but chose to ignore it. Lately, it was some member of Mom's family or Dad's work with a crisis. You would think our house was a hotline for people with problems. Minutes later, I heard Dad yelling for me to get the phone.

"Hello," I said and clinched my eyes shut from the shooting pains.

"Hi, Krista, this is Mrs. Loveless from down the street. Cindy gave me your name as a possible babysitter. I've met your mother several times."

"Oh yes," I said.

"I was wondering if you could babysit this coming Saturday. I have a little girl who is three and a one-year-old boy. We would need to leave for our party around six o'clock." I was struggling to focus on her words and not the searing pain. "Do you think you could be here around five so we could go over some schedules and have you meet the kids?"

"I'd be glad to," I said in my most responsible voice.

"Great! See you then," she said.

At least the day ended on a positive note, I thought. It was nice to make money so I did not need to beg Dad for any. It helped, too, that Cindy would be sending more business my way because she spent so much time at the hospital.

The daily throbbing continued. *How was a teenager supposed to survive with all these issues?* I walked toward the bathroom; and even before

I got there, I could tell my period started. Perfect! Katie was in the bathroom when I rounded the corner. I didn't say a word. I pulled her by the pigtail, pushed her out the door, slammed it shut, and locked it behind me. I could already hear her running down the hall to get Mom, but I didn't care. I got in the shower and scrubbed. My period always made me feel gross. I wanted to soak in the bathtub all day.

After turning off the water and feeling my way around for a towel, I stepped out and began drying off. I put on my glasses and used my towel to wipe away the fog from the mirror. At first, it looked like I'd hit my head during the night or something. I leaned into the mirror to get a closer look. Right in between my eyebrows was this huge red spot. I reached up to feel it and winced. It felt like fire under the skin, but I could also feel a huge knot. The area had been sensitive to the touch for a while, but this was different. *Great, now I have some kind of tumor on top of everything else! Will I die from some sort of rare between-the-eyebrows tumor?*

I threw open the door and yelled for Mom. Katie was standing there ready to pounce.

"She wants to see you right now in her room," she said with a pleased smirk. "You are in so much trouble!"

I gave Katie another much-deserved shove on my way to Mom's room. Mom was sitting on her toilet with her hands on either side of her face. When she heard me approach, she reached up and grabbed the towel bar to pull herself to a standing position. She reached down and pulled up her panties before she turned to face me. She looked sick.

I hated seeing her like this. My friends all said Mom was so pretty, and she was younger than most of my friends' moms. She maintained, thanks in part to boxed color, a gorgeous light brown hair streaked with blond. Her skin had gorgeous golden undertones, and she had the longest legs. I hoped to look as good as her when I got that old.

"Why for the love of God can't you just leave your sister alone?" she pleaded. She turned and made her way to the sink to splash water on her face.

"Sorry, Mom," I said, "I started my period."

"Just leave her alone, okay?" she said, dismissing me.

"I really need you to look at this thing on my face."

She came toward me, and I started to back up. There was a mad-woman look in her eyes. She touched the sore and noticed me jerk.

"I think it's some sort of cyst. We'll definitely need the dermatologist to look at it," she said.

"But, Mom, it's giving me a headache."

"Have you been having those again?" she asked. She took my face in her hand and tilted it up to examine the knot more closely.

"My head hurts all the time."

She kept examining the knot, twisting my face from side-to-side. "Let me try to mess with it a minute, Krista."

She turned to her medicine cabinet and started sifting through the silver cup that held all her beauty supplies. She brought out this thing that looked like tweezers but without the tweezer prongs. It was silver with a flattened end that had a hole in the middle. I didn't like the looks of it, and I wondered if Mom had been visiting the same torture store as those darn dentists.

"This is what you are supposed to be using on your face instead of your fingernails," she said and began to press the tool into the swollen area.

At this point, I was gripping the countertop. The pain burned so intensely tears were streaming down my face, even though I was not quite at the point of crying yet.

Have you ever seen a greyhound shoot out of the blocks after a rabbit? Or a tennis ball machine shoot out a tennis ball to someone practicing his or her swing? Well, neither one of these images would have prepared us for what happened next. Out of nowhere, the knot growing between my eyebrows shot a nasty pus missile right into my mother's face. It not only hit her dead on but managed to spread its slimy white curdle on the mirror as well.

At first, I thought she was going to cry, but then she started laughing. It was an eerie sound, one I didn't like at all. I turned to look in the mirror at the horror. Where the cyst had been, there was now a large bubbling, bleeding hole in the middle of my forehead. Like I could cover this up with my bangs or a cover stick.

"I AM NEVER GOING TO SCHOOL AGAIN!" I yelled.

"Oh, yes, you are," she said, no sympathy apparent in her voice.

"Fine, I'll go. But, I just want you to know you are RUINING MY LIFE!" I screamed, running out of her room. It was obvious my parents had children to torture them and then use them for free labor around the house and in the yard.

When I have kids, I am never going to make them do anything they don't want to do! Yeah, that'll teach Mom and Dad.

Abby

After dinner and helping with homework, I sent the girls off to take a bath. It was the first time since arriving home from the hospital I had taken time to sit down. I picked the bottom step as my perch and began rubbing my neck. A clear image of Mama surfaced. Seeing her swaddled in her sweater and witnessing the terror in her face when I announced I was leaving, opened up a file of recollections of her, moments forever etched in my memory.

During my youth, in rare, serene moments, I would find Mama curled up in her haven, an upholstery-nightmare chair in the kitchen she had shoved between the refrigerator and cabinet. It was a place she could hide, for a few minutes, until we scented her out with our childlike intuition. As a teenager, I would go on a search for her, find her, and often leave when I saw her curled up, perched like a bird on a nest, one arm tightly wrapped around her knees and the other one holding a book. Her head would be leaned against the wall, her eyes focused on the page, going to other places and getting to know new friends.

Once, she was reading aloud when I found her. "Mama, are you okay?"

She turned a dreamy face toward me, still caught in the moment of the words, the breakdown of sentences, the strut of a character. Her face lingered soft and beautiful until her eyes opened wide and the yellowed kitchen walls yanked her back to her world and me, "Yeah," she yawned, "I was just reflecting on the woman in the story. She wanted to write but felt, in her day, she couldn't."

"Why couldn't she, Mama?"

"She didn't know how to break tradition." Mama hopped up, rushing by in her red top and black pants, leaving me stumped. But also, feeling there was a part of me that understood. I reached over and picked up the book she had left behind, *A Room of One's Own* by Virginia Woolf. Flipping through the pages, I saw lined parts, circled parts, and handwritten commentary on the sides. I turned another page and saw the word *WHY?* written across it in bold red letters. A searing discomfort caused me to drop the book. Not understanding, I left it there smoldering, calling for help.

"Mama, are you going to come up and kiss me goodnight?"

I looked up. Chelsea was leaning over the stair rail waiting for an answer. It was evident she had dressed herself, a pink top with dancing brown bears in tutus and red flannel leggings.

"Get in bed. I'll be there in a minute."

I stretched deep, wanting to feel my limbs elongate and expand. Ten things ran through my mind to do; but, in one quick instant, I decided nothing meant more than going upstairs and getting everyone down for the evening, including me.

Bob arrived home later than expected and, by then, the girls were asleep. Chlesea had GloWorm nestled in close, Katie was snuggled in the middle of her growing collection of bears, and Krista had finally gotten to sleep by putting a cold rag on her forehead.

I lingered at Krista's door. It seemed she had gone from baby to thirteen overnight. A flood of memories washed over me—her first tooth, her first word, her first step—all the way up to now and her suffering with adolescence. I knew there was no cure or shortcut for her rite of passage. The awkwardness, the onset of puberty, the precipitous decline in body image, the sudden surge of interest in the opposite sex and shifting peer alliances were part of the maturation process, each stage pivotal and traumatic. The mother in me wanted to help make life saner and more settled. I knew the upheavals she was going through were heightened by the move, prospects of braces, pimples, and glasses. Not being able to help, and hearing about it daily, made life miserable for both of us. I walked over to the bed, removed the cloth from her head, and switched off her lava lamp. The pale light of night, peeking between her pulled drapes, illuminated the soft fringe of lashes on her cheek. Reaching down, I touched the silky ends of her hair, running my fingers down the length of her arm to rest in her open hand. Before I could pull away, she squeezed, and then released to go back to her dreams.

When I walked into our bedroom, Bob was hanging up his suit. Judging from his stance, I decided to bypass conversation and change into my nightgown. On the way to brush my teeth, Bob reached over, touched my arm, and asked, "How's your dad?"

I cradled the words before releasing them in a whisper, "Scared. I think we're all scared."

"Did the doctors give you any idea what they'll do next?"

The weight of the day came crushing down, immediate and unsuspecting, catching me off-guard. My knees buckled, causing me to stagger.

Bob reached out to steady me, "Are you all right?"

"I'm fine. It's been a really long, emotional day. I need a good night's sleep." I reached over and hugged him, "Thanks. I'm sorry. How was your day?"

"About the shittiest day I've ever had. The corporate guys flew in on the company jet and handed us a new proposal for downsizing. They don't have a clue how to run a business. It's all about what looks good on paper and cut costs. Trust me, they couldn't find their ass with both hands, if they tried," Bob fumed.

Holding on to the doorframe, I managed to stay focused. What I wanted to do was go to bed and sink into oblivion. Not to be rude, I responded, "I'm sorry. Looks like both of our days need to be shut down. I think I'm going to bed. How about you?"

"Not now. I need to unwind and catch up on some ball scores. I've got a long day tomorrow, too. I won't be long." Bob reached down to kiss me, "Go to bed. You look exhausted."

Howls of laughter filled the bedroom. Any thoughts of sleeping late ended. In the middle of dragging myself out of bed, Bob announced, "Abby, phone's for you."

I stumbled over to retrieve the receiver from him, fearing something had happened to Dad. I was mistaken. The revelry earlier should have clued me; it was Teddy. "Is something wrong?"

"Only my back! Have you ever tried sleeping in a hospital chair? I'll be in traction if I have to do this another night."

"Yeah, I feel real sorry for you, little brother."

"Just wanted you to know I'm hanging around today and leaving late tomorrow afternoon. Why don't you come for the weekend? Bob doesn't need the hassle of you leaving right now. Plus, I've got some business to take care of later tonight. You remember Joanne."

"Joanne?" I responded, before remembering his high school fling. "Teddy Bower you're a scum bag! How do you live with yourself?"

"When you have a good thing, you spread it around!"

I didn't even respond, hoping his laughter would choke him. Eventually he continued, "Whew! Man I can still get you, can't I?"

"I'm not even going to grace you with an answer. Let's be serious for a minute. Are you sure it's okay with Mama for me to wait and come this weekend. I did tell her I'd be back today."

"The weekend is fine. Why don't you lighten up? I don't know how Bob puts up with your hysteria and lack of humor."

"Teddy, why don't you go straight to hell!"

Dr. Gay's office called early, wanting to see if I could come in concerning my emergency room visit. They said my records had been faxed over about the same time a patient had called in to cancel their appointment. Without thinking of the calendar, I agreed.

Rushing back into the bedroom to change from sweat pants to slacks and a pullover shirt, I ran into Krista. She looked horrible. Her face blanched white, her eyes lined in dark circles.

Studying her further, I asked, "What's the matter?"

When she answered, I heard a very frightened voice, "I really need for you to look at this thing on my face."

The mom-signal went off. Drawing closer to see what she was pointing to, I almost gagged. To subdue a reaction, I drew my eyebrows tightly together. Right between her eyes grew a bulging, swollen, festered growth. I reached over to feel the tightened skin covering the ugly nodule. The heat emanating from her skin caught me off-guard.

Taking her face in my hands, I twisted it from side to side. The protrusion was enormous. I reached for the bump popper, as Katie called it, and proceeded to relieve the pressure. As soon as the edge touched the sore, it erupted. No, it exploded, going everywhere, including my face. I had never seen so much infection and blood. What was left was a horrible deep gaping hole. To my utter shock and dismay, I started to laugh. *Oh God, what was wrong with me? Why in the world would I do such a horrible thing? I must be mad!*

Krista slingshot her head toward the mirror, pursed her mouth, blinked back tears, and yelled, "I'm never going to school again!"

Without thinking, without deliberation, without rationality, without consciousness, I told her, "Oh, yes you are!" How coldhearted could I be? I wouldn't have left the house in a million years looking like that; and yet, I expected my child to go to school and face a barrage of quick-tongued middle schoolers.

Krista stormed out of the room screaming at me, each word a searing branding iron. I had failed again. Insensitive did not come close to describing my behavior, and repair would not be easy.

I struggled to smile at Chelsea and Katie, offering only light comments on our drive to school. As I pulled out of the driveway, I saw Krista out of my rearview mirror, walking toward the bus stop. I wanted to scream for her to climb into the front seat, to not go to school, and to spend the day letting me make things better for her. But I didn't. I let my eyes glide back to the road and drove away.

At eleven o'clock, I arrived at Dr. Gay's office. I dreaded the visit but knew I had to keep the appointment, for me and to fulfill my promise to Drake.

A nurse greeted me when I opened the door, "Good morning, Mrs. Simmons. Dr. Gay is waiting for you." I couldn't help but think, *What service!* Normally I had time to peruse several magazines, observe incoming and outgoing patients, and scan the news on the TV. Looking around the waiting room, it dawned on me that it was empty. On my previous visit, the office had been packed and the staff behind the glass partition hustling around. Now there was no one.

"Hey, what gives? Where is everyone?" I asked, picking up the pace to catch up with her.

"Thursdays are for emergencies. It's kind of our time to catch up on filing or whatever else Dr. Gay comes up with. Today, it's a small crew—me, the lab tech, and the doctor," she explained. Stopping beside an empty room, she ushered me inside, asked me several generic questions about my health, told me to undress, and threw me a flimsy paper gown, before exiting the room.

I had just sat down on the examination table when I heard a soft knock on the door. "Come in," I replied.

Dr. Gay opened the door, his balding head crowning his entry. I was glad he was my doctor. Not only had he come highly recommended, he was renowned as an infertility specialist, speaking overseas at several symposiums. I liked him the moment I saw him. He reminded me of my grandfather, short and rotund, acting every bit the part of a quaint, old-fashioned country doctor.

"Well, Abby, what is this all about? Are you trying to scare your husband?" Dr. Gay never looked my way, his thoughts buried deep into the pile of paperwork attached to the clipboard in his hand. It was comical watching him tip it, first this way and that, in order to adjust reading with his bifocals.

Laying his hand on my knee, he continued, "When did the pain begin?"

"A few days, maybe weeks, ago, I'm not sure. It got worse when I went to visit Dad in the hospital. He just had a heart attack."

"I'm sorry. Will he be all right?" Dr. Gay put the clipboard down to stare over his spectacles, "Have you been under a lot of stress lately?"

"Well, sure. We're still in the process of getting settled. My husband's working terrible hours, and now the scare with Dad. I know it's not his fault, but it couldn't have come at a worse time for me." When I realized how self-pitying I sounded, I added, "Sorry, I didn't mean to whine. I'm really fine. I've been tired a lot. I'm sure it's the aftermath of moving."

"Well, let me make that determination, young lady!" Dr. Gay returned to his reading. The only sound came from the chart pages flipping and his feet shuffling. A comforting feeling was settling in when he interrupted, "I agree with the report the hospital sent over. You probably had a cyst. Did they talk to you about this?"

"Sort of. The ER nurse tried to explain it, but I was in such a rush to get home; I honestly wasn't listening."

"They took a sonogram to determine the size and location of the cyst. The radiologist who read the sonogram thinks the cyst ruptured. Cysts are completely normal. They occur as a result of ovulation. There are usually no symptoms, but in your case that didn't hold true. They also ran several other tests, but most of them have too many variables. I suggest we see you again in three months and redo some of the tests. However, if you experience any more pain or anything out of the ordinary, I want you back in here, ASAP. I'm also going to put you on an antibiotic as a precaution against infection. Do you have any questions?"

"You mean I got undressed for nothing?"

"I promise not to charge extra!"

I left the office with a lighter lilt to my step. I needed this break. When I got into the car, I picked up my to-do list and crossed off the top item.

When Bob called later that morning, I was taken aback by the worry in his voice, "Abby, is everything okay? You've seemed out of sorts lately."

"I don't understand what you mean?"

"I'm worried about you. You haven't been yourself lately. Is everything okay? I mean, I know you're upset about your dad and all, but there seems to be more."

Bob rarely called me during the day and rarer still to ask how I was doing. I frankly didn't know what to say.

"Abby, are you there?"

I cleared my throat and responded, "I'm fine. I've been tired. The last few months have finally caught up with me. Thanks for asking though. Is your day going better?"

"You cannot imagine. I've been in meetings since seven this morning, and I'm already late for another one. It is nonending! Oh, by the way, the staff got together and decided to do a work-out session on Friday and Saturday. It's all a pile of bull, but I don't have a choice." His voice trailed off in resignation.

"I don't know what I'm going to do about going home this weekend. Krista's giving me a hard time. She wants to have a sleepover. By the way, did I tell you Cindy's Dad is in the hospital? It sounds pretty serious. I don't know what to do. We'll discuss it when you get home, but please see if you can change the meetings."

"I'll run it by the group later today. By the way, I won't be home tonight. We're putting together the charts for tomorrow's S2 review. Sylvia's staying over to type. By the time I red-pen all her mistakes, we'll probably be here late. Take care of yourself. I'll see you when I get home."

Thanks to the sudden change of plans, I decided to order pizza for supper. We needed a treat. I ran upstairs to check on Krista, who had gone straight to her room after school. She was propped up in her bed surrounded by scrapbooks, magazines, and school books.

"How about pizza tonight?" I asked.

"Fine," she answered, without looking up.

"Are you okay?"

"If being a freak is okay, then I guess I'm okay."

"Krista, please." I walked over and hugged her, "You know I love you, don't you?"

She hung her head and shrugged.

"I promise I'll call the doctor tomorrow."

A slight nod and a quivery lip was all the answer I needed from her.

"I'll order pizza and grab a movie. Does that sound good?"

"*Clash of the Titans?*" Krista blurted, knowing it was my least favorite movie, and banned by her father for being the all-time worst movie.

"Sure," I said, "but don't you dare tell your father!"

I took Katie and Chelsea with me to get the pizza and movie. There was a soft autumn breeze. Leaves swirled in a cyclonic pattern before falling abruptly to the ground. I had the windows down and the radio turned up high. My mood lightened as I pulled into the Pizza Hut lot. Katie spotted some girls she knew from school and went to hang out with them while Chelsea and I went in for the pizza.

Getting back into the car, I switched my attention to Katie, "Looks like you were having a good time. We need to have the girls over for a slumber party." Katie turned away and hugged the door.

"What's wrong, Katie?"

"Nothing."

I looked at her, staring, realizing she was not going to elaborate. I was perplexed. On one hand, Krista told me too much; and on the other, Katie said nothing. Now I realized the solace in knowledge. I did not quite know how to take Katie's aloofness and reserve. For now, I decided to do nothing and continue observing her behavior. I put the car in reverse, preparing to leave, when I heard, "Mama, do you think I'm fat?"

"Fat?"

There was a long pause. "All the girls I know are so skinny."

"You're the same size as your friends. I don't understand why you think you're fat. You're beautiful."

"No, Mama. I am fat. I am fat and ugly."

I held onto the steering wheel to keep my emotions in check. I stared straight ahead. High above the earth the sky was as murky as quick sand. In the distance it met the earth, and I was driving straight into it.

Chapter 9

Krista

The last few weeks had been a blur. Usually Christmas seemed to take forever to arrive. Not this year. It was only a week away. It didn't help that we spent Thanksgiving in Williamston, rushing from relative to relative.

Plus, the last two months we traveled from doctor to doctor with increasingly bad news. I was seeing the dermatologist frequently for acne surgery. Somehow, when an overpriced doctor popped my zits, it was surgery. But God forbid I pop a few at home. Then, I was ruining my skin. The doublespeak of adults was confusing, to say the least. However, the big beautiful bow on the present of my life came last week when I found out I needed jaw surgery to correct an overbite and a severe jaw joint problem (TMJ).

Mom, who had been absent a lot dealing with her father, was now caught up in something surrounding both Katie and Chelsea. I couldn't put my finger on what it was, and nobody was sharing. I was lost in the shuffle. Not a new feeling for me. To make matters worse, constant nausea compounded my horrible headaches. As part of my acne treatment regiment, I was taking Tetracycline four times a day, and it was wreaking havoc. All I kept thinking was the teenage body must be one of the most burdened systems in the world.

Thank God school was running smoothly. I had met a few more friends—enough so I had people around even if Cindy and Kerry were in another class or busy. Still, school life mainly revolved around those two. We were making so many plans these days, especially with only one more year left in junior high. To any adult, I'm sure this came across as premature planning or simply unimportant, but high school was life's dénouement—a term I am proud to say I learned in my favorite English course. If I screwed up high school, I might as well find my trailer lot and a McDonald's job. Sure, colleges looked very carefully at high school grades and achievements, but the hard-to-masterfully-maneuver social scene could also make or break you.

In the afternoons, we hooked up with ninth graders we determined were destined to be popular in high school. Our plan was to go in knowing these people in order to get into the popular groups like Civinettes, and later the dance squad Vikettes or cheerleading.

As luck would have it, several of the popular eighth grade boys lived in my neighborhood. They were so easy to manipulate when you knew anything about sports; and thanks to my football sessions with Dad, I knew quite a bit. I enjoyed hanging out with *the guys* although Cindy and Kerry cautioned me against becoming just another one of the boys.

"Krista, it's very important that we be seen with these guys and all, but you do NOT want to be one of them," Cindy said during one of our bus summit meetings.

"What are you talking about?" I asked.

"You talk too much football with them," Kerry said as Cindy nodded her head in approval.

"Please, you guys are just jealous because you don't know the difference between offense and defense," I retorted.

"You're right, I don't. And I don't care to," Cindy said, frustrated. "The point is you'll never have a boyfriend this way."

"I don't want a boyfriend right now, dork," I said. And then in a quieter voice, so as not to draw attention to myself or God forbid have Jeff hear anything from his seat only feet away, "I don't like Jeff like that." This was not entirely true because I did have a little crush on Jeff, and they both knew it.

"Sure," they said in unison, "we didn't say you did." They erupted into a fit of giggles.

Finally, we were at my stop. I waved and exited the bus. I walked up the long driveway to the safety of my home. I did not dare look back because I knew Jeff got off at the same stop.

I walked in, but I didn't see anyone, so I shuffled upstairs. Laughter escaped from Katie's room. I poked my head around the corner and found Mom, Katie, and Chelsea playing. *What a treat—laughter in this house again.* There hadn't been a lot lately, especially from Mom.

"Hey, hon," Mom said, "Helen called. You really should call her back."

"I will, I will," I said, agitated. She was one of those backup friends; and though I really liked her, Cindy and Kerry didn't care for her. Actually, they had warned me to keep my distance because she would not be an asset to us. And lately she did seem to be pulling on me too much.

"I need to make some dinner. Do you want to play with the girls?" Mom asked.

"Sure."

Sometime later, Mom yelled for me to get the phone. Immersed in playing with the girls, I had not heard it. *Helen could be a little less pathetic in her attempts to reach me,* I thought as I walked down the hall to my parent's room to grab it. Of course, Mom and Dad didn't see the necessity of a phone in my room no matter how much I begged for one.

"Hello."

"Krista. It's Cindy."

"What's up?" I asked, flinging myself on Mom's bed. I made a mental note to make sure to iron out the wrinkle I created. Those people noticed everything.

"My dad just died."

I sat up straight and clutched the phone tighter. There was an awkward moment of silence. I didn't know what to say. We were on the bus laughing and talking about boys an hour or so ago. Cindy broke the silence.

"I'm at the hospital. Do you think your mom could come pick me up and take us to the mall?" she asked.

The mall? I thought with wonderment, "Cindy, how are you doing? How's your mom?"

"I'll be fine. She'll be fine. We knew this was coming," she replied with scarce traces of emotion, "So, can you ask your mom?"

"Okay," I said, still bewildered. "I'll be right back." I laid the phone down.

"Mom!" I hollered from the top of the stairs.

"I'm cooking, come down here if you need me."

I ran downstairs and turned the corner into the kitchen. "Mom, Cindy's Dad just died." She halted for a minute, and then without saying a word to me, walked over to the phone and picked it up.

"Cindy, dear, how are you and your mother doing?" she asked her. "Yes, oh, honey," she said and then paused, "Sure, the mall, ummmm, how about I come pick you up at the hospital and bring you back here for supper? If that's okay with your mother," Mom said and hung up the phone.

"Krista, I'm going to get Cindy. Supper's almost done. Just watch the girls for me."

What had just happened? Cindy didn't sound upset at all. I could only imagine how I would have acted if my dad had just died, and this was not it. *It would be easier for her to tell me about it once she got here,* I thought and went back upstairs to watch the girls. You couldn't leave those hooligans alone for a minute.

A short time later, Cindy nonchalantly walked into the bedroom where we were. She sat down on the floor and started playing Barbies with us. I felt embarrassed to be caught playing with them, but Cindy melted right in.

"I'm staying with you guys tonight," she said without looking up. Nothing else was said about her dad. Nothing.

Abby

The next few weeks stacked on top of one another, blending until they became a unit of time. Dad came home from the hospital then returned two weeks later for open-heart surgery. The whole family scrambled, clearing their calendars to help.

My contribution was driving home three days a week, leaving when I dropped the girls off at school and rushing back for afternoon pickups. I'd get there in time to fix Dad a late breakfast, get his noon medications down him, and help him tackle items on Mama's to-do list. The pace was insane and utterly exhausting.

Sometimes I would run into Mama on her way out the door or coming in from work. New lines etched her face. During Dad's first hospitalization, she convinced her boss to let her work odd times to get in forty hours. Now her employer was pushing back, recommending a leave of absence. Dad, Teddy, Mandy, and I were suggesting she retire. I think it caused her to work harder, determined more than ever to keep some control when everything else seemed out of control.

Dad's surgery proved successful. His prognosis: six weeks of downtime, then on to a full recovery. Mom eventually quit her job. Even she had to face the facts. It seemed every day I didn't come, Dad had a health issue; and she couldn't go to work. Her boss finally told her to go home and take care of Al. Face etchings were now highlighted by deep shadows under her eyes.

The hardest part was watching her slowly shatter in confusion. I listened to her fears—struggling to attend to Dad, physically and emotionally, while making the necessary adjustments to being home and trying to define their postsurgery relationship. My hands sat limp, not knowing how to console her. My words sounded strained, stagnant. Staring at uncertainty opened up caustic wounds that festered daily—regret, remorse, and repercussion, but more importantly, for Mama, an eruption that caused her to reevaluate her life.

On the long drives home, Drake would slip into my thoughts. I would have never voiced it, but I was disappointed I hadn't seen him during Dad's last hospital stay. I would catch myself looking up and down the sterile halls, hoping to spot him, to see him lumbering up with that quirky grin of his. When I realized where my mind had drifted, I would scold myself and work even harder to squash images and feelings I had no right to have. Other times I would blame it on Mama's soul-searching, jangled nerves that had infected me with discontent.

It didn't help that the upward, exhaustive spiral I had been on didn't end. Weeks of being absent were taking a toll on the home front. Krista was seeing three new doctors—the dermatologist, the orthodontist, and the maxillofacial surgeon. Her acne was worse. The medicine to correct it made her sick, and her headaches were constant. The orthodontist wanted to put her in braces, but informed us she would also need a jaw expander and bands. The maxillofacial surgeon, who was addressing the headaches, found a Temporomandibular Joint Disorder, or TMJ, which would require jaw surgery.

With each doctor visit, Krista became more agitated and angry, and I became the direct object of those feelings. I tried to talk to her, but she seemed determined to fight me every step of the way. The doctors tried to patiently discuss options and answer questions with no success. She would screw up her face and called them all quacks. Every visit put a distance in our relationship. Sometimes, I would sneak a surprise gift in her book bag, hoping to make up for all the anguish she was going through, but nothing worked. She got more and more openly hostile with me. The only laughter and charm she doled out were to Cindy and Kerry.

Katie proved to be a different battle. I decided to devote more time talking to her. But getting her to open up became a challenging guessing game. Every afternoon I would walk into her room and try to make small talk. How was your day? What are your friends up to? Do you like your teachers? Occasionally, I would see a tiny light of response; and then it would fade, leaving me parched to figure out how to reach her.

Chelsea posed even another tap on my slim reserves. She had trouble concentrating and staying awake at school. She was up to six Phenobarbital tablets a day to control her seizures. By the time I gave her the last pill, she was flying high. Some nights I would be curled up on the couch half asleep, and she would still be playing. In the morning, not only did I have to fight to wake up, I had to struggle to get her up, too. We looked like the walking dead.

Now her teacher was giving me a hard time about Chelsea drifting off to sleep at school and not getting her work completed. After wringing our hands and working overtime at night on uncompleted schoolwork, Bob and I decided to approach her neurologist for help. Of all his recommendations, the only viable one was a slow withdrawal from the medication. He warned us it could take four or five months and her response could go from negligible to painful. During this time, we would have to be careful to monitor her for increased seizure activity, which meant taking away recess. Bob and I left the appointment numb.

Bob started traveling more. Some weeks he only came in long enough to change out his laundry and run to catch another flight. Our conversations were strained. He didn't want to know all the details of our life, and yet he felt left out when I failed to inform him. Most of the time, I couldn't remember what I had told him. Thank goodness one thing was off my plate, the tests Dr. Gay had repeated came back normal. He told me I needed to rest. REST? The man must be crazy or delusional.

I was sitting on the floor in the living room, plotting where to put Christmas decorations and what to bake for all the events I'd signed up for, when Katie came rushing in the front door. She never turned to look at me but ran upstairs at full throttle. I flinched when I heard her door slam. The thought of dealing with another crisis made me cringe. *Lord, can I please sit this one out?*

Realizing I had no choice, I stood up and proceeded upstairs to Katie's room. I stopped at her door and counted to one hundred to steady my nerves before calling out, "Katie? Can I come inside?"

No reply. Nothing. Now I counted the lapse of response time, feeling my nerve endings prickling. When she did reply, my heart was beating so loud, I barely heard her.

Entering tentatively, I said, "Hey, I noticed you tearing up the stairs and wondered what was on fire?" I tried to sound funny but quickly realized it was a futile attempt.

Katie turned a tear streaked face to me, and my heart plummeted. I rushed over, putting my arms around her tiny shoulders, "What's the matter?"

"I hate myself!"

"Katie, please. You know you don't mean that. Tell me what's going on."

She pulled away, not to get away, but to straighten up and face me, "I'm so ugly. All my friends are beautiful. No one likes me."

Reality altered. My whole being wanted to scream! What nonsense had taken hold of my daughter? How could she feel this way? Then guilt rode over my questions, pointing its thick black finger toward me. This was my fault. I had been so caught up in my own turmoil and Mama's I had put blinders on to hers. My initial reaction, to shake some sense into her, caught in my throat and burned with regret. I was sinking, slow and sure, and she was watching me. I couldn't afford to panic now or we'd both be lost. Taking a deep breath I began, weighing each word as if it were a life preserver. If I failed we would both sink, and I wasn't about to let that happen.

"Katie, I know this move has been hard on you. We haven't talked about it, but I know you miss Stephanie and Allison." I scrambled to remember a time when we laughed and joked before Katie had these absurd feelings—feelings that were beginning to scare me. I wanted to take her back to the past and have us regroup in happier times, then move slowly forward out of these treacherous waters.

"Remember when you, Stephanie, and Allison decided to go down Adam Hillard's pool slide while holding onto each other's feet? Who toppled off the slide first? Wasn't it Allison?"

Katie sat very still. I didn't realize I had been holding my breath until she announced, "No . . . it was Stephanie. She tried to flip me, remember?" A slight smile appeared at the corner of her mouth.

I rushed in, not wanting to lose the momentum, "No, I think you've got it all wrong. I'm sure it was Allison. I remember because she cut her leg. That's when you started screaming for me. Don't you remember, I was standing by Adam's mother on the sun porch?" I was hustling, hoping to draw us back into some sort of safety net.

"Mama, you never get the story right," Katie giggled. "You were serving ice cream to Adam and his stupid friend, Jacob. Allison went down the slide first, grabbed my ankles, and pulled. That's when I reached for Stephanie's ankles, but she tried to fool me. We all got messed up. Stephanie went over my head and landed half in the water and half on the cement. It was awful! Stephanie was really scared. She was the one who cut her leg, remember?"

"Boy, do I ever. I think I was the most scared person there."

"You scared?" Katie's eyebrows pulled together in disbelief.

With all the sincerity I could muster, I answered, "I get scared a lot. I get scared at night when Daddy's not here. I get scared when I drive by myself to see my parents. I get scared when I realize I'm heading down a one-way street the wrong way, as I do quite often when I venture downtown. And, I really get scared when I meet new people."

I paused a moment before continuing, "Katie, we're in a new town, a new house, and you're in a new school. We left a town we loved where we knew a lot of people. We felt protected." I stopped, suddenly feeling drained, wondering if words had the power to heal. "It's different here. Meeting new people is hard. I miss my friends, too. Is this why you've been upset?"

Katie sat, looking down at her clasped hands. "I miss my friends." Huge tears rolled down her cheeks. She reached up, swiped at her runny nose, and raised her head to look at me. "I'm scared, too, Mama."

Her body pressed into mine, and I rocked her back and forth. Silence wrapped its arms around us, too, bathing us in its treasure of time. A time to let go of fear and sink into the softness of earth sounds, the rustling of a limb on the window screen, and the dulcet tune of the wind whipping around the corner of the house.

When Katie pulled away, I studied her features for understanding, but she offered none. She just smiled. And for now, I accepted it. I felt we had propped the door open, and I firmly planted my foot in it to keep it from closing.

Before we had a chance to speak, Chelsea rushed into the room, "I'm bored!"

She had taken off her clothes, except her panties, and put on the angel wings Mama had given her the Christmas before. On her head sat a bright red ski hat with a huge white pom-pom on top.

Katie and I started laughing. That was all it took to get the girls jumping around, acting foolish. When Katie hit the radio button, music splashed the room in vibrant tones. When the "Lovin' Spoonful" came on the airway, we were all singing:

Did you ever have to make up your mind? Pick up on one and leave the other behind? It's not often easy and not often kind. Did you ever have to make up your mind?

I joined in the dancing, swirling both girls around and around. When the song ended, each of us fell to the floor.

"Hey there," I said, noticing Krista had come into the room. "Oh, before I forget it, Helen called."

Glancing at my watch, I excused myself, leaving Krista to play with the girls. Maybe the jovial moment we had been able to capture would rub off on her, too. I continued singing the song as I walked to the kitchen.

Did you ever have to finally decide? Say yes to one and let the other one ride. There're so many changes and tears you must hide. Did you ever have to finally decide?

I stopped in midstream and sang the song over again. How many times had I sung it and never realized its import? When I sang it again, for the third time, I was in high spirits determined to face life with a more positive outlook. I pulled out my writing tablet and began to jot down ways to get more organized, carve out more family time, and also some individual time with each daughter. On another sheet, I began a grocery list, and on a third, things I could do to help Mama more effectively. I was deep in concentration when Krista rushed into the room.

"Mom, Cindy's Dad died. She's on the phone. Would you talk to her?"

Krista was visibly shaken. I nodded, telling her I'd handle it. But what could I say to comfort a child who had just lost her father? With caution and reserve, I began telling her how sorry I was, asking how I could help. After a moment of silence, which I took as overwhelming grief, she said I could pick her up at the hospital and take her to the mall. Still processing the word *mall*, I almost missed her next statement. She said she needed a dress for the inurnment, saying her mother was tied up with funeral preparations. I hung up the phone shell shocked.

Leaving Krista in charge, I went to get Cindy. She was waiting for me at the front entrance of the hospital. When she slid into the car, she said a quick "thank you" and turned away to look out the window.

Several minutes lapsed before I intervened, "Cindy, why don't you spend the night with us? I'm sure your mother will be busy until quite late. Does she have someone with her?"

"Yes, ma'am. Her sister's with her. I think she'll stay the night. My sister arrived home earlier from college," she added.

"I'm really sorry about your dad. I wish I could have met him. Is there anything I can do for you?"

Cindy sat very quiet, her hands folded in her lap, her head bowed. When she looked up, a strange expression crossed her face, "No, I'm okay. We've been expecting this. Daddy's been sick a long time."

I could not imagine the grief and heartache this young girl was going through. The thought caused my chest to tighten, remembering my fear when I heard Dad had had a heart attack. But when I looked over at her, she appeared cool and collected. I knew there must be a deep, brooding sadness below the surface; there had to be. Most little girls loved their daddies.

I reached over and put my hand on hers, "If there is anything I can do, please let me know. You're welcome to stay with us as long as you need to."

"Okay, Mrs. Simmons. Right now, I need to get a dress. That will be a big help. I'll call Mom from the mall, if that's okay? She won't mind me staying. She's got a lot to take care of."

I couldn't help but watch Cindy. We stopped at the mall and went to Belk's to get her a dress. She laughed and carried on with the sales clerk and flirted with a boy outside JC Penney. Nothing about her body language or demeanor suggested this was someone who had just lost a dear loved one. An odd sensation rippled up my spine; something was not right, not right at all.

Chapter 10

Krista

Christmastime. *What kid doesn't live and breathe for the holidays?* I couldn't hide my enthusiasm. I wanted to continue the traditions I started way back when I was a kid. Ever since I could remember, all three of us slept in the same bed on Christmas Eve. We'd cuddle in after Dad read us the Christmas story from the Bible. After he finished, we'd try to fall asleep, but I'm sure Mom and Dad could hear the giggles well into Christmas morning. Don't get me wrong, I wasn't going to my friends and announcing how much I looked forward to these traditions, but I certainly did.

I had long since dispelled the notion of Santa Claus, but it was such an enjoyment to keep the fantasy alive for Katie and Chelsea. When I first went to Mom and Dad with my revelation, they told me I had a great responsibility to maintain the idea of Santa for my little sisters. Then Katie came to me when she figured it out, and I let her join me on the pact. So far, Chelsea was none-the-wiser. I took my job very seriously. Every time I heard a noise, I would wake Chelsea up and tell her Santa was on the roof, coming down the chimney or setting up the goodies. Katie and I shared a knowing wink, and Chelsea didn't catch on. We were thankful Chelsea was so gullible. We knew Christmas wouldn't be the same once Chelsea knew it was all a sham.

Around three in the morning, I quietly got out of bed and tiptoed downstairs to be the first to look at our Christmas haul. My parents always split the room into three sections and laid out all the goodies according to the recipient. First I looked over my outfits to make sure they met specifications; I was not disappointed, and there were even thrilling surprises. My Guess jeans were accompanied by an aqua sweater with a gold thread running delicately through it. I had seen it in the store but didn't want to push my luck. *Way to go, Mom!*

Next, I surveyed Katie and Chelsea's piles. As usual, Mom and Dad never fully assembled anything. They had caught on a few years ago how much I loved to come downstairs and put my finishing touches on everything. This year, it was putting the stickers on the Barbie van Chelsea got. Afterward, I went about setting up the rest of the toys. When I finished, I made my way upstairs to wake the girls.

"Katie? Chelsea? Santa's been here," I whispered as I gently shook each girl. Their eyes popped open, and they sat up. The next few minutes were a combination of thrown covers, squeals of delight, and the sound of footsteps running down the stairs. This commotion woke my parents who quickly joined us.

"Krista, come on, it's four in the morning." Mom smiled and shook her head at something Chelsea was holding up.

"Yeah, I know. But I've been up since two-thirty."

Mom headed to the kitchen to make her life-force concoction of coffee, cream, and sugar mixed until light brown. Dad, in a rare moment, sat Indian-style in the middle of the formal living room, a room we only used this time of year. Ever since I could remember, we set up a racetrack around the Christmas tree. Dad had already challenged me to a match.

The girls ran into the den to grab our stockings. This was the one treat I truly left for us to experience together. We dumped the contents on the floor and filtered through the candy to find the hidden treasures. Katie and I both had some earrings and a couple of cassettes so we sat and compared. I knew it made her feel very grown up.

After hours of messing around with new stuff, we sat down to a nice big breakfast. Then Mom went to her desk and pulled out the well-stocked box of thank-you notes. She'd made a detailed list of everyone we needed to thank—a short list, mind you, because our grandparents and aunts and uncles never sent anything. Still, this year, in what could only be called a shocking move, our great-grandmother sent us each five dollars. Mom had finally managed to stay off her "shit-list" (Mom's term, not mine). Yet even though this grand gesture only amounted to five dollars, a thank-you note still had to be postmarked the day after Christmas.

After the forced thanks, most every word of our letter dictated by Mom, we went back to gawk at our holiday haul. It was then I noticed an item conspicuously absent. I looked behind the tree, thinking I missed it.

"Mom?" I asked, perplexed, "where's the book?"

"What book are you talking about?"

I knew I was going to get an ingrate lecture from my father if he heard, so I tried to keep it quiet.

"Come on, I only asked for it a zillion times," I whispered.

"Huh?" she said.

I realized she was trying to avoid me by moving into the next room. I followed.

"You know, *Forever*," I said, putting my hands on my hips. This was the newest, must-have book at school among teenage girls. Judy Blume had long been my idol from *Are You There God, It's Me, Margaret*, to the Fudge series, and *Tales of a Fourth Grade Nothing*. This latest book was the talk of the school.

"Krista, you should focus on the nice presents you DID receive," Mom huffed and walked out of the room.

Cutting my losses for the time being, I went into the kitchen and called Cindy. I dreaded the phone call in a lot of ways because I knew this had to be a depressing first Christmas for her without her dad.

"Hey," I said when Cindy answered the phone.

"Oh my God, did you get the Esprit outfit we were looking at? I got it in purple with the matching jacket," she said with giddy excitement.

She clearly wanted normal, so I'd give her normal. "I got it in the cream and red," I said.

"Did you get the book?" she asked.

No, my parents want me to be a baby forever, I thought, "No, did you?"

"Yeah, and I've already started reading it."

"Well, I guess I'll have to borrow it when you're done."

"Sure," she said. She hesitated for a moment, "Can I come over? It's depressing around here."

"Sure," I said. I knew Mom wouldn't mind. I could get away with anything concerning Cindy right now; Mom felt so bad for her. Cindy was getting everything lately from what I could tell. So I figured I could benefit from it, too.

I hung up and called to Mom on my way back downstairs, "Cindy's coming over."

"Okay," Mom said walking into the kitchen. She then laid a hand on my shoulder and continued, "I want to take a minute and explain to you why your father and I didn't get you that book."

Oh, good God, she HAD involved Dad!

"Your dad and I thought the book was too old for you."

"Too old?" I started in on her, "It's Judy Blume!"

"That's our decision, Krista. I'm going to go bake some cookies if you want to help. Better yet, let's wait for Cindy." And, with that, she started assembling the ingredients.

What did I care what kind of conspiracy Mom and Dad had going? I would read Cindy's *Forever*. I wasn't shocked about their decision, but I didn't think they would actually look into a Judy Blume piece. I should have known, though, especially considering my parents pulled me out of my sixth grade sex education class. Now that was embarrassing.

Christmas came and went like it did every year. The beautiful lights and shiny paper could not deflect returning to work and school. The winter break came to a screeching halt with the shrill sound of the bus pulling to a stop and the slow creak of its door.

We hadn't been back to school two weeks when it snowed. Even with the level of importance I had attained recently, it was always nice to have a day off. As soon as I woke up and saw the dusting, I rolled over and snuggled in. Here, if it was forty degrees and raining, school was called off for fear ice might form. Pulling the covers over my head, I began to look forward to staying in pajamas all day and hanging out with Mom and my sisters. I knew Cindy would want to come over, but I was sick of sharing my family with her; so I was already busy coming up with a good excuse. I knew Mom would make hot chocolate and a big breakfast, followed by a hearty lunch of tomato soup and grilled cheese sandwiches. I didn't want to share.

I rolled out of bed around ten and waddled downstairs in my big fluffy slippers. I was looking forward to Mom's hospitality and family time. Mom turned the television on around eleven so we could watch the shuttle. We had planned on viewing the historic event at school; but with the snow, I was sharing it with everyone at home. It would have been easier to watch it with a teacher, though. Mom enjoyed these educational things way too much. Whenever Chelsea asked why the sky was blue, Mom would enter into an hour long dissertation about chemical reactions, etc. I always promised myself, while drowning out all the jibberjabber, I would tell my kids the quick and honest answer, "God made it that way."

While Chelsea droned on about seeing aliens, I watched the shuttle take off. Not easily impressed, I sat in awe wondering exactly what it took to get that rocket in the air. I decided I'd investigate later and avoid asking Mom.

The shuttle was slowly making its way through the sky when, suddenly, there was an explosion and a burst of white. It happened in slow motion. We sat stunned, thinking we missed something. *What had happened?*

"Uh, I think it blew up," I said, still in a daze. The gravity of the situation still had not hit me. *How many times had I watched a movie where all seemed lost and then the hero magically walked through the front door?* I was waiting for the Hollywood ending. My whole school had made a huge deal about a teacher going into space. This couldn't be how it ended. I continued staring at the screen.

I looked over to see Mom holding a crying Chelsea. Katie sat beside me, frozen. The television was overtaken with chaotic scenes of reporters trying to explain everything. They looked drugged and zombielike. They knew the world was watching but were not sure what to say.

We jolted when the phone rang. I was the first to recover and answer it—the only one seemingly capable of moving.

"Hello," I said.

"Krista, it's Cindy. Did you see it? I just can't believe it," she said.

"Me, either," I replied. *What more was there to say?*

"But, I do have some good news," Cindy said. "You know how Mom is always involved with the Azalea Festival."

Not really, I thought. Not only did I not know what the Azalea Festival was, I had no clue what an azalea even looked like.

"Oh, yeah, that's right, you weren't here last year. Well, every year...," she went on to explain the annual festival and how connected her mother was. "Anyway, this year Kim Zimmer is the Queen, you know the lady from *Guiding Light?* Mom says I'll get to meet her."

I wanted to ask how this affected me, but I played along. "Cool, Mom watches *All My Children*, so I don't know anything about the other shows," I said. Now, if Cindy was going to be meeting Adam Chandler or Erica Kane, then I might have been excited.

"My sister will be in the parade, too, as Miss Topsail Island," she continued.

It was surreal. In the middle of this tragedy, we were talking about queens and rich people.

There were times when popularity was a burden. Cindy and Kerry were both quite superficial and caught up in the newest issues of *Seventeen*. I still curled up with the latest Nancy Drew novel at night. Not that I would admit it to anyone.

A blossoming headache was building, so I said my good-byes and hung up. The mood in the house, in the nation, was sorrowful. I wanted to escape, go to my room and lie down. The headaches were becoming overwhelming. When they finally subsided, the nausea would start. My jaw surgery, the supposed miracle cure, was still several months of orthodontics away. Between taking Tetracycline pills for the acne and the hormones of being thirteen, there were days I dared not even look at food. Funny, lately, I could feel Katie watching every morsel I did or did not put in my mouth. I loved food; but on those nauseous days, I couldn't touch it.

Our moods were still gray at dinner. Thankfully, Dad managed not to assign some sort of political importance to the event. There was no talk of those damned Democrats tonight. I think he felt the need to lighten the mood, though, because he told one of his many jokes.

"So, I had this friend," he started as he always did. The man had some weird friends. Mom shot him a tired look.

"Anyway, he went to the doctor and said, 'Doctor, I have this problem with gas, but it really doesn't bother me too much. They never smell and are always silent. As a matter of fact, I've farted at least twenty times since I've been here in your office.' The doctor told him to take some pills and come back to see him the next week. The next week the man went back and said, 'Doctor, I don't know what the heck you gave me, but now my farts, although still silent, stink terribly.' The doctor said, 'Now that we've cleared up your sinuses, let's work on your hearing.'" Dad's laugh had always been infectious, and tonight was no different. The laughter was therapeutic.

Sometimes Mom seemed jealous because Dad could so easily lighten our moods. He had told a joke recently about a man in heaven and his wife, Fanny, to which Mom had not laughed either. She needed to lighten up; but with the events of the day, I figured now wasn't the right time to suggest it.

At school the next morning, the tragedy was all the talk—with anyone who was normal anyway. My lunch group at the *in* table instead filled my head with the latest fashion and who was dating whom. I noticed Helen sitting by herself, so I walked over. She and I actually had an intelligent conversation about the space shuttle, so I asked her to come over to my house soon.

Turning to go to class, I ran smack into Cindy.

"Why were you talking to her?" she asked.

"I like her," I said and continued walking.

"She's not part of the group."

"Why not?" I never thought to ask before.

"Her parents aren't even American! They must be from Mexico or something. I know they don't speak freaking English," Kerry said.

"Krista, I'm sure you'd rather hang out with us, but if not . . . ," Cindy said, leaving the threat to dangle.

I did the incredibly brave thing and looked her square in the face and said, "All right."

I was going to see Helen anyway, but I'd be smarter about it. They never talked to her, so they would never know.

Abby

In the middle of holiday preparation, the neighborhood stopped and came together for Cindy's dad's funeral. The subdivision divided up to provide food for his life celebration service and extra to send to the home for out-of-town guests. Cindy stayed at our house for a couple of days while her mother made prefuneral arrangements. Then later, she spent several postfuneral nights when their house was full of family. Krista kept questioning me about how nonchalant Cindy seemed about her dad's death. Every time she mentioned it, I reminded her that grief was personal and expressed in different ways. From her expression, I knew it fell on incredulous ears. I wouldn't admit it but agreed.

During this hectic time, I thanked God for sending my guardian angel to tend to me. I bought gifts, got them wrapped. I managed to mark off every line on my to-do list. I made several trips home to check on the folks and had the annual Christmas party for Bob's work team. In the middle of everything, I made several phone calls, setting up doctor and dental appointments for the New Year. And beginning with the first day of the year, marked the calendar off in two week increments to begin Chelsea's withdrawal from Phenobarbital.

The only one who managed to remind me I fouled up was Krista. When I didn't get the book she requested for Christmas, she acted insulted. Explaining my position seemed fruitless, so I stopped. I had no sympathy for someone whose gift pile was higher than ever.

January started out mild; but by the end, it was downright cold. I pulled out the old gray jacket I'd been wearing for the past ten years. Each year I swore it would find a home at the Salvation Army, but every spring I would hang it back up in the closet without one thought of disposing of it.

"Mom, you said you were going to get rid of that old thing," Krista said, entering the room.

I looked down at a dangling button, a ripped pocket and tenderly brushed fuzz off its front, "Yeah, I know." No one, not even me, knew why I loved this jacket, "Maybe I'll go later this week and look for a new one."

"Wear it. Who cares?"

Yesterday, her acrimonious reply would have startled me. Today it did not. Her tactics to break me were working, and the thought of coping with this through the teenage years was daunting.

With the same acrid tone, she added, "Can I have the girls over for a spend-the-night this weekend?"

I turned to face her, my mind snarling. *Who does she think she is? She is curt one minute and the next demands something.* I cautiously responded, not willing to give in, "I don't know. I'll discuss it with your father."

"Why? He doesn't care." She folded her arms over her chest and glared at me, cool steaming under her gaze.

I stood there tapping my fingers on the counter like a brush on a snare drum. I wanted to wallop the daylights out of her, but instead, channeled the negative energy to pound harder on the counter. Taking a deep breath, I countered her remark, "Krista! I cannot believe you said such a thing about your dad!"

The stare down began. My left eyebrow went straight up to my hairline. Krista's mouth tightened. Several seconds went by before I made my final threat, saying each word slow and deliberate, "Do not start with me. I promise you will not win."

Our standoff was momentarily interrupted by the phone. Without taking my eyes off my insolent daughter, I answered. It was Mama checking in.

Krista turned to leave the room but shot one last if-looks-could-kill glance before retreating. Exhaling, I pushed out the excess tension, trying to focus on Mama.

"Are you coming home?" she asked.

"Yes, but I'm not sure when. Why?"

"I'm at my wit's end. Your father's being so demanding. I told him this morning that one of us had to go back to work. You would have thought I'd hit him over the head with a frying pan. He jumped up, spilled his breakfast all over the table and himself, and walked right out of the room. What in the world has gotten into him?"

Still managing my broiling temper, I carefully selected my words, "Mama, Dad's been through a lot the last few weeks. He's . . ."

"Well, so have I! Everyone wants to focus on him. What about me? I had to quit my job from taking off so much." Her voice broke, frustration guiding her tone.

Letting a moment lapse, I backtracked my strategy to stroke her hurt feelings, "Mama, you've both been through a lot. But, you know yourself the male psyche is rather . . ."

"Childish?"

"That, too, but what I meant to say is the male ego gets bruised very easily. Dad's probably feeling very insecure right now. He's still healing, trying to gain his strength back. I'm sure he's got a lot of questions about his future. He's worried."

When I heard her puff out a wallop of air, I got fretful knowing I had upset her. Then I heard a sniffle, "You're right . . . I know he's got a lot on him. This really isn't about your father. It's about me. I can't stand watching him sleep all the time. I've always depended on his strength, and now he can barely get to the table. I'm not ready for this in my life. I thought we would have more time to do the things we talked about."

My voice softened, "That's just it. You've been given a second chance. Give Dad time to get on his feet. Start talking to him about the future. Start planning. Give him the motivation to get well faster."

Our conversation lasted long enough to shift my unsettled, pounding head back to my new coping strategy—no thought. That was the name I had given to my new emotional management plan. I had gotten through the holidays with it; and by golly, it would get me through a belligerent teenager, my parents, and an overtaxed life. Caring for everyone else's needs was beginning to feel like a disabling burden.

Right then, I became determined to become a modern day troglodyte, to quietly deal with any hurt or undue pressure from my family. I imagined myself as the poster mom for missing souls . . . a loveless being roaming the halls of her home, taking insult after injury without response, marking each day as one less to endure.

I stalled the mental movie and snickered at my theatrics, "Damn, I'm good!" But quickly clasped my hand over my mouth when I realized how pathetic I sounded.

Chelsea came up and tapped me on the back, "Mama, look what I made for you." She held out both hands. Cupped inside was a beautiful construction paper butterfly. She walked over and gingerly placed her precious treasure in my outstretched hands.

"Oh, Chelsea, I love it," I said, pulling her into my lap.

"It's to make you happy. See all the pretty colors?"

I turned the butterfly from one side to the other. I could tell she had used her new crayons on the wings; they burst in an array of neon colors.

"Wow! This is the prettiest butterfly I've ever seen."

"Katie taught me how to make them. We're going to do a play. She said I could be the star. While she's writing it, I have to make a million butterflies."

"A million butterflies? That's going to take you a long time, isn't it?"

"Nope. I'm faster than anything," she said, smiling. "Katie told me I'd better hurry up 'cause I was uninvited to her room until they were all made."

"Oh well, you'd better hurry or Katie will be grown before you see her again," I could always depend on smiles and a lot of surprises from this kid.

The next morning, we woke to a light snow. I knew school would be cancelled but got the girls up anyway. I wanted to treat them to a nice breakfast and watch the space shuttle. For the first time in history, a teacher would be going up on a space mission, and I wanted to share this thrilling experience with them.

By eleven o'clock, we were sitting in front of the television. I didn't want to miss any of the preshuttle events. As we waited, I opened up a discussion about what it would be like to go into space and what we might see.

"I think I'll see Martians!" Chelsea's excitement was evident.

"That is so stupid. The Challenger is not going to Mars, dimwit," Krista chided her.

"Krista, have you ever been to outer space?" I asked.

"No."

"Okay, can you at least admit anything is possible?" It didn't take much to stir my ire with her. Her indifference incited me; so instead of blowing up, I decided to turn my attention elsewhere, "What do you think they'll see, Katie?"

She didn't respond. That's when I realized she had moved away from us and was peering out the window. "Hey, dreamy eyes, what's going on in that pretty head of yours? We're talking about going into outer space. What do you think you'd see?" I inquired again.

Katie faced me, a detached expression veiling her face. Fear resurfaced, "Katie, are you okay?"

"I want to go to my room."

"Why? Don't you want to stay here with us and watch the Challenger take off? Remember Christa McAuliffe, the teacher? She's going into space today. Don't you think that's exciting?"

"No. Can I please go to my room?"

My fear turned to anger, "No, you cannot go to your room. We're going to have fun watching the space shuttle, we're going to have lunch, and afterwards we're going to play games, maybe make cookies. Don't you want to do this with us?"

Watching her and waiting for a response was like witnessing termites eating away at scaffolding, painfully slow. I continued searching her blank expression until I felt exhausted.

She must have sensed my desperation because she announced, "Fine." Nothing more. No elaboration. No further comment.

Chelsea quietly scooted over next to Katie, touching knee to knee. Krista threw herself back on the floor and with utter disgust said, "Kids!"

Suddenly, I heard the announcer start the countdown. "Girls, it's almost time!"

Our eyes were glued to the screen. Even Katie managed to move closer to watch. The camera was squarely positioned on the plumes of gases coming from under the Challenger. My heart was thumping, "Oh, girls, I can't believe we're watching history in the making!"

With great aplomb, the Challenger lifted off the launching pad. The steel guides fell away, and it soared. I turned to watch the girl's reaction to such an important event. Chelsea's mouth hung open in total astonishment, Katie's neck strained forward not to miss any of the action, and Krista looked totally perplexed.

"What's the matter?" She looked so comical; I threw my head back to laugh.

"Uh, I think the thing just blew up," she announced, her eyes fixed on the screen.

The import of what she said only took a second to sink in. I swung my head around in time to see the Challenger start to disintegrate. It was horrible. We sat in quiet reserve not believing the catastrophe before us.

When Chelsea began to cry, I reached over, pulling her in close. I didn't know what to say. How do you begin to explain to your children, who have just witnessed something so horrible, that death is a natural part of living? Not to be feared but embraced as part of the natural order. Knowing this and being prepared to discuss it were two different things. I found no words of solace.

Katie moved closer and put her arms around both Chelsea and me. Then Krista did the same. We sat and cried, not wanting to watch the terrible events of the morning, but not able to withdraw from the never-ending reruns. At that moment, the four of us were one entity, sharing a common experience that would forever change the way we viewed life and faced death, or so I imagined.

For days, we went about our tasks in quiet order. It was a sobering time for each of us. I had an overwhelming need to call everyone I had neglected. I wrote cards and letters to friends I had not contacted in months, maybe years. I tried to cook everyone's favorite meal. I prayed every morning. I even remembered to be kind to the dog, not forgetting her treats or to scratch her ears on demand.

And all these things shall pass. Who said that? *Did I make it up?* No matter, the need to live each day as if it were the last fell by the wayside, and the distractions of life came hurling back.

The aftermath led to a crippling tiredness. Tired had become a character in my life, managing me physically and emotionally. I would find myself dozing at inappropriate times and places. Failing to differentiate between wake and sleep was scaring the hell out of me. I felt like an accident victim holding on to my falling-out insides. Days would go by before I realized I hadn't brushed my teeth or combed my hair. I couldn't tell if I was spiraling up or spiraling down.

One morning, I surfaced long enough to realize I had been sitting in the family room recliner staring, for God knows how long, at a countryside painting on the wall. When I couldn't even remember the act of sitting in the chair, I panicked. *Had I gotten the girls ready for school? Did they have breakfast? Had I spoken to Bob before he left?* And then the crack widened. *Was Bob home or traveling?* I bolted upright. *What day was it?*

I stumbled to the small makeshift desk in the kitchen and grabbed the calendar. It was strange to see days marked off in clean, straight red lines. In my handwriting, I had written daily activities, appointments, and memory joggers. The calendar was a testimony to my daily life and interactions, but the memory of it stuck to the roof of my brain and refused to be jarred.

I had an overwhelming need to talk to Bob, to make sure life had not left me behind. When he answered, I heard the business voice and suddenly felt guilty for bothering him, "I was wondering . . . did the girls seem okay this morning?"

His impatience was evident, "What are you talking about? The girls were fine. Listen, Abby, I hate to cut you short, but I'm late for a conference call. Is this something we can discuss later?"

I could hear his fingers tapping on his cherry desk and imagined his pinched mouth and scowled forehead. There was no point continuing, "I'll talk to you later."

Returning the phone to its cradle, I walked toward the stairs. With each step I took, my head began to clear. I glanced at the hall clock, realizing it was already past noon. Determined not to lose another minute, I bounded two stairs at a time, hitting the top landing with a thud before breaking into a run for the shower.

The hot water washed away the last vestiges of mental cobwebs. I rubbed my skin raw, smearing baby oil from one extremity to the other. I emerged on guard, ready to do battle with myself and life, refusing to have a nervous breakdown or lose control. Dressing carefully, I hoped to inspire myself and impress the girls when they came home from school. It was alarming when my pants fell around my hip bones; but I quickly shrugged it off, blaming any weight loss on an elevated amount of stress during the last few weeks.

When Bob's sister, Barbara Jean, called later, I was stunned. She rarely called but did make a point of checking in on us every few months. Of all his siblings, she made the most effort.

"Mom and Dad have been begging me to bring them down and see your new home. I've been holding them off for weeks. I know how hard these moves are on you guys. By the way, how're you? I heard you've been sick," she inquired.

"No. No, I'm fine. Had a little female problem, it cleared right up. You know how that goes," I tried hard to sound nonchalant and upbeat. I really liked Barbara Jean, but threw up my guard, not wanting the in-laws to glimpse any deficiencies, "Hey, we'd love to see you. Believe me, we're as settled as we're going to get."

"Great! I'm going to call Mom and Dad and see if this weekend will work for them. You know how they are. They whine to go places, but as soon as you plan something they have one excuse after another—the dogs, the social security check getting deposited, and the list goes on and on." She laughed, "Sometimes I'd write them off if they didn't belong to me."

Her laugh had always been as contagious as Bob's, "I know just what you mean," I chuckled. "Why don't you call me back, and let me know your plans."

Good Lord, the Simmons for a weekend? Did I really sound enthusiastic? Okay, I'm a rotten daughter-in-law on top of being a loon. The worst part about an impending visit would be Gabby, Bob's dad. He was a cantankerous old fool who hated my cooking. The last time I cooked for him, I thought we would end up in the emergency room. He belched and farted the whole visit, claiming I had brought on the first attack of indigestion he had had in ten years.

Rose, Bob's mother, bounced up and down the whole time wringing her hands and wondering if we should insist Gabby see a doctor. Bob chided her hysteria, saying his dad belched and farted after every meal; it was just a good colon cleansing as far as he was concerned. I walked around with an aerosol can, trying to spray the massive, lingering fog of foul odor out of my home. Every time the girls walked in the room, their eyes welled with tears. I couldn't tell if it was from the odor or the spray that hovered in the air.

Shaking my head, I grabbed the grocery list off the counter. Now, I would need to add more items and come up with a revised menu. What in the hell was I going to cook? Maybe I should ask Bob. If he helped plan the meals, I could blame him when Gabby's sensitive gut exploded. His family would never dream about ridiculing their successful son.

The thought of them almost pushed me over the edge. *Oh crap, beam me up, Scotty!* I would have to survive a weekend of conversation about Bob's success, talent, and smarts. Then, I would have to smile and nod as they reminded me how fortunate I was to have landed their son. I could hear my grandmother tell the story about old Lucy Coddler—a girl from the wrong side of the tracks who snared a banker and rode all the way through life on his coattail and his money. I felt assured the Simmons thought the same about me.

Turning around, I tripped over the dog and slipped into a large puddle of pee. "Shit!" I exclaimed, grabbing the dog by his scruff and chucking him outside. "I cannot believe this!" In the middle of cleaning up the mess, I received a call that literally capped the rest of my day.

"Hi, Mrs. Simmons, this is Gloria Banting. I'm the director of the preschool at the Christian church near Monkey Junction. I was talking to Rev. Hayworth the other day, and he mentioned you had worked in a preschool before moving here."

Confusion and skepticism signaled, causing defensiveness to surface. I did not give Rev. Hayworth my resume. In fact, we'd only been attending his church for a few Sundays, and I had barely said more than a hello to anyone.

"Mrs. Bunting," I started.

"No, it's Banting," she corrected me.

"Mrs. Banting," I overemphasized. "Yes, I have had prior experience as a preschool teacher, but I don't remember mentioning it to anyone."

"Rev. Hayworth saw my advertisement in the church bulletin for a teacher and called me. We've been friends for years, and he knows I'm always looking for good teachers. He said your husband thought you might like to go back to work."

Her concern and compassion seemed genuine, but every word caused my blood to boil, "Look, I don't mean to be curt with you, but it has been my practice to search for my own jobs. We just moved here, and I don't feel the girls are settled enough for me to get engrossed in a job right now. I'm sure you understand," I said with finality.

"Trust me, I do. We've moved several times with my husband's company. I tell you what; I don't need an answer right now. Can you come over to my house on Sunday so we can talk? I would love to meet you and bounce around some ideas I have about expanding our services," she said.

I agreed but was fuming! How dare Bob go to Rev. Hayworth; I was mortified! I felt exposed and hurt. But I did not cave. I did not give in to whispers of insecurity or seek the pit for comfort. Instead, I opened the cabinet drawer in front of me and began cleaning, organizing, and throwing away clutter, wishing my name was on each piece.

CHAPTER 11

Krista

An autograph. With all her supposed power and ability to pull strings, Cindy's mom had only managed to wrangle me an autograph. She did not have enough power to get me an invitation to any of the Azalea festival events, but I was supposed to be on cloud nine about a soap star's autograph. Even Kerry's mom was involved enough to get Kerry into a few events. *Why couldn't my mom have cool connections like that?* I really didn't know who Kim Zimmer was, but I hated Cindy and Kerry having fun without me. And after all the big fanfare, I realized azaleas were not all that impressive.

Today was a cleaning day. Mom had been on one of her tirades. She started acting weird yesterday at dinner, and thankfully we were ushered out of the kitchen and didn't have to witness anything firsthand. Katie and I saw the demolition ball starting to swing and knew it would take its pleasure room by room. When Mom came upstairs to my room, I knew the destruction about to occur. Mom started by opening each drawer and throwing out all the clothes not folded properly. I watched her as she flung things and ranted. I had seen the *Exorcist* once, against my parents' better wishes, and stayed up with nightmares all night. Right now, Mom was the one I imagined projectile vomiting after her head finished spinning.

"How many times have I told you to fold your clothes and not just stuff them in your drawers?" she yelled as she threw more clothes on the floor.

I ducked to miss a sweater flying through the air and knew this was not the time to answer her questions.

"Answer me!" she yelled.

Okay, clearly I didn't know much, so I proceeded, "You told me to put them up, and I did." If only those drawers could produce a tight fitting suit of armor. *That would certainly come in handy right now.* Both my sisters were cowering in their rooms hoping not to be next.

"Do I look like your maid?" she questioned.

I knew the correct answer to that one. "No, Mom," I said.

"I swear, all I do all day is cook and clean for you people and what do I get? A bunch of ingrates that stuff clothes in a drawer and . . . God help me, what the hell is this?" she asked.

Since I was standing fairly close to the door, I considered bolting. When I finally worked up the nerve to make a run for it, I realized both sisters were blocking my escape route. *Come to watch the killing?* I wondered.

Then I heard the worst noise come out of my mother. It was guttural and animal like and immediately followed by shrieking, "Is this a sandwich in your bedroom? For God's sake, Krista, a SANDWICH? No wonder I smelled something. When was the last time you even attempted to clean up?" She continued with the onslaught. "I will not spend one more minute in here. You have until bedtime to get it clean, MY version of clean, or you won't do anything this weekend, and that includes babysitting!"

"But, Mom," I started.

"Don't you dare but-Mom me!" she hurled. At this point, I heard Katie snickering at the door. "You want to be next?" Mom asked, shooting daggers her way.

"No, Mom, please," Katie said, and she and Chelsea quickly disappeared.

"Okay, Krista, get to work," Mom said and stomped out of the room and down the hallway.

My sisters owed me big time for this one. I had thrown myself over the land mine for them, and now they knew to walk carefully through the remaining war zone. I began to pick up all the discarded remains of the battle, and I happened to notice my diary flung into the corner. Knowing I hadn't made an entry in a while, I stopped and opened it up. *Maybe I should keep a log of Mom's psychotic episodes in case the doctors ever need the data*, I thought.

"What ya' doing?" Chelsea asked peeking in. I think she was checking to make sure I survived.

"Cleaning my room," I said and laid down the diary to get back to work.

"Is Mommy mad?" she asked.

Master of the obvious, I thought. "Well, you know, Dad's parents are coming."

"Grandpa's funny. He says I'm his toot-toot," she giggled.

On top of suffering the crackdown of the clean police, I now had to spend my weekend putting up with Dad's parents. Dad was the youngest of six kids and the only one not to produce sons. Girls were second-class citizens in their world. Plus, Dad's parents made Mom crazy, which always directly affected us. I would definitely not bring Cindy or Kerry over while they were here. The last thing I needed was for anyone to know I was only one generation removed from pure redneck.

"Will you play with me when you're done?" Chelsea asked sweetly, bringing me back to the task at hand.

"Where's Katie?" I asked trying to politely get rid of her.

She picked up one of my stuffed animals and started cuddling it, "She never plays with me no more."

"Yes, I do!" Katie protested and joined us. "I'm just getting too old to play with Barbies and your other dorky toys."

"They're not dorky," Chelsea said, "All you want to do is play with makeup, and Mom says I'm too little. She says you aren't supposed to be wearing it either. Krista, neither."

"You make a mess with it, because you're a baby," Katie said. By now, both of them were sitting on my bed. "Krista, can I borrow a sweater for school tomorrow?"

"A sweater? It's spring. Why would you want to wear a sweater?"

"Just do, can I?"

"Sure, I guess," I said and then realized the value of my situation. "That is, you can if you help me clean my room." She fell for it, so the rest of my cleanup was a cakewalk.

At dinner that night, Dad was also the beneficiary of Mom's hissy fit. Katie, Chelsea, and I were clueless as to what exactly was happening, and we were trying to remain incognito until Mom decided to throw me in front of the firing squad.

"Is that room of yours clean now?" she asked, drawing a bead on my head.

Why was it always me?

Dad's attention immediately focused on me. He was so insanely fastidious, a grown-up term I had heard Mom use about him. He would clean off the bathroom counter every time he was in there even if he just used the toilet. Freaky! Plus, he was glad the attention was off him.

"Yes, Mom," I said.

"Krista, is it too much to ask for you to help your mother? Do you know the money we've put into your face and your mouth? All we ask in return is for you to help keep this house clean," he said.

"Oh, so sorry that I have acne and need braces. It just makes me SO happy to be running around with braces, on top of having pimples all over my face!" I said with mocking defiance.

"Don't you back-talk me, young lady!" he said, putting his finger a breath away from my nose. "You girls have ten times the stuff I had when I was your age, and you just throw it around. I didn't have a pot to piss in. I work like a freaking dog, and so does your mother."

"Thanks a lot, Mom!" I yelled and turned to leave.

In a flash, Dad jumped from the table and pinned me against the wall. I was stunned and held my breath. "That's enough!" he yelled. "I don't want any more of your smart mouth. Do you understand?"

After what felt like an eternity, he put me down. I could feel the burn where his hands had gripped my arms. "Sit down!" Dad said through clinched teeth as he returned to his seat.

Steam billowed from my nostrils. I was ready to go clean my damn room. If it passed freaking inspection later, I was going to ask Mom if I could stay at Cindy's house this weekend. I needed a break from this mental ward, especially with the grandparents coming.

There were moments in life when you just wished time away.

Abby

The next few hours, I was on a mission like Moses parting the Red Sea. No, I wasn't retreating from my enemies or saving a nation; I was devising a clearer life directive plan. If I proved faithful, once I reached the goal, I hoped to have a few pre-earned installments on future course directions. Looking at my record of failed attempts, I needed help. The only thing I seemed to excel at was brainstorming alternatives and making new starts; but the pressing problem, focusing on the walk, always kept me from entering the promise land.

The thought of a new start was exhilarating. I listed each part as headers on a writing tablet. Part I, gain back control of the house, the children, and Bob. Part II, deal with my own issues. Part III, achieve a blissful state of self-actualization for the whole family. I picked up my pencil and began writing out the details for Part I.

The house would be easy; I would go on a cleaning, throwaway tirade. The children would naturally push back, but they could be scared into submission. As for Bob, the sneak, the squealer, I would use my feminine wiles to melt his heart before thrusting in the dagger. There was beauty in knowing the heart of your victim. Naturally, I would not fire all my shots at once. I needed to reserve a fair amount of shock factor in case it was needed for a second round.

I worked all day like a banshee, going from room to room cleaning, beginning to lay the groundwork for my plan. When the girls arrived home from school, I captured them for room inspection. It had been long overdue. I wasted no time and few words on them; my demands were clear. *Do it or else!* I hated ultimatums, but sometimes, had to execute them to keep the troops in order, even if it was by hook or crook. My renewed confidence must have shown, not one of them balked.

Bob was next on the list. More planning was needed; feminine wiles would not be enough. I headed straight to the kitchen and put a John Lee Hooker cassette in the tape recorder and opened a bottle of Chardonnay. I poured myself a brimming glass full and toasted my new surge of energy and ingenuity. The sultry blues music and wine began working their spell as I created the potion.

I was in the middle of pulling bottles of this and cartons of that from the refrigerator and cabinets when Chelsea tugged on the back of my apron, "Mommy, are you happy?" She stirred the air with her well-meaning question.

"I am the happiest ever!" I fudged a reply.

Pouting, she whimpered, "You've been sad a lot."

"Well, today I am happy!"

"Mommy, will you give some of your happy to Katie?" Chelsea spoke in a sad little voice.

I reached into the candy jar and pulled out six Hershey's Kisses. "Chelsea, do you think this would make Katie smile?"

"It makes me smile, Mommy." She grabbed the sweets and ran as hard as she could, shouting Katie's name the whole way.

Dinner was ready right as Bob came through the door, "How was your day, sweetie?" I asked.

"Same old bullshit. I honestly don't think I can do this job for another fifteen years. I've got to figure out a better way to make a living. To top it off, did I tell you Sylvia was leaving?" Bob carefully draped his jacket over the back of the kitchen chair, "Something sure smells good. It's nice to walk in and see things looking normal again."

Normal? A ten-inch cast iron frying pan lay within arm's reach. I almost grabbed it and wailed the living daylights out of him. Instead, I reached for the wine, poured him one, and refilled mine. "You'd better get started. I'm already one up on you."

A gleam came into his eye. "Boy, do I feel lucky tonight."

"Oh, you never know," I said, giving him a luring smile. It was false; it was calculated, but I could not stop the darkness filling my heart and vision. "Why don't you tell the girls to wash up and hurry down to eat? Dinner's up in five."

The table was laden with country fried steak smothered in gravy and onions, mashed potatoes, fried okra, a green bean casserole, homemade biscuits, watermelon pickles, and iced tea. Before I sat down, I pulled a deep dish apple pie out of the oven to cool.

"What a feast! Let's praise the Lord for the bounty before us," Bob said with gusto. "Abby, this looks sinful!"

Everyone concentrated on their food, passing bowls and refilling plates. I sat and watched. My stomach churned with anger. I had to keep reminding myself to slow down and carefully execute my plan.

"Did everyone enjoy their meal?" I asked.

They all chimed in at once, "YES!"

"Don't forget that beautiful apple pie over there. Who wants ice cream on top?"

Hands shot up around the table. I got up and shoveled large pieces of pie topped with vanilla ice cream onto four plates and passed them around. It took several minutes before anyone noticed I wasn't joining them.

"Abby, aren't you going to have any dessert?" Bob asked, his mouth stuffed with apple pie.

"No. I don't want to get into a habit that will be hard to break." I don't know whether it was what I said or how I said it, but all four faces turned toward me.

"Are you going on a diet or something?" Bob asked, shoving another piece of pie into his mouth.

"No. I wanted to do something special for my precious family knowing I won't be able to do these things once I go to work."

"Mommy, are you going to work?" Chelsea looked worried.

"Yes, as a matter of fact, I am. Your father has arranged a job for me." No sooner had the words left my mouth, Krista, Katie, and Chelsea shot Bob a fitful look.

Bob almost choked. Ice cream and pie blew out the side of his mouth. He grabbed his napkin and tried wiping away the excess, but he was coughing so hard it was a useless task.

"Dear me, did I say something to upset you?" *Oh how sweet revenge could be*, I thought, but stood by wringing my hands in feigned concern.

"Are you serious? Why are you going to work? Who's going to be here when I get home?" Krista demanded.

Chelsea started to cry. "Please, Mommy, please don't go to work. I want you home."

I didn't answer. I only stared at Bob who struggled to regain his composure.

"By the way, Krista," I asked, feeling in full control of the situation, "is your room clean?"

Seeing a way to remove himself from imminent destruction, Bob perked up, "Krista, is it too much to ask you to help your mother?"

At this point, I closed them out and let Bob handle things. I had more important fish to fry, namely Bob. The whole affair unfolded perfectly. I didn't say a word, letting my mind trail back over my rehearsed speech while Bob tore into Krista. Electricity filled the air.

Wanting the sparks to continue flying, I reiterated, "Yes, Chelsea, I'm going to work."

Bob was quick to recover, "I'll tell you what. I think you all need to go upstairs. We'll talk about this later."

There was an eerie quiet as the girls pushed their chairs back and timidly left the room. I did not speak until I heard every one of their bedroom doors close.

"Bob Simmons, how dare you!" I blared.

"What?"

"Does the name Rev. Hayworth rattle your memory?"

"You mean the reverend at church?"

He did not deserve an answer, and he didn't get one. He lowered his head, picked fuzz off his shirt, and moved the fork on and off the plate before facing me, "What did you expect? You've been walking around here like the living dead. You won't talk to anyone. You're barely civil to the children, and you act like I'm the scum of the earth."

"Bob, did you ever once come to me and ask what was wrong?"

"Yes, many times. You never responded. What was I supposed to do? I discussed it with Sylvia, and she said I should talk to our pastor. I thought it was excellent advice. I did it because I'm worried about you," Bob declared.

My head pounded and my eyes burned. How double-dog-dare him go to a pastor I didn't know and confide in his assistant whom I'd never even met. I felt like picking up my pocketbook and slamming it into his head before getting into my car and driving to anywhere, USA.

"Care?" I fumed. "You don't know the first thing about caring. You only care that I keep your little world carefree and *normal*, as you put it. It's all about making your life easier! You don't give a flying pupshit about me or what I do on a day-to-day basis!"

Bob got up, came toward me, and reached out for a hug, "Abby, you're being totally unfair."

The game began. He knew he was wrong, but he thought a hug would erase everything. No way! Not this time. I crossed my arms, ratcheting them in close to show defiance, or was it to steady my nerves or to squelch the burning tears threatening to spill?

"Unfair? You discussed me with two people I don't even know. I mean, giving out personal data and telling them God knows what. Bob, do you know how that makes me feel?"

"Look, Abby, maybe I was a little premature about talking to the pastor and Sylvia, but I was honestly at my wit's end. I really am sorry. You're just so hard to reach sometimes. I never seem to say the right things." Bob took my arm and pulled me close. "I'm sorry. Will you forgive me?"

As mad as I was, I did not pull away. His touch felt good. I was still mad as hell, but I also had to admit how close-mouthed I had been. I truly thought he didn't care about me anymore. Maybe, if I had opened up about my fears and hurt feelings, things would not have gotten out of hand.

We stood this way for a long time. Before I realized it, we were swaying to the background music. Bob slipped his arm around me, and the tension melted in slow swirls and gentle dips. Before I knew it, we were laughing.

Bob leaned down and brushed kisses up and down my neck, stopping only to whisper, "Abby, I'm truly sorry. You know I love you, don't you?"

Right then, I diminished. My feelings, now corralled, accepted his words wholeheartedly. I looked up into the face of a man I had known since I was fifteen. Deep in my heart, I knew who he was and what he was about. He had a fierce determination to succeed, which, in his mind, had nothing to do with his family and would never interfere with how much he loved us. And in that moment, nothing could be more satisfying and intimate.

Bob pushed me away, ever so slightly, "Abby, I want our life to be right. I work long hours to be able to do for you and the girls. Selfishly, I need home to be my haven, my safe place. Work is just work. I know any minute I could be pushed aside or let go. Here, at home, I don't want to feel I have to prove myself on a daily basis. What rejuvenates me every day is seeing you, Krista, Katie, and Chelsea. Do you understand?"

Bob's words were perfect, the most perfect words he had ever said to me outside the "I Do." At that moment, I felt cherished, needed, and loved. I also felt like a total jerk!

"I'm sorry, Bob. There's no excuse for the way I've been acting." I started to cry. "I can't fully explain what's been going on with me, but that phone call today ruptured whatever was about to explode. But that doesn't change the fact that I'm still hurt you chose to discuss me with two people I don't even know. It felt like a breach of confidence and a violation of trust between us."

Bob leaned down and kissed me, long and luxurious. When my knees began to give way, he firmly planted his hand in the small of my back to steady me, drawing me closer and closer.

"Disgusting!" Krista entered the kitchen pretending to gag, her finger poised in her mouth for optimal effect. "I'll come back later!"

We quickly pulled away from one another, overheated, flushed, and feeling sixteen-guilty, "Caught again!" Bob said, before roaring with laughter. "If that had been twenty years ago, your dad would have boxed my ears!"

"You wish that's all he'd have done!"

The evening continued on a softer note, but I was still guarded. It was hard to wind up and suddenly stop spinning.

The day the Simmons were due to arrive, I went into a total tether. I wanted everything to be perfect. Funny, I never went into this kind of frenzy when my own family visited. But Mrs. Simmons had a way of making me feel like the worst mom, daughter-in-law, and wife without saying one word. It was her tight pinched look that gave her away and the snide comments she just happened to let slide out as Bob entered the room. I always had a sneaky feeling she would love to see me poof out of the picture, leaving her the house, kids, and Bob to lovingly manage. I'm sure she thought Bob deserved better. It was common for her to brag about some of the girls we went to school with—a lawyer; a doctor; and no, not an Indian Chief, but close, a consultant for the Bureau of Indian Affairs, working with the Goshute Tribe in Nevada. If I hadn't heard it a hundred times, there was no way I could remember it or care. But this woman, who claimed she couldn't remember my birth date, managed to roll off each syllable perfectly!

I had my head halfway in the oven, checking on a pot roast, when Gabby burst into the kitchen, "Damn it to hell, are you trying to kill someone? I almost broke my ever-loving neck over that last step. Someone's going to sue the hell out of you one day."

It took a second for me to recover from being startled and from slamming my head against the oven door. Rubbing the growing, stinging knot on the crown of my head, my greeting came out doubly strained, "Nice to see you, too, Gabby. Where's Rose and Barbara Jean?"

"Who the hell knows? That woman tried to kill us twice. She can't drive worth a damn. I need some tea, dammit, and not that instant shit you fixed last time." Lastly, Gabby barked, "Where's Bob?"

I went over and planted a dry kiss on his cheek, "Settle down. Bob'll be here any minute. Why don't you go in the den? I'll have iced tea in your hand before you get comfortable. I'm going to check on Barbara Jean and Rose first."

Gabby pulled a handkerchief out of his back pocket and honked several times into it. As was his style, with each blow, he kicked out his left leg like a mule, "Whew! The only way to get out the last booger is to add a little extra umph!" And then, he would laugh until tears flowed down his face. It didn't take long to realize Gabby enjoyed his own crude antics. With each repulsive act, he seemed more charmed with himself. Squeamish squeals and shrills of "GROSS!" only incited him to try additional abhorrent tricks.

Bob once described his dad as the inventor of the "Disgust Factor." As a young boy he would plead with him to be on his best behavior, especially when friends came over. His dad's response was always the same, "I've gotta tend my garden!" Bob did not have to be told what he meant. He had observed it on too many occasions: picking his nose, digging in his ears, and tilling his crotch.

Before I ever met him, Bob took me aside and told me his father enjoyed embarrassing people. He pleaded with me to ignore everything about him, which was near to impossible. And even though I had reservations about Rose, my regrets went out to her for having to endure him on a daily basis.

On impulse, I backed away from Gabby and hollered for the girls to come downstairs. When Krista walked into the kitchen, I thrust a glass of tea into her hand and told her to give it to her grandpa. Before she could complain, I ran outside to help the ladies.

Barbara Jean was struggling to get Rose out of the car. Reaching down to help, I cupped my arm under Rose's and hoisted her up, "What happened?"

Barbara huffed with exertion, "Mama twisted her shoulder trying to reach in the backseat to get that dang box of food she brought."

"Rose, why did you bring food? How many times have I told you it's a privilege to cook for you and Gabby when you're here?" I tried to sound earnest and sincere but could not manage the sour look that pasted itself on my face.

"Abby, you know how Bob misses my cooking. I wanted to bring him some of his favorites. And Gabby's been having those gas attacks lately. He said he didn't want to eat any strange food."

I clenched my teeth, struggling not to say the hateful, rotten things going through my mind. It would only start a war I could not win.

As we struggled into the house, I heard Gabby, "Kursta, you got any boyfriends hanging around I need to mess up?"

"My name is Krista, Grandpa, and no, I don't have a boyfriend," Krista did not even try to conceal the disdain in her voice.

Gabby twisted around as we entered the room. "Hey, Rose, I think we got one of them leasbans here," he chuckled.

"Now, Gabby, don't insult the child. She's shy like our Bob. And, I don't think she likes boys because she's got such a boyish figure. She'll get more interested in them once she develops."

Krista whirled out of the room with a stomp, her eyes fixed in a death glare. As she retreated from embarrassment and insult, Gabby whirled one more jab, "Girl's in a hockey trot if I ever saw one." He pitched forward in a belly laugh before adding, "Abby, you still feeding those kids that foreign food? Believe me, that mess will rot you from the inside out."

My angel returned, and I made it through the Simmons' visit. Thank God for Barbara Jean. She, at least, was not out to insult and enrage me. We genuinely enjoyed one another's company. Because I stayed out of the firing line during their visit, I think the girls received the brunt of Gabby's torment and Rose's off-colored questions. I hated it, but there's something to be said about self-preservation.

Chapter 12

Krista

We survived Dad's parents! Now we were looking forward to summer. Since Cindy and I lived just a few blocks from each other, we had lots of plans. Plans that did not include Kerry—her mom worked, and she lived about ten minutes away. My mom could have picked her up so we could all be together, but why would I want to encourage that?

Since Cindy and Kerry were so involved in the Azalea Festival, Helen and I grew closer. She lived near Independence Mall, so it was easy for Mom to pick her up and take us there. We recently went to see *Top Gun* at the mall theater. We both agreed Tom Cruise was dreamy. Unfortunately, I still saw Helen on the sly. I felt like a politician married to the woman who advanced his politics but more interested in the hot little honey on the side. Of course I wouldn't think like that if Dad didn't talk politics 24/7.

A lot of pressure was building for this summer. Come fall, I would be heading into the ninth grade, the last year of junior high. My next door neighbor, Annie, was one year ahead of me and quite popular. I often went over there, but I kept Cindy away. I wanted some assurance of making it in high school without complete dependence on Cindy or Kerry. I spent quite a bit of time thinking about these things, plotting different scenarios and strategies in case I needed them.

I continued to expand my babysitting business as well. Cindy thought I should be concentrating more on boys on the weekends, particularly Jeff. *Like that was up to me.* At least, I was earning money and not idly waiting by the phone like she was. Since her dad died, she had packed on about ten pounds to a body already ten pounds over the edge. I was ninety pounds soaking wet, and she seemed quite envious. Not that being thin guaranteed anything. At this rate, I would be lucky to have my first date sometime in college. Mom said I couldn't wear makeup until I was fourteen, so this August would be freedom. Finally some way to attempt to cover up the acne.

Mom was weird. I think she sat down after she had me and came up with this screwy list based on nothing. Krista must be eight to have pierced ears, fourteen to wear makeup, sixteen to go on a date, etc. This rotten system would be another item on my never-to-do list as a parent.

The only rule Mom had relented on so far in my lifetime was shaving. Right before we moved here, I told her all the other girls were shaving. It didn't matter until I was throwing the Frisbee with Dad one afternoon and he noticed my armpit hair. He marched me into the house and told Mom to teach me how to shave. He said it was not ladylike to be so hairy. *He always did know how to make me feel special.*

As if the hair incident wasn't brutal enough, Dad had come into my room recently holding up a multicolored polo shirt. As soon as I looked up and saw him standing there, the hair on my neck rose. Dad never just dropped by to have a chat. The fact that he was holding the shirt I wore yesterday, stretched as far as possible from his body, clued me into trouble.

"Krista, is this your shirt?" he asked.

Of course it's my shirt, you dork, or you wouldn't be in my room, I thought, but knew this was a "Yes, sir" moment. "Yes, sir."

"Krista, your mother sent me up here to . . ."

Crap, I thought, *this was the worst of the worst.* Mom only sent Dad to handle things too embarrassing for her or issues she was too emotionally wrung-out to handle. *Seriously, I was already fourteen, and not one adult in this house had approached the "birds and the bees" discussion.*

"Well, to discuss your personal hygiene," he continued. The look on his face was priceless.

"What about it?" I asked, hoping to prolong his agony. After all, he was putting me in the horrible spot to actually discuss such things. This was a man who puked if you accidentally left a sanitary pad in the bathroom. I could understand how it might gross someone out, except Mom had emphatically showed us the proper technique of disguising it. We were taught to roll it up in tissue paper until it resembled a small football. It wasn't even recognizable. Still, he would run out of the bathroom screaming about someone "leaving their personal items" like one of us had the plague and he could catch it.

"Krista, I just don't know how to say this, but I was walking by the bathroom and smelled something awful. I went in to see what it was, and I swear the hamper almost greeted me at the door."

I wasn't getting it, and I think he could tell by my expression, because he continued, "Krista, are you bathing every day?"

I admit it; my jaw dropped. "Of course I do, Dad. Well, I skip every now and again, but not often!"

There are moments in life I wish I had made up as a joke, something funny to tell my kids someday to scare them into submission. This wasn't a made-up moment. Dad marched over, pinned my arms above my head as if he was going to tickle me, and then actually sniffed my armpits. I knew this episode was going to be the highlight of one of my future psychiatric sessions. How embarrassing and disgusting to have your father smell you like you were in a military school being inspected.

"Good God, child. That's just gross! Are you using deodorant?"

"Mom!" I yelled. Someone had to save us from this.

"Krista, seriously, your mom is giving Chelsea a bath. I swear I will have more of these BO checks if you can't take better care of yourself. You've got to use soap and water, deodorant, and whatever else your mother can find for you. Do you understand?"

"Yes, Dad," was all I could muster. Thoughts were exploding in my head. *Was I stinking at school? I had just ended my period; was that causing me to smell? Could the boys at school tell?* I was totally freaked out. While my mind was racing, Dad stomped out of my bedroom.

As I was sinking into despair, Chelsea came in. She had radar for sadness; she sensed need and sought out the desperate. She shimmied up to my bed and crawled up to play with my tattered stuffed unicorn. Absorbed in my own melodrama, the next sound I heard was a thump.

Panicking, I quickly looked over the side of the bed. Chelsea was thrashing about, not making a sound.

I ran into the hallway. "Chelsea fell, something's wrong." Mom was already rounding the corner.

"I didn't do it, I promise," I said.

Abby

Bob's parents left early. Normally, I would have started cleaning right away, but I had promised Gloria Banting I would see her at two o'clock. I dreaded the visit, still fuming over Bob's interference and indiscretion. Still, some part of me was curious. If nothing else, it would be interesting to hear how her preschool program was set up and discuss educational philosophies. That's exactly how I was going to approach this, an enlightening afternoon of adult conversation.

I arrived early. Before I could ring the doorbell, a perky little brunette, barely five foot tall, with a full, beaming smile met me at the door, "You must be Abby," she said in a singsong voice. Her pale green eyes were striking and friendly. "I'm so excited to finally meet you."

She motioned for me to enter. Stepping in, I almost fell over a squeak toy. Before I could recover from the near miss, a huge black Labrador charged around the corner, his teeth barred.

Mrs. Banting grabbed his scruff, "Stay, Mason!" The dog stopped on command but continued to glare at me, twitching his upper lip. He wanted to make sure I was uncomfortable, and he was doing a fine job of keeping my knees shaking.

"Abby, come on in the kitchen. I'm cooking for a shut-in who goes to our church. The boys are finishing up their lunch. By the way, please call me Gloria. Surely I didn't hear you say Mrs. Banting. I'm not old enough for that!"

"I'm sorry. Have I caught you at a bad time?" I asked, dodging first one toy then another.

"Heaven forbid, this is my life! If David Banting would ever stay home long enough to see this chaos and pitch in, my life might be more organized. On second thought, I haven't been organized since I left home. And only because my mother made me!" When she laughed, it was contagious. I couldn't help being drawn into her magnetism.

Gloria never stopped talking, her hands flying around her torsos like wings. When we entered the kitchen, I saw twin boys sitting side by side in high chairs. They were completely covered in macaroni and cheese. The orange goo was in their hair and hanging off their long black eyelashes.

"Grab a towel over there and help," Gloria instructed.

Without thinking, I reached for the kitchen towel, wet one end, and started wiping the boy nearest me. The more I wiped, the more I spread the disaster. Food was everywhere. I'd never seen such a mess in my life. I knew I had lost the battle when Gloria grabbed the rag out of my hand and suggested I fill their bottles with juice. Her charge was running around, smeared with cheese and happy as a lark. Mine, half-clean, jumped down to follow.

"Hurry, Abby! The only thing that will calm them down now is a bottle," she shouted.

I flew into action, filled the bottles, and handed them to the boys without a moment's hesitation. I watched in wonder as they ran into the adjoining room, fell on the floor, and raised their bottles in unison. I didn't realize I was sweating until Gloria told me to sit down and handed me a glass of tea.

I looked around at the pile of dirty dishes in the sink and pots on the stove. Then I looked back at my hostess. Her hair was coiffed. Her nails painted a pale coral. And she was smiling.

"How do you manage?"

"You mean with those two?" she laughed. "I have two more upstairs."

"Another set of twins?" I gasped.

"No, silly. Two more kids. Jonathan and Jeremy are sixteen-months-old, Terrence is three, and Thomas is five. Then, there's Lilly and Samantha and Tiny and Mason. You met Mason at the door."

"Oh my," I exclaimed. "How do you manage?" I began to laugh, realizing I had repeated myself. "I'm sorry. I was just caught off-guard."

"Please don't be sorry. This is all my choice. Okay, let's get down to you. I need a preschool teacher, and you come highly recommended."

"Really, Mrs. Banting, I mean Gloria, I don't know if I'm interested in a job right now, and I don't know if I would be a good fit for your program," I straightened my back along with my resistance, but it did not take long for Gloria to infuse me with her enthusiasm for teaching and igniting mine. I left with a promise to think about her proposal and call back within a few days.

When I got home, everyone lingered around making chatty talk. I knew they were curious about my visit and my decision to pursue the job, but I didn't want to share my thoughts with anyone, especially Bob. I knew he wanted to ask, but he was smart enough to avoid another ass-chewing session. Frankly, I was glad they were backing off. In some demonical way, I liked lording it over them, having an air of mystery, even if it only lasted for a few hours.

I glanced at the calendar to see how my week was shaping up and realized I had forgotten the girls' upcoming recital. "SHIT!" I said aloud. As much as I didn't want company, I had to phone and invite Mom and Dad.

"Mama, how's Daddy doing?" I asked, trying to slow down and prioritize my conversation.

"Well, he doesn't have a choice; he has to keep doing," she retorted.

I did not have it in me to dignify her remark with a response, so I plunged forward, "Do you think you and Dad can come down for the weekend? Krista and Katie have their dance recital. I can call Mandy and see if she can drive you," I offered.

"Oh, I'm not sure. Let me check with your dad, and I'll call you back later tonight. Mandy's dating a new man, so I doubt she will want to break away from him for the weekend."

"A new man? What happened to the last one, or do I want to know?" My recently divorced sister decided to make up for six bad years of marriage by courting every available man in the county.

"You don't, and we didn't ask. All we know is we have Connor here almost every night now. Then again, I'm glad he's here. He's a lot of company for your dad. You know, I think he loves that boy more than he did his own children."

It didn't matter how many times I heard that statement about my sister's ten-year-old son, Connor, it still stung. It resurfaced feelings I harbored as a child of being totally insignificant. Even though I knew family issues were often hard to swallow, even after chewing on them for years, certain issues threw me off balance. And the only way I knew how to deal with the hurt was to embrace the tactic of memory-duck-taping, hoping it would eventually stick in order for me to forget. Most of the time it worked. In weak moments, however, the tape would loosen at its edge, and the hurt would slowly seep out. Today, I chose to punish myself and stall on the love quotient part of her statement, but swiftly decided to find solace in knowing Connor was providing joy to Dad. Recovery happened, and the tape tightened before any mentally draining relapses occurred.

"Abby, are you still there?" Mama's voice sounded irritated.

"Sorry. Hey, you know I don't mind Connor coming with you. He and Katie get along well together. It would be nice to have someone for her to pal around with."

"I'll check with your dad. A trip might do us good. It will be fun to see the girls. I'll call later and let you know what we're doing."

I shifted quickly to the tasks at hand: more cleaning, more grocery shopping, and another confrontation with Krista. She would have to give over her room to my parents; no doubt a battle would ensue.

I was totally immersed in this line of thinking when I heard screaming coming from upstairs, "Mama! Mama! Please hurry!" Krista sounded scared and frightened.

I ran. Before I even entered Krista's room, I saw Chelsea lying on the floor beside her bed. Katie knelt next to her, looking small and pale. Chelsea's arms and legs were sprawled in an unnatural position, her torso wringing from side to side. She was having a seizure. I gently turned her on her side to prevent her from biting her tongue. Intuitively, I began rubbing her arm, hoping to calm her and my rapid heartbeat.

"Krista, go get your dad, now!"

Krista was bug-eyed and scared. For once, she jumped up and did exactly what I said without one rhetorical comment.

"What happened?" Bob asked as he entered the room. His eyes went straight to Chelsea, lying limp on the floor.

While I checked vital signs, I barked orders, "Bob, I need for you to get Chelsea downstairs and lay her on the couch. I want to keep a close eye on her while I'm cooking." I turned to the older girls, "Krista, bring down Chelsea's baby blanket and pillow and put them beside her. Katie, grab her GloWorm."

Bob gently lifted her off the floor, cradling her in his arms. Krista and Katie followed behind him. I continued to sit, staring. My insides churned and bile filled the back of my throat. I fought the urge to vomit and scream at the same time. But as usual, I willed myself to do what life demanded.

The rest of the night was quiet. The girls sat vigilantly beside Chelsea, rubbing her arms or touching her head. Bob sat in the chair next to her, monitoring. I busied myself with routine household chores, pausing to check on her often. When I brought in sandwiches and soup, we sat near her at the card table and ate. We knew we had to be together. Somehow together, we felt we could combat anything. But when Bob's eyes met mine, one question sat between us. Had our decision to withdraw Chelsea's medication been in error?

Chapter 13

Krista

After numerous discussions and scheduling sessions, it was time to get my braces. Mom picked me up from school because the orthodontist requested the honor of my presence. I was not keen on having a mouth full of metal and hearing the snickering comments from friends and sisters but wanted the ordeal over.

It only took a couple of hours for the nightmare to unfold. You'd think a procedure as life altering as this would take longer than a movie.

The process to outfit me with the latest in metal fashion left me sore from one earlobe to the other. Prior to this, I had no clue teeth could ache. Yank, stretch, and pull would forever have new meaning. Before dealing with everyone's comments, I ventured up to my bathroom to examine my new look.

I stared into the mirror first without opening my mouth. *Looked like the same me.* Then I forced a smile. All I could see was metal. It covered over seventy-five percent of my teeth. Any show of joy on my part would bring a gray metallic glow to a room. *Wonderful!* I pulled my bottom lip down to expose the rest. I did not recognize myself. I moped downstairs when Mom called for dinner.

"What's wrong?" Katie asked, noticing I wasn't eating much. "Why aren't you eating?"

"Krista's really sore from her orthodontist's appointment today, so she's having a hard time right now. But, Katie, I made one of your favorites. Eat up!" Mom said.

"Oh, yeah, she's a brace face now," Katie said and stuck her tongue out at me.

Not willing to be outdone, I shouted, "Yeah, well you're a brat."

She pointed her finger at me and spat, "Everyone at school will think you're a nerd." *Katie was pushing her luck.*

"Oh, yeah," I said a volume higher than was allowed at the table. I knew I needed to come up with a good retort to shut her up. I didn't need more of her crap right now, because my teeth were sore and my head was pounding. Suddenly, I remembered Mom had recently bought me a smaller size in pants than Katie during our last trip to the mall. *I had my angle.*

I wrapped a smug I-got-you look on my face and with devilish delight said, "Well at least I wear a smaller size than you do, and you are three years younger!"

I regretted saying it as soon as it spewed out of my mouth. The look on her face was heartbreaking. The smugness fell off of mine and was replaced with shame. She ran from the table. I followed her to her room. Rounding her doorway, I saw her just standing there, facing her bed. I didn't know whether to stay or go, but her voice broke through the chilly silence.

"Have you seen the tap costumes for our recital?" Katie asked. She held up the blue and yellow sequined monstrosity.

We had been taking tap lessons together. I didn't tell Cindy and Kerry about it; they thought it was a very elementary school kid thing to do. I kind of liked it, but I was never going to admit it aloud. If they found out, my excuse would be Mom forced me to do it to spend time with Katie. I couldn't dance, which was evidenced by my role in the recital. When the teacher heard I had taken gymnastics classes before and was quite good, she put me in the center doing front walkovers while the rest of the girls danced around me.

"I don't even know why I'm up there. I'm going to be doing so many flips in a row, I'll probably puke," I said. Plus, I didn't know how

I felt about getting up on the high school stage in front of Mom, Dad, and as rumor had it, at least one set of grandparents.

"Do you know how tight they are?" Katie asked.

I was not following her. "Tight, what do you mean?" I hadn't really thought about the outfits except how hideous they were. Maybe we had the wrong sizes or something. "Well, let's try them on and get it over with," I said.

I took the bigger size and started putting it on but could tell it was too big, so I handed it to Katie. She held it up, threw it down, and ran out of the room crying. Mom, whose extrasensory perception must have been clued in, was by my side in a moment.

"What did you do?" Each word Mom said fell like a hammer.

"Nothing. I swear! We were just trying on the recital outfits."

Mom continued to bore a hole in my head with her intimidating scowl. The government should recruit her to extract confessions and read minds.

"Mama, seriously, if she cries at these, just wait until she sees the giant raisin costumes we have for the grapevine song."

Still in interrogation mode, Mom lifted her left eyebrow and said, "You'd better be telling me the truth." She gave me one last cold hard stare and then went racing after Katie.

I could feel sweat beads on my forehead. It was a workout trying to face Mom. Thinking back on Katie's behavior, I noticed a trend of weirdness. Maybe she was ready to start her period or something. Everything set her off, and I always seemed close enough to get blamed.

Days became a blur. Summer was so close even the teachers were winding down. At lunch, Cindy mentioned she wanted to have a big pool party for her birthday, which was in mid-June.

"And, we need to start planning this now?" I asked with more sarcasm than they ever seemed to absorb.

"Krista, you can be such a dork," Kerry said. "So, who are we inviting, Cindy? We should get invitations out before school ends, so people will know."

"Why don't we get together this weekend and plan everything?" Cindy actually jumped up and clapped. I thought we were over that kind of behavior in fourth grade for heaven's sake.

"I can't do it this weekend, because I've got that rec . . . ," I stammered. I quickly recovered, "I've got plans with my family."

"Oh God, you're really going to do that recital?" Cindy asked in horror. "Isn't it at the high school? The one we will be attending next year?" Okay, maybe I didn't recover.

I should have known it was a mistake to admit to her recently about the recital.

"Krista, come on, you're kidding," Kerry chimed in and then turned to Cindy, "We've just gotta go watch."

Great! Mom's parents were coming, which meant another round of Mom's insane cleaning; I couldn't dance, and now my friends were coming to watch the humiliation. Shoot me now. *It couldn't get any worse!*

"Hey, guys," Cindy and Kerry said in unison to the group of guys we usually ate lunch with, a group that included Jeff. "Want to go see Krista dance this weekend?"

"Funny, really funny," I said and left quickly. *I spoke too soon. This was worse.*

When I got home, Mom told me to move my things to Chelsea's room. My lucky streak continued. I hated being kicked out of my room for people who just ignored me anyway. I dug up my diary and added yet another item to the list of things I would never do to my children. After some therapeutic writing, I went to bed. This weekend would be grueling.

The weekend began like any other, except I was cramped up on a twin-size bed with stuffed animals all over my face. Mom greeted us early and started doing hair and makeup, because at times like these, you had to have makeup and three foot high hair. Mom teased our hair until it hurt, and we gathered in the living room for pictures.

Mom had vacuumed the formal living room, you could tell by the perfect pattern marks in the carpet. Its dark gray color served as the perfect backdrop for the flashy costumes. A sequined zigzag pattern ran all the way down and around the blue spandex leotard. Then, to top it off, we had a skirt made of tassels. Like sequins and spandex weren't bad enough! Mom had found me thick panty hose to wear since I complained about how white my legs were in the costume. Katie was lucky; she had a tan and didn't have to wear the hose. They looked like something old women in the nursing home wore.

Cindy and Kerry were the only ones my age there. I thanked God for small favors. My stomach flipped and turned as we waited behind the curtain for the signal to go on stage.

Katie must have noticed. She reached over and touched my arm, "You'll be okay."

She had a huge smile plastered on her face and her hands on her hips, striking a pose. She looked like a professional.

Suddenly the curtain went up, and we were sashaying out into formation. I was the last one. The girls formed a semicircle around me. When the music started, I began the front walkovers with the other girls dancing around me. I focused on the lights and kept going. Just as I feared I would spin myself into the next universe, the song stopped. I stood up to get back in line to dance our way out. Or rather I tried. I was so dizzy I could barely walk, much less dance. The girl behind me had to reach out and grab my shoulder to steer me backstage.

Katie and I ran to change into the raisins outfits for our next performance. The dance instructor got this hilarious idea to use black tights and black garbage bags stuffed with balloons to make us look like the California Raisins commercial. She also found huge plastic sunglasses for us all to wear. Katie was so into it, an actor about to take the lead role in a play. I had never seen her so excited.

"Well at least, I don't look fat in this," Katie said, stuffing in her last balloon and tying up her head hole.

"Are you kidding me? We all look fat in this," I said and off we went.

This time, I led the procession. Halfway across the stage, my plastic glasses broke. I was holding them up with one hand while trying to strike a predetermined pose. Then I noticed another girl starting to lose her balloons. Giggles erupted throughout the audience. As soon as we got home, I ran up to my room to retrieve my diary. I added the following entries to my list of traits a GOOD parent would have: make guests sleep in a hotel or on the pull-out sofa and NEVER make my children wear a garbage bag with balloons and sing *I Heard It through the Grapevine*.

Abby

Beyond today's choices lies an uncertain path.

The words kept turning round and round on a mental turntable. Each time the word "choices" came back into focus, I would stop and contemplate its import. After the tenth time, I was a wreck. There were so many choices to make each and every day. Some were careless, others required more thought. Still, it was the careless ones that haunted me. What repercussions would those bring? How many had I made today? Were uncertain paths made worse because of my decisions?

There were times, like this, when I struggled. It was like tiptoeing through a minefield. Choose the right door and win a car or the wrong one and drop through a trapdoor. Normally, I reveled in the beautiful complexity of the female psyche; but when I had to fight to unravel it, its intricacy seemed far too complicated, too dimensional, resulting in an explosion of insecurity bombs. When this happened, no-thought strategy came into camp and temporarily wiped the slate clean, giving me time to regroup.

Walking downstairs, I suddenly found myself in a lag between dinner and the next casualty. Feeling somewhat guilty, but not enough to deny myself this moment, I stretched out on the brown sectional in the family room. What a luxurious feeling! I pressed my toes into the arm of the couch and pushed hard, feeling my ligaments pull and tighten. When I flipped onto my side, I grabbed a pillow, pushed it under my head, and looked out the sliding glass door, admiring the new flower beds.

The first bed I spied was full of Brazilian verbena, already beginning to spread. The early signs of summer and heat had made the wiry, shrubby leafed plant flourish. Before long, its purple blooms would be profuse and showy. Behind it, cone-shaped Stella D'oros, from my grandmother's garden, bushed a lush green, readying itself for vibrant yellow flowers.

Next to it, I filled a second bed with society garlic. Known for its pungent odor, most gardeners avoided it. But I loved its finely textured clusters of narrow grasslike leaves and its bursts of star shaped pinkish-lavender blossoms. Bob hated the plant and threatened to tear it up every time he mowed in the area. He sneezed; his eyes drained and swelled. He thought the plants were the culprit for his allergy attacks; but to his indignation, I pointed to the true villain, the smog of gas fumes emanating from behind his dilapidated push mower.

Admiring my garden, I nodded in affirmation, "Those choices were certainly good ones."

"Who ya talking to, Mama?" Chelsea asked.

"Hi, baby," I responded, automatically, "Just talking to myself about the pretty flowers I planted." I patted the couch for her to join me.

She snuggled in; and we stared out the window for several minutes before she sat up, placing a paper object in my hand, "Mama, I found this upstairs. Read it to me."

"Well, what do you think it is?" I turned it over and over, a tightly wadded square folded and tucked several times. "Wow, whoever did this worked very hard, didn't they?"

"I think it belongs to Katie. She loves to write notes and fold them this way. She won't show me how. Will you?"

I slowly started to peel each corner up and over, revealing isolated words printed in dark, bold black. FAT UGLY HATE GROSS. Internal alarms blared.

I lowered the note, keeping it tight in my fist, "Chelsea, why don't you go see what Krista's doing?"

Chelsea ran out of the room, thrilled to have permission to check on her sister. I reached up and rubbed my aching forehead. My chest tightened, straining with each breath. Immediately, my hands flew up in a prayer pose, placing my index fingers on each side of my nostril, and clamping tight. *Maybe,* I thought, *if I don't breathe another breath, life would stop here and not progress to scary and complicated.* If there was such a thing as the weight of the world on your shoulders, I felt every ton of it right now.

Rather than opening more of the note, I forced myself toward Katie's room. I didn't need to read anymore. What I had been trying to deny for weeks had come to fruition.

Entering her room, I held the note in my outstretched hand and presented it, "What is this?"

She sat on the end of her bed, painting her toenails. Nonchalantly, she glanced my way offering no comment, looked back down, and proceeded to finish her task.

I tried not to scream, but fear pushed it out of me, "Kathleen Ellen Simmons, answer me now!"

"It's just a note," she announced dryly.

I was coming undone. Desperately I tried to regain composure, but I was past the point of no return. Fear offered its assistance, helped me raise my hand, and gave me the extra push I needed to slap her across the face.

Katie didn't cry, instead she stared at me with a full blank face. I was mortified, "Oh my God, Katie, I'm so sorry!" What had I done? I lowered my hand and reached for her. Her stiff body was unresponsive and unyielding. At that moment, I didn't care. I needed to touch her and somehow transfuse the love I felt for her through my arms and into her body.

"Mama, I'm sick and tired of Chelsea coming in my room without my permission," Krista said, then stopped to stare. I sensed her examining the situation, then in an asinine voice she added, "Hey, what gives? You guys having some tender moment or something?"

Fury has no wrath worse than mine, and I struck out, "I am sick and tired of your blame-name-and-shame mentality. Go to your room, now!" I snapped. She was smart enough to hightail it without one comment.

It was all I could do to refocus and not go spouting off or foaming at the mouth, my typical reaction. From past experience, I knew emotionalism, in that fashion, never worked on anyone. I sat quietly, watching Katie resume painting her nails.

If someone had looked in the window, the picture before them would appear serene, Katie intent on her nails and me watching. No one would hear my heart pounding or witness fear's stranglehold. This was indeed the hour of lead as Emily Dickinson penned it. I vividly remembered parts of it, learned in a high school English class:

> *After a great pain, a formal feeling comes-*
> *The Nerves sit ceremonious, like Tombs-*
> *The stiff Heart questions . . .*

> *This is the Hour of Lead—*
> *Remembered, if outlived,*
> *As freezing persons, recollect the Snow—*
> *First—Chill—then Stupor—then the letting go—*

Letting go. I repeated it over and over until it became a soothing prayer, bathing me in a comforting euphoria. Deep in thought and meditation, I almost missed the small voice beside me, "Mama? Have you ever been fat?"

Sighing, I lifted my eyes and looked at Katie. "I remember thinking I was."

"Did you hate yourself?"

"Yes." No reason to lie. No reason to deny the sound of my heart's knowledge and push lies toward ears that thirst for the truth, "I remember when I was in high school, maybe ninth grade, Sarah Jane Griffin told me I was curvaceous. I heard the words, but interpreted them to mean I was fat. Dragging myself home that day, I decided I would never eat again."

I stopped to think and reflect before continuing, "Mama got so mad at me. She said I was being foolish. I had the Bower body and needed to accept it. She probably thought her words were supposed to make me feel better, but you know what the Bower girls look like—tall, wide, shapeless, and let's face it, fat!" We both laughed out loud. She knew exactly the image I was describing. She had seen the whole slew of them.

Katie threw her arms around my neck, and I squeezed back. "I'm sorry, Katie," I whispered in her ear. For now, there didn't seem to be anything else to say. After one more hug, I left her room trying hard not to let my mind settle into dark places.

The rest of the day, I willed myself to be happy and eager to please. Guilt was the driving force; but if guilt could keep my mind centered and focused, then so be it. I challenged myself to have a noncatastrophic day, one that rode easily on the wind instead of always pushing against it.

"Hey, girls," I said as three sleepy heads entered the kitchen for breakfast.

"Mornin', Mama."

"I hate to rush you, but we've got to get going. Chelsea you have show and tell, don't forget to take the book you picked out. Katie, remember to go over your spelling words for your test today. And, Krista, we need to be at Dr. Murray's office at nine o'clock."

"Just kill me and put me out of my misery," she hissed.

"Your headaches are getting worse. The jaw alignment and braces might be the only thing to ease it. Doing nothing means more pain."

Reading the cues, I could tell Krista was working up to a full-fledged dissertation. To cut her off at the pass, I ran in with, "Tonight, I'm going to make everyone's favorite meal, fried chicken and mashed potatoes!"

The visit with the orthodontist went smoothly. Thinking I would be there half the day, I had brought along three magazines and two books. I could never tell the kind of mood I would be in, wanting to idly turn the pages of a magazine or concentrate on something literary.

I chose the magazines, pulling out pages featuring new recipes and fashion tips. I skipped over the "How to please your man in one hundred ways" section, but not before wondering why they published such nonsense in a female magazine in the first place. Where was the "One hundred ways to please your woman" section? I'm sure a survey would confirm that women were tired of seeing only one side of the erotic picture. Don't the editors realize who their readership base is composed of? When the headlines are about pleasing a lady, she will shove it in her partners face or insist on reading it to him. How else would he find out about the art of seduction? None of the car, business, or sports magazines received at our house even touched on love, unless it's defined by scantily clothed juveniles. I made a mental note to explore the possibilities of writing my own article to submit. Men needed to be enlightened; I married the proof!

"I'm ready," Krista said, blowing right past me without one glance in my direction.

Huffing, I gathered my stuff, ran out to the car, and opened the door, "Hey, that didn't take long." Her door slammed before I put one foot inside. I knew if I said another word, there would be a major confrontation. I quickly turned on the radio, sat back, and hoped the raging fury beside me would not unleash.

When we got home, Krista jumped out of the car and marched inside, reminding me of Hitler's youth's goose-step. I decided to leave well enough alone, knowing she could not keep her feelings to herself. I'd hear about it soon enough.

Before supper, I warned Bob to be kind about the braces. I had already talked to the younger girls, reminding them about Krista's sensitivity. Both of them had promised not to say anything at all, if they couldn't say something nice. I had them repeat it twice. The strategy worked for Bambi's friend, Thumper, it might work for us, too. But even the best laid plans fail. Hell blew open before I could even sit down to eat.

"What's wrong?" Katie asked Krista. "Why aren't you eating?"

Choices. I messed up again, talking before thinking. Rather than have them recite something from a Disney movie, I should have grabbed their necks and threatened them with bodily harm, Terminator style.

After a horrible exchange of bruising words, Katie ran off and Krista followed. I stood up to follow, but Bob caught my hand.

"Let them be, they need to work this out. My brother and I use to fight like dogs. It's normal."

"Bob, you don't understand. Katie has some real problems right now. She's ready to go over the edge. She has this insane idea that she's fat. Krista should have known better," I said, still ready to sprint.

"You're too hard on Krista, Abby. Katie openly started that battle. I don't blame Krista for name-calling. She didn't mean it. Kids understand that. Why don't you sit down and quit interfering all the time."

I did, not to agree with Bob, but because I did not know what else to do. It was clear by his declaration he did not understand his daughters' needs. But I bit my tongue and said nothing.

Later that evening, as I passed by Katie's room, I heard the older girls talking. Wanting to follow Bob's advice, I walked on by. While I was filling up the bathtub for Chelsea, I heard a commotion. Poking my head around the corner, I saw Katie rushing out of her room crying. I flew in before thinking. Krista, wide-eyed, immediately professed she did not know what had happened.

Steam blew out of my nostrils, but I turned and followed Katie. When I caught up with her, she was outside, swinging. Rather than charge over and get involved with her emotions, I waved and yelled, "I'm fixing dessert, wanna help?"

No answer. Instead, she pumped her legs harder, pushing the swing higher and higher into the air. I watched not knowing what to do. Did parenting have to be so brutal?

When God gave Moses the Ten Commandments, why didn't parents get a tablet, too? Doesn't it make sense that rules for effective parenting should have been a preface to the Commandments? Couldn't we avoid a lot of chaos right from birth by raising children astute in laws on how to behave? Every time a child acted out, he'd be firmly guided to the applicable rule; and at the end of each, it would read:

> *Do not disagree with thy parents for I, your Heavenly Father, have endowed them with the knowledge of child rearing. Follow their advice and you shall profit, and your heirs shall profit, and their heirs shall profit by your eminent wisdom. Fail to comply and you will die.*

Considering this, I had to wonder if God made a mistake from the get-go in creating Adam first. If Adam had been fashioned from Eve's rib, she would have had a bigger voice in the wider picture. Trust me, naming all those damn animals first, would not have happened. More importantly, she would have been the contact point to understand exactly what God said about the rules in Eden. Getting the WORD first, rather than secondhand, might have prevented the BIG screwup that lead to ten rules verses one. That taboo fruit, unless donned in Macy's shoe department, would have remained off the beaten path. She would have seen to it.

Trust me, unlike Adam, Eve would have delivered the complete directive from God to him from precise dictation, eliminating the need for more rules down the road. Frankly, it was apparent right after the nuptials that man was totally incapable of hearing or delivering the right message. So when she sat down with him to announce, "This is important!" she knew he would immediately tune her out and tune into his latest obsession, chimpanzees throwing coconuts at one another. And he would do this while nodding confirmation to each word she said. To test him, she often demanded him to repeat what she had said. To her dismay, he recited it verbatim, frustrating her to no end. So when the inevitable "fall" happened, man stood totally clueless, pitching his usual excuse, "No one told me," while woman stood poised with pen and paper in hand, writing down all the tips God could spare. When He said, "With pain you will give birth to children," she winced, writing without ceasing. Eve knew she only had to multiply Adam's dependency to understand the tedious parenting job ahead. God's heads-up helped, but more data would have been appreciated!

Laughing at my daydream did not help my dilemma but did ease the muscle constrictions in my neck. I paused, thinking back over Bob's comment. He was right; the girls needed to handle their own battles. I guess God got it right after all. He designed a plan to put two people together with different viewpoints and strategies, somewhat like opposable thumbs, to be able to work together or independently toward a goal, common or not. I quickly said a prayer not wanting my latest mental-run-off to be counted as blasphemy and be added to my growing list of sins.

The next few days were a whirlwind of planning, housecleaning, and grocery shopping for my parent's visit. The preparations turned out to be easier than hauling everyone to the auditorium for the performance. I almost missed it. Mom brought Bob's mother to surprise the girls. I ended up the one surprised. At the recital, they flanked me on either side and talked continuously, to me, over me, and around me. It was beyond annoying.

When the song "I Heard It through the Grapevine" began, I grabbed both women's wrists and insisted on their attention. The dance troupe traipsed onto stage, and the fun began. Within minutes, balloons were flying out of costumes; and laughter erupted in sweet chords. Rather than getting embarrassed and messing up, the team continued with their rehearsed format. I was laughing so hard I thought my side would pop. Or maybe it was a swelling pride at seeing my girls onstage, having fun together. It really was the small things that made a mama proud.

CHAPTER 14

Krista

Finally it was summertime. School had wrapped up without a hitch. My weekends were filling up with social obligations and babysitting. I had pretty well conquered the competition for jobs in our subdivision and was even branching out into the adjoining neighborhoods.

"Mom, guess what?" I yelled excitedly as I walked into our front door.

"Krista, you're home," Katie answered, "wanna come help decorate the clubhouse?"

"You finally got Mom and Dad's okay to do that?" We had asked about decorating it as soon as we moved here, but we weren't sure Dad would ever okay our big plans. The other day, he had given Mom grief about hanging clothes to dry in the backyard. He was afraid what the neighbors would say. After Mom explained the money she was saving, he offered to carry the clothes basket outside himself.

Katie skipped toward the garage door. I followed. Edging around the garage into the backyard, I couldn't believe my eyes. They had already started painting, and Mom and Chelsea were still working on it. On one side, in bright pink letters, was *Cool Cats*. Katie, who loved *Grease 2* and the song "Cool Rider" decided on the name. She took great pride in taking me on a tour inside and told me I could visit anytime.

Mom looked amused as Katie and Chelsea settled in. "The girls love it," she said. "I know you are busy, but maybe you could spend some time out here with them. They miss you."

Katie then rushed up and squealed, "Hey, Krista, remember the jackets the Pink Ladies wore? I asked Mom if she'd make us some that said Cool Cats. Want her to make one for you, too?" Their excitement pulled me in, and I picked up a paintbrush to help. I also took this as an opportunity to corner Mom.

"I promise to spend time with them, Mom, but I'm not making any promises about playing dress up. Hey, by the way, Mrs. Osborne told me her neighbors are looking for someone to keep their thirteen-month-old daughter two days a week for a few hours. It would be like forty guaranteed dollars a week."

"Krista, you know I'll need to meet them."

"But, Mom, Mrs. Osborne from church knows them. I don't understand the problem." I told her.

Mom was so insanely protective. She always demanded to meet every new person in my life. She had embarrassed me so many times with friends who invited me over for slumber parties. I could never go until she personally met the parents. I wondered what kind of childhood made Mom so skeptical of people. We didn't live in New York City for heaven's sake.

"Krista, this is not up for discussion. You know that," she said simply.

"Fine, Mom, it's embarrassing, but fine," I said. I wanted to say, "Whatever," but I'd seen how that hadn't benefited me in the past.

Mom smiled, trying to soften my bristling temper. "I'll call Judy tomorrow. I'm sure she'll introduce us. If you'd like, we could go over together and meet the little girl and her mom."

Now, wait, this had promise; it would probably help me get the job. Mom had an amazing way with kids, and they always loved her. With dollar signs floating through my head, I joined the girls as an official *Cool Cat*. This was yet another time where having younger sisters afforded me the opportunity to play and goof off without worrying what my friends thought. Even if they found out, I would blame Mom for forcing me to play with them. It was a perfect setup even though I enjoyed every minute of hanging out and singing with them.

It wasn't long before Mom called us in for dinner. The time had flown by. I really hadn't spent much time with the girls since school ended. Katie seemed back to her old self, and Chelsea was always an energy ball.

Dinner was ready; but as usual, Dad wasn't there yet. So we sat and waited. It was such torture to smell food and not be allowed to touch it. I imagined hell had some kind of similar torture where you could see and smell a slice of chocolate cake; but every time you reached out to get a bite, it would burst into flames.

When the phone rang, Mom got a disgusted look on her face, assuming Dad was calling to say he was running late. As for us, our mouths started to water; we figured she might get ticked off at him and give us our food. But it was only Cindy. Mom handed over the phone with a look of half relief and half murder.

"Hey, you guys eating?" Cindy started off.

"We would be," I yelled loud enough for Mom to hear, "if my Dad cared about his starving children. What's up?"

"Kerry and I have been working on the pool party, where have you been?"

"Working?" I asked. "How could it possibly take days to plan? You have a pool, and I know you have an *it* list to invite." Annoyance was starting to outweigh my hunger.

"This is big, Krista. How many times do I have to tell you? Anyway, we decided not to invite Susie, because she's been borderline for a while now, but we added Paige."

I was staring intently at the door, focusing my super powers, the ones I was praying to have right then, to will my Dad home. Mom had made fried chicken with mashed potatoes and gravy, and I wanted it! Now!

"So, what do you think about Gary? Or should we invite J.B.?" Cindy continued.

"Oh, um, I'd say Gary," I tried to cover up my lack of interest and honestly couldn't remember who J.B. was. I did not notice many boys beside Jeff, other than the ones who liked the Dallas Cowboys. In this area of the country there were a lot of Washington Redskins and Pittsburgh Steelers fans. Neither team was acceptable according to Dad, so I usually hung out with the guys who liked Dallas. "He likes the Cowboys."

"The Cowboys?"

"You know the football team."

"I swear, Krista, if you don't start listening to me about this stuff, you'll never have a boyfriend. No matter how well you fit in with the girls, you are never going to make it into the right group in high school without the right guy," Cindy lectured.

Just then, I heard the distinct sound of the garage door opening. Oh, the sweet sound of the dinner bell.

"Cindy, I gotta go. Dad just got home. I'll call you tomorrow." Freedom from annoyance and hunger. Thanks, Dad!

The sweet words of a little ditty Dad used to sing when I was little played in my head, "You take a chicken and you kill it, you put it in a skillet, and you cook it to a golden brown, and that's Southern cooking, and it's mighty nice."

"Let's eat!" were the sweetest words I had heard all day.

Abby

"Judy, I'm not sure Krista should be babysitting for people I don't know," I said, remembering my own experiences with Henry. The thought of Krista getting into a mess like that made my blood boil. Also, I did not want to endure the same fiasco we had in Kentucky with me constantly being the go-between in Susie and Krista's battle to win the babysitting kingdom rights. Alert lights were already blinking.

More and more of Cindy's contacts were calling. I figured the day of reckoning was approaching and war ready to ensue.

"Abby, I've known the Capshaws since they moved here a few months ago. For Christ's sake, he's an executive with the Boy Scouts," Judy argued.

I didn't answer right away. I was trying to balance the phone on my shoulder, the responsibilities of a parent, and a leg of chicken coming up out of hot oil.

"I tell you what," I conceded, "call the Capshaws and set up a time for us to go over and meet them. You can do the introductions."

"Come on," Judy interjected in disbelief.

I didn't care how my response sounded. "Set up the meeting or Krista will not babysit for them."

At supper, I told Krista about my conversation with Judy and the request for her to set up a meeting. Krista furrowed her brow, huffing, "Mom, this is so embarrassing."

"I agree with your mother," Bob interjected. I couldn't tell if he was agreeing to show a united front or trying to antagonize Krista. Frankly, I was surprised he offered a comment at all. His nose had been buried so deep in the sports section of the newspaper I expected to see black smudges when he surfaced.

Krista assumed a wounded stance, "I don't see why you guys always have to meddle in my business."

"Well, remember this and you won't ever forget it. My mama taught me logic every time she said, 'Because I said so, that's why.'"

Bob reached up and gave me a high five. Never in my life had I reacted on the Q-tee with a quip. It was elating! We rejoiced over my comeuppance, but it was a short-lived victory. Krista ran out of the kitchen yelling, "Fugeddaboddit!"

"Oops!" Bob declared, a smirk still lining his jawline.

"Oops is right. I'd better go and check on her."

On the way upstairs, the phone rang. "I'll get it," I called out, counting my lucky stars for a reprieve.

"Hey, Abby, is that you?" a male voice asked.

I could barely hear over the thrashing of my heart, but I didn't have to ask who it was. "Drake?"

"Yeah, it's me. I was just calling to check on your dad. Dora and I saw your folks at the Oyster Bar the other night, but didn't get a chance to talk to them. Your dad didn't look so good."

Feeling the heat rise in my face, I swallowed hard and stammered, "It's nice of you to call. I've been worried about Dad, too. Mama thinks he's going through some kind of depression. She said he won't let her leave the house without interrogating her."

Bob called from downstairs, "Who's on the phone, Abby?"

I cupped the phone hard and yelled, "Bob, I've got the phone. It's for me."

I pulled my hand away from the receiver then went over and gingerly closed the door. "I'm sorry, Drake. You know how it goes, as soon as you get on the phone some kid starts yelling for you." I tried to laugh, but the sound erupted as a strained mew.

"Earth angel," Drake said softly, "How're you doing?"

"Earth angel, my foot!" I chuckled, feeling the tension loosening. "When's the last time you said that?"

"June 3, 1967, to be precise. It's in the journal, my last entry."

"Send me the journal, you forger," I countered.

"No way! I got the journal when you said good-bye, remember? Hey, before I forget it, Reed Jenkins died. He was up toward Asheville in that old rig of his, and it went over the side of a mountain. Don't know if he had a heart attack or what. Dora and I are going to the wake tomorrow night. His wife's in pretty bad shape. He left four youngins behind. Thought you might want to know. Reed was kinda sweet on you back then." Drake paused, before adding, "But who wasn't?"

He was always a teaser. "I'm not going to comment on that last lie." My attention turned quickly to Reed. "Oh my God, Drake, I'm so sorry. I know you two were close. I'm not sure if I can drive up or not. I'll talk to Bob when I get off the phone. If I can't come, would you please give my condolences to his wife?"

Grief hollowed my tone as the past expanded into view. Clips, full of memories, flooded my vision; and the world created and erased itself in one second, mimicking a life's unpredictable journey.

Before I could respond, Drake interrupted, "Well, guess I'd better mosey on. Take care, kiddo. Don't forget this old fool."

Just as I opened my mouth to answer, the line flattened into dead tones. Holding the phone tight, like a lifeline, I sank to the floor. A part of me wanted to go to a quiet, alone place and wash myself in memories, but that was an indulgence I didn't have. I was a mother who needed to repair a rift created earlier. Standing up, I ran my hand through my hair, stiffened my shoulders, and headed toward Krista's room.

I spilled an apology as I entered. Krista sat on the side of her bed, giving the dog belly rubs with her feet. I eased over, remaining quiet, trying to give her space to open up. Finally in a hurt voice she asked, "Why did you make fun of me? You and Dad always treat me like a baby."

Daggers pushed in a little further. "I'm sorry about the snide remark. It was uncalled for." I paused before continuing, "Krista, I know you can't understand this right now, but I'm looking out for your best interest. People you think look normal, our very neighbors, may actually be bad people. That's how bad people get away with so much meanness; they put on false faces."

Krista looked up in admonishment, "So, you think you can just look at people and tell if they're good or bad?"

"Heavens, no! But I do think it's my responsibility to be involved in decisions that could put you in harm's way. I never feel I have all the answers; I make plenty of mistakes. Hopefully you can learn from mine and not make the same ones."

Her face went sullen as she alleged, "You never give me the option to make mistakes. You always take control of everything."

Feeling the discussion might go nasty, I intervened, "I'm sorry you feel that way, but I'm going to go with you to meet the Capshaws. I'm sure they're very nice and will understand my concern. As you gain more experience in judging people, and I gain more experience at parenting, maybe you can have more say-so in these decisions. It would certainly be a burden off me."

For once, Krista did not have a rebuttal. I hugged her and walked back downstairs. As I approached the kitchen, I heard Bob talking.

"No kidding," he said, laughing out loud. "Sure, I'm free. It'd be great to catch up on old times. Call me at work tomorrow, and we'll set a time and place to meet. You have my office number, don't you? Great! I look forward to seeing you."

"Who was that? I didn't hear the phone ring," I asked.

Bob whirled around so fast he had to catch himself from falling. "Oh, that," he sputtered.

"No, not that," I proclaimed, "who?"

I've never seen Bob act so meek. "Well, it was an old girl, I mean neighbor, from Williamston. Her family lived next door to us for years. You've heard me speak of the Strawbridges, haven't you?" I could tell Bob was trying to recover, getting all matter-of-fact with me.

"No, I don't think I recollect any conversation about a Strawbridge family. So who is this girl, why is she calling you, and what does she want?"

Bob's business face slid on. "Mary's in town for a convention. She left a message for me at work, and I returned her call. She wants us to meet tomorrow and catch up on old times, our families . . . you know, stuff. I told her to call me at work tomorrow to set up a time."

While I continued to stare, he coughed, a strategy he used to buy recovery time. "Look, I feel obligated. Our families used to be very close. She called out of the blue. Mother probably gave her my phone number."

"Well, I certainly believe the last statement. Your mother would undermine me in a New York instant! Yes siree, why don't you go meet Mary tomorrow and catch up on old times. I could care less!"

"You're taking this all out of context, Abby. Mary is just a neighborhood friend."

Walking over to the table, I picked up dishes and glasses, slamming items almost to the point of breakage to further state my point. Irrational jealously crippled my thinking. *Just a neighborhood friend, huh?* Bitter rage hovered right above my uplifted left eyebrow; and as long as I didn't speak, I could contain it there. But, oh no, Bob couldn't walk out and leave well enough alone. "Get off it, Abby!" he snarled.

Before I could fling an answer at him, Katie came to the door, "Mama, who was on the phone earlier? Did Madison call?"

I halted, remembering Drake's phone call. "It was a wrong number," I lied.

Katie walked out of the room, and I turned to face an angry, indignant Bob. The facedown began, and I refused to give in. Finally Bob huffed, turned, and walked out the room. Stomping my foot in a tantrum, I seethed, wanting to scream and throw something at him.

That's when guilt rode in like the headless horseman, holding my head in his bloody hands. I shoved the image aside. *I've done nothing wrong*, I thought. I am not going to feel guilty over some chance meeting with Drake. It's not like I planned any of them, and I certainly never phoned him. Bob's meeting with Mary had been prearranged and calculated. But the image refused to disappear. Instead it reared its ugly head, mocking me.

My temper slowly defused, and a new vision emerged. Mary was probably beautiful and a size two. All of a sudden, I imagined Bob sitting across the table from her with moonstruck eyes. What if this was just the beginning of many trysts? What if he fell in love with her? Why wouldn't he? Lately, I'd been a shrew. I buried my head in my hands as my stomach gnawed.

Chapter 15

Krista

The Capshaws passed Mom's scrutiny. They had one thirteen-month-old daughter, so the job was a piece of cake. Most of the families I babysat for had at least two kids who ran me ragged. Everyone seemed to love the Capshaws. Mrs. Capshaw worked part-time during the summers and was a school teacher the rest of the year, and Mr. Capshaw was some bigwig with the local Boy Scout chapter.

Becky, their little girl, was precious. She was very social and took to me right away. My second day there, she and I were playing blocks on the floor when I heard the front door open. A little startled, I gathered Becky in my arms and peeked around the corner. Thankfully, it was just Mr. Capshaw.

"Hi, Krista, I hope I didn't startle you," he said entering the room. "I decided to take the day off and get some yard work done. I'll be outside if you need me."

"No worries," I said. He disappeared outside, and I settled back in with Becky. I was getting paid either way, so I wasn't going to argue.

Around ten, I laid the baby down for her morning nap. I sat down on the couch and turned my favorite soap opera on. This was definitely an easy way to make money. Her Mom was going to be home by noon, and Becky would probably sleep until then. I began to envision the new pair of Guess jeans I'd soon be wearing.

I had just walked into the kitchen to get a drink when Mr. Capshaw walked in the back door and asked, "Do you mind pouring me one of those?"

"No, not at all," I said, reaching for a glass and filling it up. Knowing my immediate compliance to get Mr. Capshaw a drink would tick off Dad made me chuckle.

Mr. Capshaw sat down and used a kitchen towel to wipe away his sweat. He was a short round man with little hair left on his glistening head.

I felt him looking at me. Before I could get too uncomfortable, he finished his drink, rinsed out the glass, and set it in the sink. He didn't look back but asked, "Did Bec go down okay?"

"Yes, sir. She's a cutie. She didn't cry or anything," I offered.

"Good," he said. Instead of heading back outside, he started circling me. "Hey, my wife tells me you're quite the gymnast."

I didn't remember saying anything about my gymnastics to Mrs. Capshaw; but then again, I tended to talk too much so maybe I had.

"I guess so. I trained with a former Olympic coach in Tennessee. That's where we lived before we came here," I said, feeling nervously chatty, "But we couldn't find a good program here, so I quit."

"I hope Becky will love gymnastics some day. Could you show me some moves?" he asked and led me into the living room with his sweaty hand at the small of my back. I could feel the heat of it, and it made me nauseous.

I was very nervous at this point, afraid I might puke. I wanted to call Mom, but I was so confused because my parents were emphatic about the need to respect adults and not question. "Ummm," I stammered, not exactly sure what he wanted. "Like a cartwheel or something?"

He sat on the arm of the couch, never taking his gaze off me. "Sure, whatever."

I hesitated, but I thought I would do a quick cartwheel and then a roundoff.

After I finished, he excused himself and headed toward the bathroom. *Adults were weird,* I thought. When he came out, he said he needed to pick up some groceries and left.

My luck held. Just as Becky was stirring, I heard her mom come through the front door. I greeted her and asked if she wanted me to get Becky. Mrs. Capshaw was definitely responsible for her daughter's beauty—both had auburn hair, green eyes, and porcelain skin. We could have been considered family, their coloring so similar to my own.

"No, no, perfect timing on my part, I'll go get her. Would you mind if I pay you on Thursday when you come? I totally forgot to get cash," she asked. "Well, unless you don't mind a check."

"Thursday is fine. I'll see you then," I said. Halfway during my walk home, I realized I hadn't mentioned her husband being home. I felt uneasy. She might think I shouldn't be paid since her husband was here anyway. It wasn't as if he helped. I hoped she wouldn't be angry with me on Thursday or think I was devious.

The wonderful smell of cookies greeted me when I finally got home. Living here was such a treat. *I guess I should tell Mom stuff like that more often.* Many of my friends came home to empty houses or had to attend camps all summer because their parents were working.

"How was your day at the Capshaws?" she asked.

"Fine," I said and grabbed a cookie. "What kind of cookies are these?"

"Chocolate chip," she responded.

"No nuts, I hope."

"I wish, but I know you girls don't like them. Heck, I don't even like chocolate, so this is all for you gals and your father," she said as she scooped cookies onto a cooling rack.

"You're so weird! Who doesn't love chocolate?" I said.

Two days later, I practically skipped to the Capshaws. I would get forty dollars today, and the weekend was coming. Perfect timing, because last night I remembered I had to buy a new swimsuit for Cindy's big pool party the next week.

This time, I wasn't shocked when Mr. Capshaw showed up an hour after his wife left. He disappeared into the backyard after a quick hello.

Becky was fussy and didn't go down for her nap until around eleven thirty. Within minutes, he came in for a cold drink. But today was different. After drinking some of the iced tea I poured for him, he asked me to look at something in his office. Figuring he wanted to show me some pictures of Becky, I followed him.

He went behind his desk and opened a drawer. Then he handed me a magazine. I had never heard of *Playboy*. The woman on the cover was holding a towel over her boobs and down between her legs. With his prodding, I started to open it, but first gave him a puzzled look. He came over and began turning pages for me. Suddenly, images of naked women screamed up at me.

"So, what do you think?" he said and began to rub my arm.

I laid the magazine on his desk. Commandeering what little bit of bravery I felt, I said, "I think I hear Becky." He grabbed my arm, but I pulled it away. "Mr. Capshaw, I really need to go to her."

My breathing labored, I ran to Becky and scooped her up half asleep. I wanted to kiss her for saving me from such an awkward situation. I held her against me like a shield and returned to the living room. I was beginning to calm down when I noticed Mr. Capshaw in the doorway. I jolted.

"Becky woke up," I said, my voice cracking. "I'm going to rock her. I'll be in her room if you need me." I tried rocking Becky for a bit, but both of us were too keyed up. I heard Mr. Capshaw leave, and relief washed over me. But only momentarily. The more I thought about everything, the worse I felt.

The sound of the door made me jump out of my skin. I only settled down when I saw Becky's face light up and realized it was her mom. Mrs. Capshaw seemed a bit surprised to find Becky awake when she returned.

"Oh, hi, Mrs. Capshaw," I said. "Becky didn't nap well today."

"That's okay, it happens to me sometimes, too," she said, picking up Becky.

"I need to go," I said, so ready to leave. To run. I was afraid she would look at me and know everything.

"Let me get you your money," she said. I was so anxious to leave I had almost forgotten my money, and I never forget about money. She went around the corner and got her purse, "How was the day?"

"Becky was great, as always. And Mr. Capshaw just left," I said, stuffing the cash into my jeans. *There, I said it.*

"Jason was here?" she asked.

"Um, yes," I said, ashamed. *Had I done something wrong?* "He usually works outside when he's here."

She seemed to fumble with some toys. Her hands were shaking, but so were mine. I wanted to be home.

"I really need to get home and help Mom with some stuff," I said. Mrs. Capshaw gave me an absentminded wave, and I was off.

On the walk home, I wasn't sure what to do. I knew I should talk to Mom and Dad about everything, but I felt stupid. Not only did I do cartwheels and look at gross magazines, but I never mentioned any of it to my parents or Mrs. Capshaw. *Was it all my fault?*

When I got home, I went straight to my room, ignoring my sisters' pleas to head out to the clubhouse with them. When Mom came to check on me, I told her I wasn't feeling good. *I wasn't.*

The more I thought over the events of the last few days, reality hit me. I must talk to Dad. Not Mom. If all of this was silly man-stuff, he would tell me. Or, if I had acted inappropriately, he would tell me that, too. He was not one to blame adults and let kids off the hook.

After dinner, I asked Dad if I could have a minute. Mom, sensing something was amiss, excused herself and the girls, saying it was bath time.

"Dad," I said to get his attention away from some mail he had grabbed. *Couldn't he sense how scared and worried I was?*

"Yes?" he asked, irritated. *Maybe this wasn't the best time.*

Wringing my hands and fidgeting in my seat, I suddenly found myself at a loss for the right words. Feeling defeated, I wimped out, "Well, I guess it's no big deal, I'm just going to go to my room." I started to leave. But then I realized I had to at least find a way out of this babysitting situation. "Well, I do have one question."

He finally looked up, "Ask it then."

"What's the best way to quit a job?"

"Krista, when you make a commitment, you stick to it."

"But, Dad, I don't like working for the Capshaws," I said.

"Well, that's not a reason to quit," he said, but he had put down the mail.

Maybe I did need Mom. This man was dense. Mom would have known something was wrong by now. I cracked. "He showed me naked ladies, Dad. Is that reason enough?" I started crying. I needed his help. I couldn't face the Capshaws again.

"He did WHAT?" he yelled and grabbed my arm. "You tell me everything he did, right now."

Suddenly, I was scared. I didn't want Dad mad at me. I told him everything I could remember: the gymnastics, the *Playboy*, how he always came home. Dad called Mom and then took her to another room.

Next thing I knew, she sat down beside me and hugged me hard. Dad was heading to the phone. I was scared.

"Where's his number, Abby? I want it now!" I had never seen him like this.

"Bob, calm down," Mom suggested and started sifting through a drawer. "Here it is." Her attention turned back to me. "Krista, honey, it's okay. Please don't cry," she said and pulled me closer.

"Oh, God, don't call him, Daddy. I don't want to get in trouble!" I yelled and turned to Mom. "Please don't let him call. I'll never get another job. I didn't mean to do anything wrong, I swear."

"Honey, you didn't do anything wrong. This man did. And your father WILL call him," she said.

I watched Dad intently, but he placed the phone down. "Nobody there," he said. "Krista, go to your room while your mom and I figure this out. You will not go back; and if you see him for any reason, stay away! Do you understand?" he asked and then gave me a hug. The kind you gave someone you were never going to see again. I was so confused. He looked frightened.

Later, Mom came up to my bedroom. She explained something was wrong with Mr. Capshaw, and the police would have to get involved. They might need to talk to me. I wanted the whole thing to go away. I had Cindy's pool party coming up, and I didn't need all this other stuff going on.

The weekend passed quickly, and Mom seemed eager to take me swimsuit shopping when I asked. I begged her constantly to find out what was going on with Mr. Capshaw, and pleaded with her not to tell anybody because I was ashamed and embarrassed.

On Sunday, I started my period. The world was closing in on me. Weird neighbors, and now a pool party at Cindy's while I was on my period. Luckily, the party wasn't until Friday. Maybe, just maybe, I'd be done by then. *Are you there, God? It's me, Krista.*

The whole week was spent with constant phone calls from a panicked Cindy about things that quite frankly bored me. But it was a nice distraction from the Capshaw mess. I talked to the police about the incident. I had never been so mortified. Mom was with me, but it was still too strange. Even stranger was the news. Nothing was going to happen. The police called, seems they had gone to the Capshaws and nobody was there. They were gone. Just gone. They had taken their stuff and left. The house was empty, and the neighbors couldn't offer any information.

We never talked about it again.

Abby

I was knee-deep in the middle of folding four loads of laundry when Gloria called. Juggling the phone and continuing my chore proved tedious, but knowing who I was talking to settled me down, "Hey, what ya doing? You sound like you're in a tunnel or something." I heard children crying and dogs barking in the background. Next came a large crash. "What's that?"

"The usual. At least everyone hopped back when Mason knocked the flower vase off the table. No one's bleeding, thank goodness! I hate blood." As usual, she seemed totally unhitched by the chaos.

"How's everything at school?" Before she could respond, it hit me. I had forgotten to call her back to accept or reject the post at the preschool. I immediately began to apologize.

"No, really, don't worry about it. I could have called you, too," she stated. "It so happens, I'm in a bind for the next three weeks. David's been transferred, and I'm trying desperately to get the school ready for a new director, and my family ready for a move. I need someone to take over my class. There's not much left to do. The summer program is not as structured, and we have lots of activities which consume most days."

"Believe me I understand what you're going through. The girls are out of school . . ." I started to roll off excuses and then changed my mind. "Yes, I would love to help. I think it would be good for me. But, I'll have to work it around the Capshaws."

"The Capshaws?"

"Sorry. My oldest daughter started babysitting for them. I'll need Krista to watch Chelsea and Katie for me while I'm working. I'm sure it won't be a problem."

"Perfect! I only teach on Tuesdays and Thursdays. I'll be in the classroom with you on Tuesday to introduce you to the children and show you the ropes. Then it's all yours. You'll fit right in."

The backdoor opened, and Bob walked into the kitchen as I finished my conversation with Gloria. "Hey, what's up?" I asked. "I thought you were meeting Mary today."

"I did," Bob said, draping his coat and tie over the back of the kitchen chair. Without looking at me or elaborating, he picked up the newspaper and walked to the den.

"So? How did your meeting go with your friend?" My comment, meant to be half-joking, sounded rudely sarcastic. Moving closer to the doorway, I sneaked a peak wanting to judge his reaction.

He sat in the recliner, opened up the paper, and answered, "Okay, I guess."

Instinctively, I focused on what he was *not* saying. "Where did you meet?"

Bob lowered the paper, answering with a snippy tone, "Abby, what's with the twenty questions? I met an old friend. We caught up on our families. That's it! I'm tired, and I'm home!"

Stunned, I pulled back. Icy anger threaded my posture. This was certainly not like him. He was the one who couldn't wait to get home and tell me all the minute, boring details of his day. Now, all of a sudden, he wanted to give the impression that it was paining him to even give me the abbreviated version. I didn't think so.

Unfortunately, God did not give me the wits to shut up; but from past experience I knew to proceed with caution. I softened my tone, "I'm not trying to play twenty questions with you, Bob. I just wondered how your afternoon went with Mary. Your attitude's making me feel there's something you're not telling me."

Bob shifted in his chair, flinging the paper to the floor, "Okay, here goes. I met Mary Strawbridge at T.G.I. Friday's for lunch. We ate from the salad bar, had a glass of wine, and reminisced about old times. We left at two o'clock. She got in her car and drove off, and I got in mine. Instead of going back to work, I decided to spend some time browsing around a couple of used car dealerships. Then, I came home. End of story! Unless you want me to make up something," he scowled, picking up the paper and snapping it open in one final melodramatic I'm-done-with-you act.

Rather than answer, I walked back into the kitchen to finish supper. I tried hard to guide my jealous mind to reason, but all I could see was Bob and Mary sitting all cozy and intimate around a little table at Friday's, munching on salad, sipping wine, and staring deeply into one another's eyes. Images, filled with critical falsehoods, edged my vision—their little tête-à-tête, droning on about old times; brushing hands as they reached for their glasses, then pulling away as if their mere touch burned one another; giggling at each other's remarks; getting closer as their speech became more animated, knees touching, offering unspoken promises . . .

I grabbed a meat clever, turned it on its side, and slammed it down on an unsuspecting tomato, spraying pieces from one end of the kitchen to the other.

"Damn!" I yelped.

Shaking, I gently laid my weapon down and scoured the enormous mess. I deserved it for jumping to conclusions that did nothing more than spike my jealousy. All of a sudden, I felt like a balloon losing air. My chest, no longer heaving from anger, was weighted and sore. Tears rolled down my cheek, making me realize how quickly I had let my emotions get out of hand. Self reproach began to bubble from deep within, making it harder for me to regain control.

While wiping up tomato pulp and seed strewn from counter to cupboards, I kept berating my behavior. I didn't realize I was murmuring out loud until Chelsea grabbed my hand, startling me.

"There're white bugs on your face, Mama."

My hand flew up, swiping at my cheeks. Lowering it, I saw the evidence of my fit. Feeling more than foolish, I shook my head and grinned, "It's tomato seeds, silly, not bugs. Have you ever seen a tomato explode?" Chelsea shook her head. "Well, mine did," I said, continuing to clean.

"Mama?" Chelsea's tone caused suspicion.

"What, sweetie?"

"Katie hates me."

"Now, why would you say such a thing? Katie doesn't hate you."

Chelsea dropped her eyes and hung her head. "That's what she said. She said, 'I hate you, Chelsea.'" Big droplets of tears ran down her cheek and fell, spotting the lavender collar on her blouse.

Gathering her in my arms, I folded my unwarranted emotions deep within my breast in order to deal with hers. I ached, not understanding why children said such hurtful things to one another. And of all people, Katie was not the one to deliberately hurt anyone. She usually thought of others before herself. More than that, she adored Chelsea. From the minute I brought her home from the hospital, Katie mimicked everything I did, even pretending to breastfeed her baby dolls.

I leaned down and whispered, "I'll tell you a secret. You can help me cook supper and bake cookies. I bet that'll win Katie over. In fact, why don't we have a surprise party? What can we call our special event?" I pulled away and watched her transform from a very sad little girl into an imaginative little elf.

She sparkled, exclaiming, "A Happy Party!"

"Okay, a Happy Party it is."

The next couple of hours we were engrossed in fixing supper and baking cookies. I turned on the radio and drowned myself in chopping, pouring, and making light conversation. By the time supper was ready to be served, we were both laughing and singing our favorite songs.

"Hey, I'm going to run upstairs and tell the girls to wash up and come down to supper. Why don't you get Daddy?"

The first door I came to was Katie's. A twist on the doorknob failed; it was locked. I tapped lightly, saying, "Katie, supper's ready." I listened, my hand raised to tap on the door again, but something felt wrong. Without knowing why, I turned to leave but stopped when I heard the door edge open, "I'm not hungry." She hid behind the door, letting her flat words slide through the crack and fall at my feet.

"Are you sick?"

The door creaked shut; her "no" barely escaping its prison.

Numbness filled and blanched me before I regained my composure and spat, "You don't have to eat, but you will come downstairs and sit with us. Do you understand?"

I didn't wait for an answer but turned and proceeded down the hall. Right now I didn't want to deal with her. In fact, I didn't want to deal with anything else today.

When I got to Krista's room, she was sitting Indian-style in the middle of her bed, staring out the window. "Are you okay?" I asked.

"I don't feel so good."

"I'm sorry," I said, trying hard to sound sympathetic. "Come on downstairs. Chelsea and I have a surprise. I think it'll make us all feel better."

When everyone gathered around the table, Chelsea made her big announcement. "We're going to have a Happy Party! I made the cookies."

In dull harmony, the girls said, "Great."

Not wanting Chelsea's feelings hurt or my elevated mood shattered, I added, "I think a Happy Party is just what this family needs." I reached over and put a large platter of chocolate chip cookies right in the middle of the table. "Tonight, I'm breaking all the rules. Eat what you want and then dig into the cookies or just eat cookies. Eat all the cookies you want!"

"Wow, this is a Happy Party," Chelsea squealed, bypassing the food and going straight for a cookie. Even Katie seemed in better spirits as she grabbed one, too.

Toward the end of the meal, I was reminded of an earlier conversation. "Bob, Mother called today. She would like for me to come home and go with them to Dad's follow-up appointment with the doctor on Friday. I'd like to leave after supper on Thursday. Can you possibly work from home or take a day off?"

Bob hesitated, "Abby, I'm not sure. I'll have to see what's on my calendar." Pausing, he added, "On second thought, I'll have my assistant readjust things. Don't worry about the girls, I'll be here. Maybe we'll drive up Saturday morning. I'd like to see your dad, that is, if you want me to."

I lifted my arm to hug him; but Bob pushed his chair from the table, grabbed the mail off the counter, and walked out of the room. I lowered my arm, watching, feeling a deep, residing sense of disappointment.

All at once, Krista jumped up from the table and rushed after him, "Dad?"

Bob kept walking and Krista stayed in step behind him, repeating his name until I heard him utter a disinterested, "What?"

Sensing Krista needed time alone with her dad, I rounded up Katie and Chelsea for baths and story time. Within minutes, Bob was standing at the bathroom door, spewing, "Abby, I need to talk to you, now!"

The intensity of his words gave no room for rebuttal. I followed him down the hallway and into our bedroom. Bob slammed the door shut and started pacing. Emotion bristled around him like an energy field. His fair complexion turned beet red; his nostrils flared.

For one brief moment, I panicked, holding my breath waiting for that one crushing statement, "I'm leaving you for Mary."

When Bob stopped and turned toward me, I wanted to begin pleading for him to stay but halted. The broiling anger I saw wasn't about me.

"I'm gonna bust somebody's ass!" He rasped. "How could you let our daughter babysit for a pervert? I've a good mind to go over there right now and kill him!"

"Bob, what are you talking about?" At this point, I was totally confused and still reeling from the panic that Bob would ditch me for an ex.

"Jason Capshaw, that's who I'm talking about. That son of a bitch pulled out a filthy *Playboy* magazine and shoved it in my daughter's face. Heaven knows what else that little bastard had up his sleeve. Killing is too good for him!" Bob continued to pace, flailing his arms like a mad man, stopping every few minutes to slam his right fist into his left palm, "I'm gonna bust his sorry ass, Abby!"

"Bob, please slow down. I don't have any idea what you're talking about. I met the Capshaws. They seemed like nice people." My hand flew to my mouth as my imagination went wild. "Oh, my God, Bob. What did that man do to Krista?"

Bob stopped, his face bending in folds of pain, "Abby, Krista's only thirteen. She had no business over there unsupervised. How could you?" He yelled, shooting an upturned fist in my direction.

The horror of his implication smacked me full force. "Bob, how could you say such a thing? We discussed this, and you agreed she could babysit. You were the one who set the parameters—a few hours during the day, until eleven at night, and only if we were going to be home."

My mind whirled. In a biting tone, I continued, "How dare you make me out to be an irresponsible parent? I'm the one who insulted your daughter by insisting on a visit to meet them, to make sure they were trustworthy. Don't you remember the church member who called and recommended the asshole? How in the hell was I suppose to know the man was a child molester? Hell, he's with the frigging-ass Boy Scouts, for Christ's sakes!"

Anger fell out of Bob's face and pain replaced it. He lowered his arm, relaxed his fist, and closed his eyes, "Krista's my little girl, Abby."

His shoulders curled in as he lowered his head, "He could have hurt my little girl."

It only took two steps to reach him and wrap my arms around his shoulders as he heaved into deep racking sobs. Realizing the import of what had happened, I went into action. "Bob, we have to call the police. The man's dangerous; he has to be stopped."

With an unexpected suddenness, Bob pushed me away and rushed out the door. I followed but paused at Krista's room to check on her. Her eyes were huge with fear, pleading for comfort and reassurance. Frankly, I didn't know what to do or which way to turn. Life seemed upside down and inside out.

From down the hallway, I heard Bob yell, "Where's that son of a bitch's number, Abby? I want it now!"

"Bob, please calm down. I'm coming!" Krista grabbed my hand and ran with me.

Papers were sprawled all over the floor, and more were being thrown as Bob scrambled through the desk drawers. His crazed look terrified me, but I edged around him and grabbed my address book. "Here it is," I said, shoving it into his hand, before turning to comfort Krista.

Wrestling from my arms, she grabbed her dad around his waist begging, "Oh God, please don't call him, Daddy. I don't want to get into trouble." Panic iced each word. Bob shrugged her loose, picked up the phone, and began dialing. Krista reached over, grabbing my arm, "Mom, please stop him! I didn't mean to do anything wrong, I swear!"

Taking her by the shoulders, I urged her to focus, "Look at me! You didn't do anything wrong, do you understand? Jason Capshaw has to be stopped. He could have hurt you. He could hurt someone else. Your father has to call."

We watched intently while Bob dialed the number, waited several seconds, and then slammed the receiver back into its holder.

"Nobody's there," he said, facing us. "Krista, go to your room. Your mom and I have to figure out what to do next. You will not go back there, and if you see him for any reason, stay away." Bob reached over and squared his eyes on hers, "Do you understand?"

My head throbbed. I followed Bob back into our bedroom, watched him go into the bathroom, and slam the door. I sat on the edge of the bed, waiting for him, begging my staggering emotions to regain control. Finally he emerged, sat down beside me, and said, "You're right, Abby; we need to call the police. If I go out of here half-cocked and do something foolish, I'll be the one they put behind bars."

I reached over and took his hand, "Bob, you're doing the right thing."

He lowered his head, saying, "I'm supposed to protect my family." Agony and hurt merged, "Abby, I feel as if I've let Krista down. I put her in harm's way."

"Bob, you did no such thing! There're a lot of evil people out there who look just like you and me. It's not like the scum had molester printed across his forehead. We did what we knew to do to keep Krista safe. Now we have the responsibility to do that for others."

Instinctively, I began to rub his back, trying to ease the ache and sorrow. With each downward rub, I kept repeating, "We'll get through this," and with each upward sweep, adding, "I promise."

CHAPTER 16

Krista

It was Friday. Time for the social event of the year. Well, according to Cindy anyway. I was still sidelined by my period, but figured I could get away with a panty liner in my swimsuit. Most girls would have worn a tampon and not worried. But my mom, terrified because a friend of hers had gotten Toxic Shock Syndrome, never taught us how to use tampons. Plus, thanks to her retelling my Aunt's harrowing story, I was convinced if I used them I'd get it, too. So I didn't have a choice. I couldn't call and cancel.

When I showed up to the long-awaited pool party, twenty friends were already there with more to come according to Cindy. I must have ignored too many of her frantic phone calls because I didn't realize there were going to be so many people. I preferred smaller groups. Otherwise, it was too much competition for attention, especially when it involved mostly girls. Still, Jeff was there, and he was cuter than ever. When he was around, a heat enveloped me that words couldn't adequately describe. The warmth, normally a welcome feeling, lately left me confused. I didn't like the lack of control over it. It was one thing when Jeff stoked it; but when the thoughts of the sexy scenes in that *Playboy* did it, I felt dirty and ashamed.

There were times at night when I couldn't sleep and would explore my own body. It always seemed okay, but now I questioned myself. I would have died if my parents caught me. The first time I had experienced such a sensation was when I borrowed Cindy's *Forever*, the book my parents had emphatically told me not to read. I had reread the same paragraph several times one night. I touched myself trying to find the heat source. I had never told anyone. *Who do you tell such things to?*

There were a lot of subjects I couldn't discuss with my parents. But the Capshaw issue topped the list. *Were they disappointed in me or thought I brought it on myself?* As long as the subject was taboo, I'd never know. Plus, their constant scrutiny over anyone new was heightened now, making me feel under a microscope.

Jeff's voice brought me back to the present, "Cindy told me you'd be here." I swung around to face him. He had sandy brown hair that curled just at the bottom. This complimented sparkling green eyes, a chiseled nose, and a great smile. There was a bit of a Kevin Bacon quality about him. But maybe my obsession with *Footloose* clouded my vision.

Blushing, I said, "Good to see you." Lame! An overwhelmingly awkward silence followed, so I continued the conversation, "I can't wait for football season to start. Not much longer now. Preseason is coming soon. I wonder if the Bears will be as good this year." I was rambling. I knew it, and he knew it.

"God, are you talking football again?" Cindy asked as she handed Jeff a pop. "I swear. You might as well be a boy." With that, she walked away.

I reminded myself she was my friend. Although I was heading the pack right now, I knew even the slightest cut would have the sharks turning on me.

Jeff and I exchanged some sports talk, and then everyone jumped in the pool. Swimming and goofing around, I suddenly felt something strange. With a huge rush of embarrassing nausea, I realized my panty liner had started moving up the back of my swimsuit. I made a quick excuse, got out of the pool, and ran inside. It never dawned on me that the adhesive wouldn't hold in water.

Thankfully, the party was almost over. I started to throw the used liner in the trash can, but noticed it was empty. I didn't want anyone to see it. This was the bathroom everyone would be using. I decided to flush it down the toilet, but it kept floating back up. After several tries, panic was taking over. When it finally went down, and stayed down, I sat on the toilet to catch my breath. I felt like I had just run a marathon. My reaction seemed over the top even to me.

I stayed after the party to help Cindy clean up, but I had to borrow some personal products. We had a nice chuckle over what happened. She couldn't believe I still wore pads and had never used a tampon.

Embarrassed beyond belief, I admitted, "I wouldn't know where to put one."

Cindy looked at me with a judging eye. "That's just stupid. Sometimes I think your parents keep you in a bubble. My mom told me she'd put me on the pill as soon as I hit high school. My sister is on them. But, she has a real steady boyfriend, so she needs them," she said, picking up various pool supplies outside. "Mom said she should still have the guy wear a rubber, though."

At this point, I felt like a kindergartener. I had no clue what a rubber was. I knew I was going to get major grief for my upcoming question, but I had to ask.

"Okay, I know you're going to laugh, but what is a rubber?"

Cindy started laughing so hard her soft drink came spewing out of her nose. I should have waited until she wasn't drinking. "Come on! Seriously?" Taking one look at my face, she blurted out, "Oh my God, you really don't know!"

I wanted to go home and yell at my parents to burst the bubble and give some clue as to what was happening every day all around me. "Never mind," I said. *Clearly I was going to have to find out another way. Or, better yet, go the library and study.*

Cindy continued laughing for a bit but then realized I was not finding anything funny about this conversation. I wanted to tell her about the Capshaws and see her reaction, but I didn't want her to know.

"Didn't you have a sex ed class?" she asked.

My parent's decision on that one was going to haunt me forever. "My parents wouldn't let me attend," I said and hung my head.

"You've got to be kidding! Whatever, okay, a rubber is a piece of plastic that covers the guy's penis so his sperm can't get into you. You know, so you won't get pregnant. My sister says guys don't like to use them, though."

I had no clue what a penis really looked like. The only penis I had ever seen was a dog's. And nothing about it made me ever want to see a man's. Red and ugly when it was out and covered with fur when not.

Still, I was curious about kissing, especially the first one. I replayed the fantasy a zillion times in my head. But I had never tried to picture actual sex. The best my fantasies got was to fall into a bed of roses with a tall, dark, and handsome man then blur out.

All thoughts of penises and sex purged from my brain when Cindy abruptly announced, "I think Jeff likes the new girl."

I thought I would cry but asked, "What do you mean?"

"It's all that sports talk, I tell you. You never listen to me," she said without remorse or consideration.

Like she had all the answers, I thought as I gave an air of so what.

Later in the evening, I saw Mom sitting alone in her bedroom. I seized the opportunity to talk with her. "Mom, am I pretty?" I asked.

Without stopping what she was doing, she said, "Of course you are, darling."

"Am I too boyish?" I continued.

My mom looked up with panic, terrified of one more teenage drama. "Okay, what's this all about?" she asked and patted the bed in invitation.

I took her offer and sat down. I didn't face her, but I wanted to be close. "I like Jeff," I started.

"Our neighbor Jeff?" she asked, sounding shocked.

I thought she had already figured this out. She always seemed to be on top of things before I could even tell her. I turned to her, "Yeah. But, Cindy said he likes this new girl. She said I talk about sports too much."

"Since when did Cindy become all knowing?"

"She isn't ALL-knowing Mom, but she seems to know more about boys than I do. I'm such a dork. As if the braces and glasses weren't enough, these zits are out of control," I said, pointing out several offenders.

"You stress too much, Krista, that's part of your problem. Relax and be young. You have years to stress and worry about things."

The standard kids-have-it-so-easy lecture was not what I was looking for at the moment. "Boys don't look at me like a girl," I said.

"Krista, you're not in a rush, are you? Trust me when I say it's better if you just wait with the whole boy thing," Mom got up and walked over to her dressing table, opened a bottle of lotion, and began smearing it on her face and hands.

Realizing this conversation was going nowhere, I decided to change the subject, "Are you still going to see Grandfather this weekend?"

"Yes. You all might join me on Saturday."

"Oh, that should be fun. Let me go to bed now and prepare for the excitement," I said, hoping to make her laugh. But she didn't look over or even seem to notice me leave.

I knew Mom loved me, but at times she had a strange way of showing it. I walked into my room and plopped down on my bed. After several minutes of staring at the ceiling, I reached under my bed for my rogue copy of *Forever*. I had dog-eared a certain paragraph, and I began to read it.

Someday, someone, is going to want me just the way I am, I thought. Then I lost myself in the romantic notions of my mind.

Abby

Pulling a bar stool up to the kitchen sink, I watched the sun break through the thicket of Wax Myrtles separating our backyard from the neighbors. The sun's yolk yellow bled around the limbs, spilling onto the brown patches of dead sod. As it rose higher, light and shadow moved in detailed increments, highlighting flower beds, forgotten toys, and fragments of broken shells used for compost. Dew drops sparkled, bouncing off each item before radiating into a prism of colors. I sat at the perfect vantage point.

When daylight made her full entrance, she curtsied; and I applauded. Within minutes, the ocean blue sky was filled with white capped clouds. Butterflies tripped and teased around the yard, their wings opening wide, then half-folding into a prayerful, welcoming dance.

Two finches ran a game of tag and retreat. Even with the windows closed, I could decipher their rich musical warble. It was mesmerizing. The female was streaked in shades of dull brown, while the male sported his feather finery, a raspberry-purple hue. At first, the female appeared shy and reticent to the advancing antics of the male; but before long, she was right in the play and fast becoming the speedier opponent.

Along the southern edge of the property, a long hedge of Burning bushes had begun to change to a deep red color characteristic of late summer. The transformation had begun early; but the weather had been unusual, cooler and wetter.

Watching nature's opening scenes lifted my spirit but only temporarily. Events layered one over another, dividing my mind, distracting me, affecting my judgment, discoloring my senses. And no matter what I did, it felt as if a dark storm hovered—ready to advance, stall, change directions, make a direct hit, or die out completely. And I knew, if the pattern held true, an aftermath of wallowing in disappointment, discouragement, and depression would follow. For days, I felt it approaching like an incoming storm, predicted, yet unpredictable. Mooring it to the back of my brain had not worked; the knotting failed at the first tug.

That's when Despair walked in and sat down, posturing herself like a streetwalker. It was always her pleasure to advance ahead of storms, to heighten anxiety. Like a strong wind, her wiles were used to destroy objectivity. I knew her, knew what to expect; and yet, it was all I could do to hang on as she lashed at my reserves. When she finally exited, she turned for one brief glance at her handiwork and offered a tasteless, winning smirk as I stumbled over the toppled damage she had left behind.

I shook my head until Reason showed up to pick over and discard pieces from Despair's effects. Even with a clearer path, a brooding sadness stirred. Was it possible that I was one of those women who could never find happiness? Was I pushing Bob into the arms of another woman? Had my judgment about the Capshaws gone awry? The thoughts repelled, coiling their slimy insinuations into a tight knot.

Anguish welled up so powerfully; my chest expanded, causing my breath to momentarily cease. I struggled to think of an uplifting song, a spirited movie, a quirky phrase, or a verse from the Bible to stop the insane path to mental demolition. My head pounded from the exertion, birthing no comfort.

Panic swelled. It began as light as rain's first droplets and increased in strength until it became a torrential downpour, a frog-choker, Dad would say. And without warning, Fear's chokehold claimed me.

She did not look the way I thought she would—tall, wistful in movement, and elegant. Fear seemed harmless until I sank under her hypnotic eyes and felt the draw of her hands, gently pulling me into the vortex of her bosom. The closer I got, the louder hell's sirens beckoned me into captivity.

Sweat droplets rolled down my face; and at the same time, a sweet coolness made me shiver. I clinched my eyes tight, willing the turmoil to stop. That's when each image, each uninvited guest, slowly walked away. Fear walked out last. She withdrew as casually as one would remove their hand from another's.

Quiet nestled in, and languor deposited itself. Its softness enveloped me, beginning a renewal. The sideshow ended; but the main event, the name highlighted on the marquee in bright neon lights was mine. It was a profound moment when I acknowledged, and not for the first time, the reason internal bashing plagued me and why; at times, it was so hard for me to acknowledge happiness or love. The truth of the matter was I lived with an addiction—Insecurity.

Insecurity owned despair, anguish, panic, and fear. It birthed every negative feeling I had and labored to destroy me. Every silver bullet it fired was meant to disable, maim or kill. As much as I sought the safety net of Reason, Insecurity had a bondsman's contract on her, too. Sometimes the effort seemed futile until I weighed the cost with the reward—a husband and three daughters.

I glanced down at the countertop. In a soap dish next to the sink there were three pieces of beach glass we found earlier in the summer. Once objects of intent, they were now irregular shards that had been shattered, tossed, and resculpted by powerful oceanic forces. The end result was something entirely different, lovely, and intriguing in its own right.

Cracked and hazy, I noticed the green one had a character of its own. I picked it up and turned it around, seeing my reflection in a faceted corner. *I am beach glass,* I thought. God had created me, stood back and announced, "It is good." And even though life's storms had altered and changed His original design, I was still the Creator's child, loved regardless. The image spun a protective web around me, causing two giant words, *I'm sorry,* to form and spill out.

Sweet perfumed incense filled the air; and I inhaled deeply, imagining it traversing throughout my body. In a matter of minutes, I had thrown myself into a concaved pit, sustained injuries, applied a soothing balm of faith, and crawled out with an apology and prayer. For now, I had ceased the dangerous drift into miry clay.

Rumblings spilled down the stairs, and I realized the household was awakening. The sounds were soothing. They were the promise, the payment; and it was worth all the cost to be here in this moment and time.

Going through the rote work of morning, I continued my internal lamenting, remaining cognizant of its slippery slope. I realized Bob's meeting with Mary and the bedlam over the Capshaw's had resurrected my she-demons, but I also knew I had the power to restrain them.

Change had to begin now. After years of conditioning, I had a lot of work ahead of me. And realistically, I knew relapses were bound to happen. Human emotion was not like gathering an accumulation of quantitative or qualitative scientific data and simply applying it. It had to be tested over and over again until the success of its application proved reliable. From experience, I knew stumbling blocks in life could be enablers or disablers. The determining power was mine. Like beach glass, I could weather the storms and come out entirely different and intriguing.

The girls rounded the table, giggling and teasing one another. Seeing them reminded me how crucial my role was in their life. The thought was staggering. If all they saw was the pathetic figure I was earlier, what chance would they have to break free? Three girls watched every move I made. I wanted to role model a course filled with confidence and security, essential elements to battle whatever came their way, rather than have them default to the pit like their mother.

I finished serving breakfast, excused myself, and rushed to dress. Before I made it to the bedroom, Krista called out for me to answer the phone.

"Abby, how're you doing?"

Not one name surfaced. The quietness must have clued the caller, "Abby, this is Mac . . . Mackenzie Manning, you know, from across the street. Remember, we met the other day at the Frank's?"

"Yes, I'm sorry. I don't know where my mind is these days. Sure, I remember. How're you doing?"

"I'm fine, thank you. I feel awkward about this, but I've got a favor to ask of you. I take Todd in for some minor surgery next Wednesday and wondered if you would catch Lane and Jeff when they get off the bus and watch them until James gets home? I really hate to ask, but I'm in a bind. Mother's out of town and Dad's in a golf tournament."

It wasn't what I wanted to do, add more kids to the mix, more traumas to the day, but I wanted to reach out. This might be a way to make a new friend and extend myself beyond the house and children. "I'd be happy to help. I've marked it down on my calendar. I hope Todd will be okay?"

"It's a minor procedure to remove a suspicious mole. The dermatologist thinks it's fine, but we're going to make sure. Hey, you're a doll for doing this. Let's get together soon for coffee or lunch."

The next few minutes were a mad rush and dash to get everyone to school. I grabbed my sunglasses and pocketbook and got in line behind the girls. My last thought, *you're a piece of beach glass,* empowered me, making me wistful and eager to face the day's challenges.

I worked hard to keep my goal intact. I did not snip at Bob for being late or chastise Krista when she announced at bedtime that cupcakes were due at school the next day. I bit my acrid tongue. I bit it, hoping the taste of blood proved more palatable than an irretrievable remark.

Later, Bob came up behind me, circled my shoulders with his strong arms and professed, "Hey, I'm sorry if I hurt your feelings last night. I don't blame you for what happened with the Capshaws. Everything I said was out of anger and fear . . . nothing more. It was like my worst nightmare coming true." He pulled me closer, "Oh, Abby, I need you. Please don't be mad." His voice trailed kisses along my neck.

I leaned back, feeling an old tinge of resentment swell and resistance rear, but I squelched them. Turning around, I saw Bob's olive eyes deepen into a smoky hue. Liquid warmth flowed, swelling, leaving my knees weak and unsteady. He saw and knew. Responding, he lifted my chin and kissed me on the mouth, lightly at first, stoking the flame until it blazed into urgency. Pushing up on my toes, I pressed in close, wanting more. He met my passion with his own, reaching around and firmly lifting me onto the counter. His moist breath trailed, tracking from hairline to cleavage before stopping.

"Wow!" he said. I could feel his raspy breath inches from my face. I didn't want to open my eyes or end this moment; but when I did, his look was searching. Before I could comment, a faint smile spread, cocking up one corner of his mouth, "To think I almost worked later."

Through half-mast eyes, I silently begged him to hush and continue. Hunger mounted. My arms, laced around his neck, fingers fanning, sliding through his thick hair. I drew him in tighter, winding my legs around his waist, demanding, needing him fully. Leaning closer, I ran my tongue over his ear, tracing the curvature from top to lobe. Bob shuttered and then let out a hearty laugh.

Pulling away, I looked at him in surprise, "Did I do something funny?"

"No! It's just the best damn thing you've done in a long time. I think I need to start coming home earlier." He reached down, kissed my throat, and whispered, "Did I tell you how much I missed you today?" He lifted his head, his eyes solemn, waiting for a response.

My heart tripled bounced, uttering soft words spoken like a hurt kitten, "No . . . not today."

Bob gave a teasing smile, knowing he was in full control of the situation. With each breath on my neck, each tingle of a kiss, all the nonsense circumventing my brain peeled away.

I tenderly edged back, wanting to face him at eye level. "Bob, this is so perfect. Why can't it always be like this?" Bob's face softened in acknowledgment, as I resumed, "Today, I've been questioning everything about myself. Am I a good mother? Is my judgment totally off? Am I pushing you away?" I gulped hard, "Into an affair? I've tormented myself with these questions. I've been in a horrible, frightening place. Last night, your accusations really threw me into a tizzy."

Bob started to withdraw, surely thinking a lecture was rising; but it wasn't, and I had to explain, "No, listen. This moment, our closeness, answered a lot of those questions. I love you so much and need to know you love me, too. I need time with you. I need for you to kiss me. I need passion. I can't wait, hoping for intimate moments; I need to create them. It's like happiness. Sometimes you have to make it happen because if you wait for it, you'll probably find yourself in a deserted terminal."

The hunger to tell him how I felt sounded desperate, but the excitement over this newfound knowledge had to be shared, "Bob, stay with me, one more minute. Do you remember the classic movie we saw the other night, *A Streetcar Named Desire*?"

Bob sighed, shrugging, "Maybe? I dunno, Abby." I could see and feel him disengage, physically and mentally.

"Yes, you do. Your favorite actor, Marlon Brando, was in it and Karl Malden. Remember you kept making jokes about his nose?" I rambled, taking forever to get to the point, but I wanted him to understand my revelation.

His voice edged in agitation, "Yeah, I guess."

"It hit me, just now, what Blanche DuBois meant. It's what I've been feeling and couldn't find a voice for. I'm not unhappy with life; I'm trying to figure it out. I need to talk about it and even act it out sometimes to see what works for me, what works for this family. Do you understand?" No holds barred, I was going all out, with nothing to lose, feeling I had found a treasure map.

"Abby, I don't understand. I wish I did, but . . ."

I motioned for him to stop, "Blanche made an historical statement for me, Bob. She made it for a lot of women. She said, 'I don't want realism. I want magic! I try to give that to people.' Bob, here I am, rushing around taking care of everyone else's needs, trying to make everyone happy, and there's nothing left. When I fall apart, it's out of sheer desperation. That's the truth. All the other crap I spout off is only half truths, just a bunch of hurtful words because I'm hurting. Until this very moment, I haven't known how to express what I need out of life."

Breathless, I stopped, thinking about each word I was about to say, spreading them out on a mental clothesline to gauge their import. "Bob, I need some magic in my life!"

Bob halted my speculation by putting his arm around me. Before I could respond, he attempted to address my deluge, "Honey, I didn't realize the effect I had on you. I mean *that* was some kind of response you gave, but I think it popped a wire." He began rubbing my back, pity filling in between each word.

Popped a wire, I thought. I pondered what he had said, trying to align it with my newfound knowledge. My task had failed. For the first time in my life, I had made a clear distinction between the terms *reacting* and *responding*, knowing it was unbiased truth. Self-disclosing had been hard. Now it had been reduced to a quip.

An awkward pause ensued as our eyes met. We stood at a crossroad. I could berate him for totally missing the point or get over myself. I chose the latter. It wasn't worth the work to dissect each line of

Blanche's speech for him and deal with his deer-in-the-headlight look. In fact, I felt sorry for his linear thinking and proud that even in tragic moments I could somehow entertain myself, especially when it came to new revelations. I loved being a woman—able to push a daisy up into a mushroom cloud and not get blown to smithereens.

Leaning over, I patted him on the back, "Bless your heart! Here I go rattling on and on. You must be exhausted! By the way, you're quite the man." I pulled him in, planting a luxurious kiss on his forehead. He truly couldn't help he was limited. More than one revelation had opened my eyes today.

A fresh, new look crossed Bob's face, "Abby, why don't we go away, just the two of us? We could get your mom or my mom to come down and watch the girls. How does Cancun sound?" Bob looked like a kid who had been offered his pick in the candy store.

I hesitated for a minute. The thought was heavenly, but was it practical right now? "I don't know. What about Dad? He's still recuperating, and . . ."

"Oh," Bob said, disappointed. "Well, we don't have to go right now. Let's wait and go in the fall when the girls get back in school. I'm sure my mom would come down. That is, unless you've decided to go back to work."

Just then, Chelsea came running into the kitchen, "Mama? Mama? I can't find my picture!" She ran over, pushed Bob to one side, and clutched me around the hips. Her eyes brimmed with tears.

Bob bent down, "Chelsea, where do you think you put it?"

"I don't remember. Daddy, it was beautiful. I made it special," Chelsea whined.

Bob took her hand, "How about we let Mommy go upstairs and take a nice warm bath, and we go look for your picture?"

My mouth dropped open, but I quickly popped it shut before Bob and Chelsea saw. I felt I had a free pass. My initial reaction was to rush from the kitchen before I drew the "Go to Jail" card. Tonight had been exhausting. Talking to Bob was as intense as listening to Krista talk about her teenage traumas. None of us could see the other's point of view. Right now, a nice hot bath sounded prescriptive. I left the pots in the sink to soak and trudged upstairs.

While the bathwater ran, I poured in my favorite lavender vanilla bubble bath. The aroma drifted through the air, sweet and floral. Closing my eyes, I leaned my head back to inhale the perfume. My shoulders slid down into a more relaxed position, and my pounding head slowed to a dull ache. Feeling tiredness wrap its arms around me, I walked back into the bedroom to retrieve pajamas. My mood felt purple, so I picked out the beautiful deep violet gown Mama had given me for my birthday.

Krista knocked and entered the room, halting my walk back to the bathroom. "Mom, am I pretty?"

Her tone and posture belied a casual teenage conversation. I excused myself, walked into the bathroom, and turned off the water. I took one deep breath before exiting and braced myself for another hurdle.

Sitting on the side of the bed, I motioned for Krista to join me. I could tell she was nervous. I ventured to guess it was something personal. Rubbing her back opened up a geyser of hurt feelings. She began her delivery reciting comments Cindy and a neighborhood boy had made. I almost laughed because it sounded so childish but quickly refrained. I knew, for her, this was serious. She thought her whole feeling of young womanhood and self confidence were tied up in what other people thought. I understood. I had a lot of painful memories of my mother's aloofness when I came to her with problems. Now I realized they were not as big as hers; but to me, at that time in my life, they were monumental.

Letting the conversation round out, it became apparent the insensitive remarks were only more additives to her already-brewing turmoil. Krista saw life as a series of disappointments she had no control over.

I tried hard to listen; but after elongated minutes of hearing her drone on, I checked out. Rather than being attentive, my responses became mechanical, too textbook. Without thinking, I berated her for creating stress in her life. When her face suddenly went sour, I halted, realizing I had done the same thing Bob had done to me. The difference was I understood, making the act seem more heinous.

How could I have been so insensitive? She had more reason than most to be stressed. I essentially told a girl who saw twenty-four hours as an eternity to wait until she was an adult to have life work out for her. *Just turn me into a pillar of salt*, I thought. The worse part? I realized my mother showed up in my voice.

Feeling awkward and embarrassed for botching up the whole conversation, I sat staring at the floor, not knowing what to say or do.

Krista broke the edgy silence. A swift change crossed her face. No longer whimpering and sad, she became indignant and spiky. She shot me a darting glance and marched haughtily out of the room.

"Shit!" I said, shaking my head. If I had tried, I couldn't have blown a conversation any better.

Walking over to my dresser, I placed the purple gown back in the top drawer and reached for the faded, torn yellow gingham one. Slowly, I walked into the bathroom and pulled out the drain plug. Sitting on the rim of the bathtub, I watched the scented water swirl around and around until the last drop descended into the drainpipe. Getting up, I put on my tattered gown and went to bed.

Closing my eyes, I prayed, "Hey, God, it's me, Abby. I need some guidance down here; I'm fouling up right and left. What I don't want to do is mess up my daughter's life by walking in reverse. Grant me wisdom and a lot of mouth control to show strength and confidence.

God, it's just You and me, and I'm outta answers. If you could spare a few, I've got pen and paper handy. Amen."

"Oh, PS, forgive me for all my sins. They're wracking up faster than I can amend them."

CHAPTER 17

Krista

As if the summer wrapping up wasn't disconcerting enough, Mom had volunteered to have Jeff and his brothers stay at our house while their parents were out of town. *Did Mom want the pleasure of plunging a knife deep into my abdomen?* I would have asked her, but she lacked all humor lately. Perhaps the serious route would do better. I'd simply explain how awkward and mortifying having Jeff at our house overnight would be. But that wasn't appealing either. One, I knew Mom wouldn't understand. Two, she had given her word to help. Cindy managed to shine an even brighter light on the situation.

"Your mom offered what!" Cindy yelled into the receiver.

"You heard me, please don't make me repeat it," I said.

"And when is this happening?" Cindy continued the tortuous conversation.

"I don't know, I'm praying it doesn't," I said.

"Krista!" Mom yelled from downstairs.

I swear! How does she always know I'm talking about her? I excused myself from Cindy, cupped my hand over the receiver, and yelled back, "What is it, Mom? I'm on the phone."

"I know. You've been on it for a while, I need to make a call, so wrap it up."

"Cindy, I have to go," I said. The conversation with her was not helping, so I wasn't totally destroyed to end it.

I went downstairs to let Mom know she had phone control. "Okay, okay," I said, handing the phone over, "what's so important?"

"Krista, I know you think the world resolves around you, but I need to call my mother," Mom said in a huff.

If only the world revolved around me, I thought and then went to find my sisters. Chelsea was doing artwork in her room. She always had a crayon or pencil in her hand. She looked pretty intent and satisfied, so I went in search of Katie. She'd been so aloof and off to herself lately. I wouldn't admit this to anyone, but I missed her.

"What ya' doing?" I asked Katie. Her room was a wreck with clothes everywhere. "If Mom sees this mess, I wouldn't want to be you," I warned her.

"I know, but I have to go through my clothes," she said as she pulled out different outfits and held them up to her body. Mom and Dad had installed a mirror and ballet bar in her room as a big surprise for her birthday. It was now serving as a dressing room mirror, and almost everything was getting flung to the side.

Assuming she was tossing clothes because of fit, I reminded her, "Mom said she'd take us shopping next week for school clothes." This was something we generally all looked forward to, but Katie did not look at all interested.

"I don't want to go," she said.

"You've got to be kidding," I responded, "of course you want to go." I had a list a mile long made with several weeks' worth of intense thought.

"No, I don't," Katie said. She sounded old and pained. I wanted to feel sorry for her, but what did she have to worry about? I mean, I just turned fourteen, and I was heading into ninth grade. She was eleven and going into sixth grade. She was heading into her last year of elementary school and heading the school.

Even next year, as a seventh grader, would not be tough because the schools were changing the system. Sixth and seventh would both head to middle school. She was darned lucky! We would both benefit by entering seventh and tenth with the sixth and ninth graders, so we would never really be *freshmen*. This kind of stuff was overwhelmingly important, but I was sure of one thing. Katie did not care right now.

"Krista," Katie said jolting me back, "I don't like how any of this looks on me."

"What do you mean? I kind of like this," I said, lifting a pair of jeans and trying to be supportive. I normally leaned to the sarcastic side, but I sensed a need for some seriousness.

"They make me look fat," she said and threw the jeans to the floor.

"That's crazy, look at you," I said and pointed her toward the mirror. What was she talking about? She was almost as thin as I was. Nobody on any planet would accuse either one of us, or anyone in our family, of having an ounce of fat. Even my Dad had chicken legs.

"I can pinch an inch," she said.

"That's ridiculous!" I said. I hoped ignoring the absurd would make it go away. "Come on, let's go downstairs and bug Chelsea and Mom."

"No, thanks," Katie said and turned back to her closet.

I wanted to run straight to Mom. Deep down I was concerned, but I had my own problems to deal with, and this insanity would surely pass.

The final days of summer went by in a snap. Mom's parents seemed needier than ever, and her draw to them grew stronger by the day. So we soon ended up in Williamston for a week. Thanks to the late notice, and an ever-approaching school year, we had to do our school clothes shopping there. Not my ideal, since there was only one decent store in a forty mile radius. And that store, quite frankly, catered to the average elderly crowd with a juniors' section the size of a walk-in closet.

When we found nothing there, Mom took us to Roses Department Store. It supposedly had name brand looking merchandise for cheaper prices. Mom brought me a pair of *Guess* jeans, but the triangle symbol was upside down. None of us really found much, but Mom picked up some handkerchiefs and underwear for Dad. We tried to find him some T-shirts, too; but every one we looked at either had a huge stain, odd stitching, or a head hole completely sewn up.

The ultimate benefit for us was Mom always felt guilty when she promised something like school shopping and she couldn't follow through or it was compromised. She would inevitably feel the need to make it up to us and soon.

Knowing guilt clothes were coming added to the recent discovery that Jeff's parents' plans were cancelled. *Bring on school. I can handle anything now!*

The first day of school finally arrived. I headed to the bus stop while Katie and Chelsea were chauffeured to their respective big days. I didn't really mind, but I did give Mom some grief so I could shore up some more guilt points for those real *Guess* jeans I still wanted.

Cindy and I were elated when we compared class schedules and discovered we had almost all the same classes. Unfortunately, so did Kerry. I was getting sick of the whole three musketeers thing. Kerry brought nothing to the table. I was starting to strut my stuff with them like a peacock dancing around all the ugly females. These days, I would even talk to Helen within earshot of either Cindy or Kerry without a care in the world.

"Krista, she's a dork," Cindy said with Kerry hanging on her every word.

Fearlessly I said, "I like her, and we are actually going to the mall together this afternoon."

"Fine, but I invited Cindy over to my house, and we thought you'd want to come, too," Kerry said with a fake smile. It dawned on me she was perfectly happy having Cindy to herself, but I also didn't care as much as I should. Cindy was really my friend, not Kerry's. Kerry was a simple fill-in for when I was unavailable.

"Have a good time then," I said and opened my locker door. With all the social importance of high school combined with my natural ability for getting grades, sometimes I forgot those books piled high in my locker were the true reason for being in this building.

"Okay, but what about next weekend?" Cindy asked.

See, I thought, *she's already making sure she wraps me up for next weekend.*

"I think I have to babysit Friday night, but we could do something at my house on Saturday. I'll check with Mom," I said. I knew I might have to use up some of my reserve guilt points to finagle a spend-the-night with Cindy.

Cindy and I headed to geometry class with Mr. Tuttle. He was insanely obese and totally favored the guys in his classes. Kerry was only in algebra, so I got a much-needed break from her. For the first time, I had trouble with math, normally my best subject. I couldn't quite imagine the three-dimensional figures. It seemed inane to memorize a bunch of theorems in order to come up with coordinates on a three-dimensional image. No wonder the guys were better at it; it was completely useless knowledge.

And God forbid, I go home and ask Dad for help with it. It took forever to solve a sheet worth of problems because he would spend so long proving the theorems.

One night, I couldn't handle his *help* anymore, "Dad, someone smarter than both of us came up with this theorem, so why do we have to prove it? Can't we just use the formula and move on? I'll just memorize it."

"No, you need to understand the basis behind the theorem in order to properly use it," he said. He scratched out some more x's and y's. I should have known better. Who would go to an electrical engineering major for a simple approach to math? He was enjoying torturing me combined with the pure pleasure of proving he could do it. I made a mental note to request Mom's help next time.

In the middle of proving what looked to be a particularly perplexing formula, Dad pulled out his handkerchief and wiped his brow. As he pulled the wet object down from his forehead, he stared at it. He turned it over several times and then called Mom.

"Abby, where the hell did you get this thing?" he yelled, assuming Mom would heed the call.

"What are you fussing about? I'm getting Chelsea ready for bed," Mom bellowed from somewhere in the house.

He kept talking to her in this manner until she got frustrated enough to come into the kitchen. "Look at this," he said, showing Mom the handkerchief.

I had no clue what was going on, but soon they were both laughing hysterically. *What did I miss?*

"What's so funny?" I asked. I could use a good laugh.

"The hem . . ." Dad started and tried to finish amidst chuckling, "is like an inch all the way around."

This was funny? I wondered.

"I got it at Roses when we were visiting Mom and Dad," Mom said. She was laughing so hard now that she had leaned into Dad for support.

Someone help these people. Calling an interpreter now. HELP! Another kid-doesn't-understand-her-parents moment.

"Seriously, Abby, I think I earn enough money to have real handkerchiefs," he said, wiping tears away with his object of scorn.

I grabbed my geometry book and retreated upstairs for some normalcy. Chelsea was sitting in her room talking to her stuffed animals. I sat down next to her, picked up Mr. Pig and Fuzzy Wolf, and started acting out the Three Little Pigs. Within minutes, Katie had joined us. Right after the wolf blew down the second pig's house, I looked up and saw Mom in the doorway.

"Mommy, Krista told me a bedtime story," Chelsea said, brightening at Mom's presence.

"I heard. Good job, Krista," she said.

"You and Dad feeling better now?" I asked.

"Much," she said. *Adults are strange. It can't be life's ultimate goal to become one.*

Abby

Summer wound down like an old Victrola, but not the heat. The full-on summer blast refused to abate, falling like a merciless sledge hammer. My grandmother referred to these as the dog days of summer, saying it caused everyone, even the hounds to bristle. The girls constantly bickered, and Bob traveled most of August. I got up each morning, not stopping until I could no longer put one foot in front of the other. And at night, I would often fall into a tightfisted, dreamless sleep.

It wasn't a completely depressing place to be, but an elevated happiness quotient would have been welcomed. My new awareness and oath to be more positive and upbeat was harder than I thought it would be. In fact, the more I gave, the more I role modeled good choices, the more animosity and contention flowed. I quickly realized my need for magic, created or caught, was not high on my family's chart of needs. It didn't matter how much addictive-packed enthusiasm I dealt out, fitful stares halted most attempts midstream.

One morning, I awoke to a storm. The bedroom was so dark I had to check the clock to confirm the time. Outside the window, a tree limb caressed the screen—a whispering, murmuring, rustling sound. Lightning flashed, highlighting the pale green pattern in the bedspread, before it tumbled off the walls and floor. It was perfect and terrible at the same time. The solitude, coupled with Mother Nature's symphony and light show, was intoxicating. I refused to move, hoping to absorb the energy that danced treacherously in intermittent strikes.

Closing my eyes only enhanced it. Between each blast of white, my eyelids responded with reciprocal flashes of multicolored brilliance.

While sinking into a world of utter bliss, a hand tapped my shoulder, and a soft voice spoke close to my ear, "Mommy?"

Not wanting to break the trance by tending to reality, I answered softly, "Yes?"

"I'm afraid of lightning."

Gently moving over, Chelsea slid in beside me—a natural reaction amid a natural occurrence. Her warm body snuggled in close. The lightning continued, and we both fell back to sleep.

When I finally woke, a deep peace settled into my marrow, time-released to maintain a steady euphoria. Not to disturb my bedmate, I slipped out of bed and went to check on Krista and Katie. Both slumbered. Even the dog snored loudly from under Krista's bed. Tiptoeing downstairs, I put on coffee and heated the oven for muffins. I was in the middle of gathering ingredients when three sleepyheads flopped down at the kitchen table.

"Good morning," I whispered.

"Mornin'," spilled out in a soft concert.

The rain continued, pouring its contents out of a heaven bound reservoir of unending droplets, each bouncing playfully off the windowsill. I reached over and turned on the radio, dialing into a classical station, keeping the volume background low. When the muffins were done, their aroma filled the room, beckoning the girls to shake off their dreaminess and come alive.

"Mama, are you going to Grandma and Grandpa's this weekend? Do we have to go?" Krista asked.

"Well, I don't know. I've got to discuss it with your dad. I thought I'd go alone this time. Dad'll be here, and there's nothing for you girls to do there," I responded.

"Amen!" Krista said. Katie and Chelsea giggled at Krista's rush of approval.

"You don't have to be so eager to get rid of me, you know. Do you think your dad will cook his famous macaroni and cheese?"

Krista looked stricken. "Mom, please cook before you leave! Dad almost burned the house down the last time you put him in charge of cooking. Forget him. I'll do the cooking myself!"

"I know how to cook," Chelsea said, looking very pleased. "I can fix peanut butter and jelly sandmiches with cheese crackers." Frowning, she added, "But I don't like that marmalee you like, Mommy. I only like grape jelly."

Everyone howled, and Katie clarified, "It's marmalade, silly."

"I don't like that orange stuff either," Chelsea continued, "but I can make the best sandmiches ever! Right, Mommy?"

I winked at her, "Yeah, you're some cook! Remember the Happy Party?"

Behind me I could hear the girls continue their debate over food and who liked what when Bob called, "Sounds like you're having fun."

"The same old stuff. How's your day?"

"I didn't call to talk about me. I miss you. Let's go to dinner Friday night."

When I realized I was holding my breath, I exhaled, "Dinner? Friday night? That's a great idea. But, I can't. Did you forget my plans to go home Thursday night?"

"Go Saturday. Your folks can fill you in on the doctor's visit. We'll do something special with the girls when they get home from school, and then go out to dinner. If you go early Saturday, I'll stay home on Monday. I'll take a vacation day. Is it a deal?"

"Well, I don't know what to say, except yes!"

"Then we have a date. I'll make the arrangement. All you have to do is be ready. And, no shop talk, I promise," Bob assured.

I lowered the phone but quickly picked it back up to say, "Thanks!" The dial tone filled my ear; and for a moment, I was disappointed until a schoolgirl tingly feeling rushed over me.

"Who was that?" Chelsea asked.

"It was your dad. He invited me to dinner Friday night."

"Mama and Daddy sitting in a tree, k-i-s-s-i-n-g."

"Okay, girls, that's enough. Put your plates in the sink and go make up your beds. We have a full day of shopping ahead of us."

Krista and Chelsea ran from the room, still continuing their rhyme-chime, but Katie stayed behind; her countenance set firm, "I'm not going shopping."

Anger struck like a lance. The room got quiet, so quiet you could almost hear it breathe. We stood testing time, never losing vision of one another's eyes. Acquiescence was not a hallmark of my temperament; so I held fast, digesting the import of her words and trying desperately to calm my response. Then I felt it. It was like a crack in the ceiling of time when the heavens open up and pour out their contents on your heart—erupting, crushing, and brutal.

I saw it in her eyes and recognized the truth for the first time. In a wisp of a moment, I saw another pair of eyes, but this set glared in reproach and disgust. They belonged to Mama; and I was eight years old.

* * *

"Abby, who filled your plate?" Mama pointed her finger at the mound of mashed potatoes, corn, and roast beef. "If you continue to eat like that you'll be big as a barrel. I want you to take half of that food and put it on another plate and save it for tomorrow. I'm ashamed of you for being so piggish."

My small hands trembled, lifting the plate and doing as I was told. All hunger was gone as I felt her beady stare, daring me to overeat.

When she finished her meal, she commented, "Abby, now that wasn't so hard to do, was it? I'm proud you showed some restraint. Your Daddy's people are so heavy; I see you have those tendencies, too. Lord help me, it looks like it'll take everything I can muster up to keep you from being fat."

* * *

Tears welled, and I tucked the memory back into its protected place. Taking Katie's hand, I nudged her toward the table to sit down. I turned a chair around to face her, sat down, and began, "What's wrong?"

Steadfast and unwavering, she replied, "Nothing. Nothing's wrong."

"I don't know what to say to you. I'm not even sure I want to go down this path, but I have to. You don't seem happy, not for a long time now. I hear you making comments about being fat and that your friends don't like you. You've become distant with you sisters. I've tried to give you some space to work through this, but I think it's time we do something."

Seconds lapsed. I hoped she would say I was wrong, maybe break down and explain, but she didn't. At this point, I drew a blank, not knowing how to proceed.

Finally I edged cautiously, "Katie, I want what's best for you. I want you to be happy. More than anything, I want you to be healthy. You cannot achieve or do anything effectively if you aren't well."

Katie blank-faced stared at me, through me, and beyond me. Fear gripped my throat with her rusty iron fist. Tears ran unabashed. My soul screamed for resolution or acceptance or denial . . . anything to put an end to this issue. But still, she sat. Panic arrived, causing me to grab Katie by her wrists, screaming, "What are you trying to do to yourself?"

Her face remained a dull white mask. Her eyes as cold and lifeless as a porcelain doll.

Feeling bruised and battered, I released my grip and sank back into the chair. She, and only she, held the key to the thousand rooms of her emotional state. I did not know how to gain access to her vault of insecurities. Never in my life had I faced the possibility of losing a child. What was I going to do? No words formed to describe the horror picture forming in my mind. At this point, I had to call a truce and get help. Maybe Bob would know what to do. Hell, maybe Bob could ask Sylvia; she seemed to know everything. The one person I wouldn't ask was Mama.

"Can I go to my room?" Katie asked.

"You may go to your room, but trust me I'm not finished with this discussion. We will find answers; I love you too much to see you hurt like this."

I reached over and hugged her, willing my arms to provide a miracle. But they did not. I felt the resolve in her body, a stubborn resignation that no one could fix her. All of a sudden, I remembered the wordless picture book she often brought me at bedtime; and there among the one dimensional characters, I saw her, wordless and frozen.

Chapter 18

Krista

We were back to school after a couple days off for teacher in-service days. Soon, it would be the holiday season again. We'd been warned, although I'm sure my parents didn't see it as a warning, we'd probably spend a portion of our Christmas break with our grandparents. I started to view Williamston the way my Catholic neighbor described purgatory. It wasn't hell, but it was a holding cell. All the souls there appeared unhappy and restless, and those feelings spread like a plague.

My babysitting business was thriving, and luckily no new incidents with any weird adults. It took awhile to convince Mom and Dad to let me expand to new families. Even I was hesitant at first, but now I had enough business to pick and choose when I wanted to work. A small cedar box in my closet safely protected my earnings. It was right on top of the diary I turned to less and less these days. Things were going too smoothly to warrant much self-motivated counseling.

Cindy and I were tighter than ever, and Kerry hung on pathetically close. On top of spending most of our school day together, we were all in the competitive choir, the Troubadours. This ensured most of our evening events corresponded as well.

We teetered on being the most popular girls at school. The cheerleading squad seemed to have a firm grasp on the title; but most of the school talked about them horribly behind their backs, throwing around words such as slut and easy. The boys only talked to them when school dances came up or the girls' parents were out of town.

My knowledge of things my parents would have loved to keep me from grew by the day. It felt liberating and secretive.

Nothing could ever be perfect, though. One day, Cindy and Kerry rushed into choir practice and grabbed me by the arm. They quickly thrust me in a corner and looked around for prying ears.

"Did you hear?" they said in near unison.

"Hear what?" I really had no clue what was going on, and I didn't like the looks on their faces.

"Jeff is dating Jenifer," Cindy said and braced for my reaction. Kerry had a look of barely hidden glee.

"Jenifer. The cheerleader Jenifer?" I asked. I hoped my voice wasn't as shaky as I felt.

"Yep," Kerry said, "and rumor is it's serious."

Serious in middle school always translated to at least second base. The thought of Jeff with his hands up Katie's shirt made me ill.

"Girls, girls, it's time to practice, please take your seats," Mrs. Haliburton said, motioning to our corner. Cindy and Kerry were altos and sat in a different section of the room than I did as a second soprano. We were practicing Whitney Houston's "The Greatest Love of All," and it seemed cruel and ironic. I was glad to be separated from them and have time to process the news.

Thankfully, choir was the last period of the day. I rushed to my locker and onto the bus with little fanfare. Kerry took a different bus, and Cindy's Mom was picking her up today for some reason or another. I closed my eyes to absorb my day. My stop came with relief.

I wanted to talk to Mom about Jeff and his new girlfriend the minute I saw her and heard her usual, "Hey, Krista, how was your day?" But her attempt at caring was halfhearted at best. So I simply answered, "Fine," and went to find Chelsea. She had an ability to force a smile on you no matter what had happened. Her cuteness was her weapon, one I did not possess.

I was avoiding Katie. I knew I should be the big sister and seek her out, but I was at a loss. Apparently, so were my parents. Just last night, I heard them talking in their bedroom after they thought we'd all gone to bed. It's amazing how parents talk over you or around you when you are a child. Or, they think you are asleep and can't hear their voices floating up the stairs or through thin bedroom walls. I wasn't completely clear on what was going on with Katie, but even I could see the changes in her. Her face looked sharper than ever, like skin barely stretched over a skull, and her eyes were always shaded a dark gray. The life and spunk that used to spill from her seemed trapped in a body of darkness.

When I noticed Chelsea had innocently fallen asleep in the middle of a pile of stuffed animals, I decided I should attempt to find Katie.

As usual, Katie was sitting by herself in her room. I timidly crossed the threshold into her sanctuary.

"Hey there," I said sheepishly.

"Hey," she said. It was clear she would have been just as happy if I had turned and left her be. But the sadness in her face pulled me in.

"What's up?" I asked, hoping for a simple response.

"Why is everyone asking me that?" she said and threw a pillow across the room.

My first thought was to retreat and go ambush the little zookeeper next door, but I decided to stay and tame this beast first.

"Okay, that's it! What is really going on with you?" I said. It was time to play hardball, and clearly nobody else was willing to step up to the plate.

A dark cloud swooped over her face, and I hardly recognized her. "Nothing's up with me! I don't want to talk about it with you, Mom, or Dad! And I certainly don't want to talk to some stranger. I'm not going to! They can send me, but they can't make me say anything," she said. Her eyes were cold as steel.

I had no clue what stranger she was talking about, and it didn't seem the time to ask for details. "Come on, Katie. I honestly just want to help, but I don't know what I'm supposed to be helping with." The hard stand was not going to work, I could see that now. And I should not have to be the heavy, parents owned that role. I changed course, "Hey, why don't we just put some music on or something?" I wanted peace. I wanted my sister.

"I don't want to," she said.

"Then what DO you want?" I asked, frustrated. The question startled and softened her.

After a moment, she said, "Okay. I want to play with She-Ra." I turned and got the plastic bin out that held her figures. We settled down and went through the motions. She brightened up a bit until Mom knocked lightly on the door.

"Girls, it's almost time for dinner," she said.

Katie tensed immediately, so I answered. "We're coming," I said. We could hear Mom moving around in Chelsea's room, so I knew we had a few minutes.

"I want you to talk to me if you need to," I said and put my hand on Katie's shoulder. "I mean that." I did. I could tell she needed someone, and I always felt the enormous responsibility to hold this family together.

"Okay, thanks," was all she said.

The next few weeks were busy. I had several singing and social events; Mom was up to her eyeballs with crafts and projects for Chelsea's school, which she occasionally commandeered my help with; and there would be another weekend trip to Williamston. Katie had started some doctor visits, and the tension in the house between her and my parents was touchable. Selfishly, it felt good not to be the one my parents were shaking their heads about.

The holidays were fast approaching, and it was time to sit down with the various catalogues arriving daily in the mail. Our favorites were still the Sears and JC Penney wish books, but every holiday season there were more and more. We had plans to spend Thanksgiving in Williamston, so I was hoping it would be the last trip there until 1987. I did have a lot to be thankful for this year, but I knew I'd be even more thankful if we could spend Christmas with just the five of us.

Helen and I were hanging out more and more, and Cindy and Kerry had given up trying to control the issue. I solidly claimed the leader position in this pack of girls, although we constantly talked about how the cheerleading group was gaining ground on us. Cindy and Kerry strongly believed we all needed boyfriends to compete, but the one guy who really interested me was still hot and heavy with Jenifer. And for all their talk, neither one of them had boyfriends either. I could only imagine the wrath I would incur from Mom and Dad on the issue, and I was more than happy for Katie to keep the spotlight right now.

Before Thanksgiving break and the long-anticipated (read sarcasm here) trip to see my grandparents, the three of us had a powwow about boys.

"We NEED boyfriends," Kerry said during lunch while balancing a Tater Tot on a fork.

"Why?" I asked in a half-amused, half-bored tone.

"The cheerleaders are all the talk now, and I think we all know why," Cindy said.

"Okay, then, we don't really need boyfriends, we just need a guy to shove his hand up our shirt or better yet, down our pants," I said. Teenage issues could be so exasperating.

"Funny, Krista, really funny," Kerry responded. "How many times have we told you how important next year is? We are at such an advantage with not having to be TRUE freshman. We all know that we will be going up against these girls for Civenettes and such.

Heck, maybe we should try out for cheerleading or the Vikettes," she finished.

"Wait, that's a great idea," Cindy said and literally started bouncing up and down in her seat, "my sister knows a class we could take this summer that would help. And, Krista, you have all those years of gymnastics."

"I don't really want to do cheerleading. It's not my cup of tea to stand up and yell and be all fake and phony," I said. Even I could sense the ultimate unimportance of these talks.

"Well, maybe we do," Kerry said and started poking at Cindy, "right, Cindy?"

"And, didn't you guys hear about Jenifer?" I asked. They both silenced at that point, and I felt some satisfaction.

The lunch bell rang, saving me from the rest of the conversation. I sensed Kerry wanted to be the leader; and oddly, I found myself less interested in the position. I didn't want a boyfriend and all the corresponding issues. I wouldn't mind a guy to turn his attention on me, and the thought of kissing seemed pleasant enough, but the other stuff was disconcerting. Maybe it was because of the Capshaw incident; and yes, I did factor in what I knew to be my parent's thoughts on the matter. Ultimately, I didn't feel an overwhelming desire to rush into the land of boyfriends. I had found a few more books since *Forever* on the subject, and reading about it seemed to be more satisfying than the reality of it. At least for now.

After all, it was only yesterday I had run into Jenifer sobbing in the girls' bathroom. The rumor quickly spread—she hadn't been willing to go *all the way* with Jeff. The fallout was cruel, and the guys were like a pack of wolves protecting their young. They had labeled her a tease. I didn't get involved because I didn't know much about it, and it didn't hurt me directly. Matter of fact, maybe it would end up helping me.

Relationships with friends and sisters were taking up enough of my life, and I wasn't even in high school yet. *How many junior high romances lasted, anyway?* Mom and Dad had started dating when they were fifteen and freshman. Nowadays, you probably couldn't make it unless you waited until at least junior year or even college. So I had a couple of years.

Abby

"Wow! You look beautiful!" Bob beamed with pride.

"Thanks," I said, blushing. "Where're we going?"

"I know I mentioned Elijah's, but then I thought it would be better if I surprised you." Bob crossed the room, came up behind me, and buried his face in my hair, "You smell delicious; maybe we should skip the meal and get a room instead."

I pushed him away playfully. "Robert Arthur Simmons, you are abominable, and I love it!"

"No, I'm horny as hell and can't wait to have a wonderful evening with my sexy wife."

Pulling back into his arms, I hugged him tightly.

"Gross!" Krista said, entering the room, "Do I need to take the kids and move out?"

Not wanting to rush out of Bob's embrace, I smiled, saying, "No, I think you're all safe."

Bob winked, "Krista, I'm depending on you tonight. You know our agreement, right?"

"What agreement?" I queried.

Krista opened her mouth to speak, but Bob interjected, "Abby, never you mind. This is between Krista and me."

She quickly clamped a hand over her mouth and rushed from the room.

The evening was beautiful. Bob got dinner reservations at the Pilot House, overlooking the Cape Fear River. Before going inside, we sat in the car enjoying the view. Anchored to our right sat the USS *North Carolina*, a World War II battleship. At the time she was commissioned, she was considered one of the world's greatest sea weapons. Now she sat idle, rescued from the scrapper torches after the war ended and brought home to rest. Perched across the river, her massive steel structure loomed skyward as a reminder of our country's strength and perseverance. Her presence seemed to give her a new purpose as the town's protector.

To our left, the bridge leading from Wilmington to Southport seemed to cut through the sunset. Ripples of pink, purple, and gray provided a backdrop of intense color while the bridge's grid work set time and space. I wanted to capture the breathtaking scene that stretched before us so I'd never forget the feeling of tonight.

Deep in thought and reflection, I was startled when Bob reached over and took my hand, "Abby, I realize this move has been difficult. I'm sorry I get so engrossed in work and myself. I don't know how to balance it all. I'm not trying to make excuses; I just want you to know I'm aware of it. Frankly, I don't know how to fix the expectations I've created."

Time rippled like the soft current flowing before us. I did not want to break my gaze from the night sky to discuss what could or could not be. I did not want this moment to end—the sound of the water lapping at land's edge; the muted voices of people strolling along the riverfront; the rhythmic puttering of a tugboat pushing upstream; or the seagulls squawking, scurrying to night safety.

In the middle of serene beauty, and with the man I adored, the only words I found were, "I love you."

Dinner was incredible, but I do not remember what I ate. The evening's magic was laughter. Laughing at Bob's animated stories, feeling we were dating for the first time, laughing when he teased, laughing when he fumbled, nervous and schoolboy shy. It made me wonder. Had I ever noticed how his eyes twinkled when he spoke? Or how he pushed away from the table, putting his arm across the back of the chair when he boasted, feeling cocky and confident? Or how he pulled in, getting close when he expounded on a tall tale? Had I forgotten or had a film been lifted from my eyes?

Leaving the restaurant arm-in-arm, we walked slowly back to the car. Bob had felt the magic and reconnection, too. Tonight, Blanche would be smiling, knowing we truly understood magic, even if only for tonight.

Leaning my head on his shoulder, I sighed, "The evening was perfect. It will be etched into my memory forever."

Bob leaned down and kissed my head. He didn't say a word, but in princely fashion, opened the car door, keeping his hand in mine until he lowered me to my seat. Our drive home was quietly sacred. Every few minutes Bob would reach over and touch my shoulder, hand or cheek. Like me, he didn't want to break the spell but needed a touch to reassure himself he wasn't dreaming.

Pulling into the driveway, the headlights bounced on and off the front of the house. Our eyes instantly traveled to the small figure sitting on the doorstep. Bob stopped the car, put it in reverse, and slammed it to a stop, spotlighting the porch area. Magic trailed into stardust as parental reality hit.

"What the hell is Katie doing sitting outside in the dark?" Bob questioned.

He yanked the shift into drive, accelerating so fast the tires squalled as he rounded the corner of the house. I reached out and grabbed the dashboard with both hands to steady myself. When he jolted to a stop, my head jerked forward, barely missing the windshield. Before I could respond to his actions, he had the car door open and was bolting toward the house. Feeling a heightened sense of danger, I followed him.

As I entered the garage, I saw Bob framed in the doorway talking to Krista. Shouting tumbled over his shoulder, landing squarely in my path, "What the hell is Katie doing on the front doorsteps, Krista?"

"I'm sorry, Dad, but she's such a pest. I tried to stop her, but she wouldn't listen," Krista retorted, meeting her dad's voice level. Instead of proceeding toward them, I slipped out and edged around the foundational bushes toward the front door stoop where Katie sat.

"Hi," I said, sitting down beside her.

"Hi," she responded.

"Boy, I never knew how dark it could get out here. Guess I don't make it a habit of sitting on the porch this late at night."

A heavy covering of clouds blanketed the sky. The moon weaved around the thickness, barely observable unless you caught a flash of light or a slip of its shape. The quiet was interrupted every few minutes by the shouting match in the garage, but we remained in our own solitude. Surprisingly enough, it was Katie who spoke first, "I come out here a lot. I love the night."

"It's peaceful, isn't it?"

"Are you going to Grandma's tomorrow?"

"Yeah, I think I'll leave early in the morning and come back Sunday. Dad's going to take everyone out for pizza tomorrow night. It'll be nice for ya'll to spend time together. He doesn't get to do that much, especially since we moved here."

Katie stood up and reached for my hand, "Come on, Mama, it's getting late. Let's go in." Her voice sounded very grown-up, very worn-down.

I took her hand but did not rise, "Katie? Was there a reason you wanted to be out here alone?"

"I was just thinkin'."

"Well, it's okay to think, but sometimes it's nice to talk about things, too. When you get through with your thinking, can we talk?" I asked.

"Maybe."

I let her pull me up and lead me into the house. Krista came flying by on her way up the stairs, sparks of anger trailing, "Thanks for getting me into trouble, Katie!"

Katie looked at me, winked and walked up the stairs. I watched her, willing myself to find the right words and strength to deal with whatever tormented her.

A scream, spilling down the stairs, jerked me back to the present. "Mama?" Krista bellowed.

I didn't rush; I recognized the tone. It wasn't, "Mama, I'm hurting or bleeding." No, it was a "Mama, life is unfair" holler.

Just as my right foot hit the first stair to check on her, Bob yelled, "Abby, come here!"

I lowered my foot, placed it back on the step, and lowered it again. *By dammit,* I thought. *Let them find me. In the middle I sit, and in the middle I stay.*

Krista rushed angrily from her room, and Bob pounded in from the kitchen. Both were ready to spew out their feelings and thrust their anger down my throat. I sat on the foyer floor, my legs crossed Indian-style, no, 1960's Bohemian style, in protest. All I wanted was to savor my wonderful evening, but clearly they had other plans.

Exploding in frustration, I hunched in my shoulders and lowered my head. Let the enemies approach. My fury was loaded and ready!

CHAPTER 19

Krista

The women in this family were finally self-imploding, I thought. Surprisingly, Dad had yet to run from the house screaming in a terrified search for testosterone. At times, I know he felt we had opened a special home for women and all their issues.

Mom dealt with one melodrama after another, each sucking the life force and any resemblance of her from her body. Doctors were making a pretty penny off our family. Katie was seeing a doctor and getting more and more withdrawn, and I lived at either the orthodontist or dermatologist. Dad traveled on more business trips lately—Mom said it must be nice to have an excuse to escape. Only Chelsea, still enveloped in her childhood innocence, remained immune to the crazy trappings of the Simmons women.

My increasingly dependent grandmother and Cindy and Kerry with their constant popularity conferences made us eligible for a new scientific study on estrogen and its effects on the female brain.

At a recent doctor's visit, my dermatologist re-emphasized her theory that an imbalance of hormones caused my acne and the best course of treatment was birth control pills. Mom again made it clear she did not like the idea of putting a fourteen-year-old on hormones, but I knew she had bigger issues than my chemical makeup on her mind. And I had to admit the thought of more estrogen in this house terrified me enough to make the acne a blessing instead of a curse.

Thankfully, the holiday spirit provided a salve to our wounded household. Everyone seemed happier in varying degrees. Katie and

I discussed all the latest fashion trends and constantly revised our Christmas lists. I put a lot of effort into avoiding any discussion of sizes. Katie was still so fragile, both physically and emotionally. Still, a new feeling of hopefulness surrounded us. Even Mom and Dad's attitude was less cautious and fearful than it had been.

School was winding down and our two-week Christmas break became more of a reality. Dad worked late hours so he could take his annual week off, and Mom was busy buying gifts and shipping them to various destinations. She was so in the spirit she gave each of us a small allowance to buy presents for friends. Cindy and Kerry wanted to spend more money than Mom had allotted, so I reached into my treasured cedar box for the extra. Soon I discovered they had decided together on exactly what present we were all getting each other. During lunch on our last day before the break, they sat me down with the details.

"Okay, this is the purse," Cindy said and pulled out a magazine made up almost entirely of advertisements versus any real articles.

"Purse?" I naively asked.

"Please, you had to have seen this purse," Cindy said, "My sister has one, and she says ALL the high schoolers who are anything have one. We could get it in different colors, but I think we should pick one color and match. What do you think, Kerry?" she asked. It was obvious why she did not pick me for that question after I had showed little interest in the whole idea.

"Let's match," she agreed, "and I prefer brown."

"That works for me," Cindy responded gleefully, "what about you, Krista?"

"How much is it?" I asked. Mom was giving me fifteen dollars to spend on each of them, and I would have to cough up the rest. It was amazing how much more I cared about my own money than Mom and Dad's.

Cindy scanned her finger down the description until she found the price, "Well, in this magazine, it's forty dollars, but I bet we could find it on sale."

Great, so now I was going to shell out twenty-five dollars of my own money for them. Still, it was cheaper than buying the purse outright; if I *had* to have it for high school, I would be ahead of the game without trying to convince Mom to buy it for me next summer.

"All right, I'm in," I said with as much enthusiasm as I could muster.

"Oh, we almost forgot," Kerry stated. For a second, I was caught up in the fact that "we" described Kerry and Cindy more these days than the three of us. "Cindy's coming over to my house the weekend after Christmas. We figured we'd exchange gifts and hang out. By then, we'll all be sick of our families and need a break. Plus, we can compare our Christmas load."

"I don't think I can. Rumor has it we might be back in Williamston then," I said.

"Good Lord, Krista, you guys go there all the time. Well, let us know if you can make it," Kerry whirled around and made an overly dramatic exit.

When I walked into our house after school, I smelled something wonderful, so I rushed directly to the kitchen.

"Hey, Mom, what smells so good?" I asked.

My mother, not the woman who had been masquerading as her for at least a couple of months now, turned to look at me. Her smile cast warmth in my direction. I wanted to throw my arms around her and never let go. I did not want to lose her again.

"Mommy made cookies!" Chelsea shouted, barreling into the room. Mom grabbed her with one hand to slow her down and used the other to expertly transfer cookies to a plate.

"Come on, girls, eat 'em while they're hot," she said. That's when I noticed Katie had joined us. I realized the significance of the occasion—not only were we being allowed to have cookies this close to dinner, but Katie was there. I wondered if I had accidentally wandered into an episode of *Leave It to Beaver* or something.

"Okay, what gives?" I asked. I did not want to cast a stone; but for some reason, my teenage mind could not sit back and simply accept the moment. Then it hit me . . . there was bad news coming! "Wait a minute! Are we going to Williamston again right after Christmas or something?"

"Nope," Mom answered, "your dad and I decided we'd spend the holidays just with our girls, does that sound good?"

Man, does it! Chelsea was even more enthusiastic with her little squeals. Even Katie brightened.

As I was devouring the warm cookie, I decided to take advantage of Mom's obviously good mood. "Hey, can we get by the mall soon? I need to get a gift for Cindy and Kerry," I asked. I mentioned the purse, decreasing the cost a bit, but not enough to avoid a response.

"Come on, Krista, that's ridiculous," Mom said.

Somehow, I managed to put an ice cube on our tea party. "I know, Mom. I'll use my babysitting money for the difference."

"No, Krista, you are not going to use your hard-earned money to buy some thirty-dollar purse. Heck, my purse came from Kmart, and it's just fine."

"But, Mom, you don't understand, I have to," I said and defiantly surrendered the cookie back to the plate. I hated to give it up; but if she was going to be unreasonable, she left me no choice. She could not bake up cookies and get whatever she wanted. No siree!

"We'll discuss it with your dad then," Mom announced.

"Oh that's going to help. You're the one who constantly says we shouldn't bother him with all the little stuff." Score one for the teenager.

"We'll just see how little he finds it," she retorted and then proceeded to clean up the cookies and start dinner. The girls shot that *thanks-for-ruining-it-for-us* look, and I truly wanted to crawl under the nearest rock.

In the end, in what could only be described as another Christmas moment, Mom and Dad relented and took me to the mall for the purses.

On Christmas Eve, Kerry called to find out if I would be coming to her house. Something in me, some sort of small rebellion, declined. I was really enjoying being with my family for a change and was afraid if I left and came back, it would be back to the way it was just a few weeks ago. But more importantly, I wanted to take a little stand with Cindy and Kerry. Kerry shrugged it off with little to no fanfare.

Right after I hung up, we left for Christmas Eve service. I have to say I was very happy. I wanted our family to feel like this all the time. I was excited for the haul of Christmas, but I was also hoping the New Year would break the spell we had been under. Nineteen-eighty-seven was going to be a good year—a year full of planning for high school and a year of mending. After all, my big surgery was planned for late July. On top of getting rid of headaches, the doctors said it would create a nicer silhouette shape for my face by bringing my chin forward. A new face for a happier me.

As we sang "Joy to the World" that night, I was truly joyful and eager.

Abby

I turned up the radio and headed out of the driveway, wanting to get to Williamston early and do some sidetracking. There was an urgency to revisit some of my old hangouts. Silly teenage feelings seemed to be erupting in the middle of my thirty-something life. Maybe it was a sign I needed to return to a more carefree time. Or maybe I needed to find a conclusion to ties left dangling between scattered memories and steamy car windows.

Entering Martin County, I turned left onto Bear Grass Road and headed west. Farmland flanked both sides of the road. On some, tractors were tilling under yellowed corn stalks and churning up a half mile dust stream. On others, dilapidated tobacco barns seemed to rear their head from the soil, their cavities rotted and splintered from years of smoke curing. Farmhouses sprinkled the scene. Some were whitewashed and repaired, others board-rotted and mildewed. Cattle lumbered, feasting unencumbered. I rolled down my window, inhaling the pungent scents of upturned earth and livestock. Without warning, an image of me at fifteen filled my vision.

* * *

"You're not listening!" Mildred said, sounding puffed-up.

I stood outside an old tobacco barn, next to my best friend, Mildred. She had spent most of the school year harassing me to quit my part-time job at Cato's, selling women's clothing, and work with her at her daddy's farm sorting and staking tobacco leaves. She insisted I'd make a lot more money, and we'd be together all summer. After pestering me to death, I consented but refused to quit my job until I checked it out.

Ten minutes into hearing her lay out my job description, I knew I should have stayed in bed. I looked at the women going about their jobs mute and robotic and realized this was not for me. But wanting to keep my friendship, I tried harder to listen, knowing by the end of the day I'd kiss the farm good-bye.

"Sure, I'm listening," I answered. "But, I don't see the purpose in sorting leaves. They all look the same."

"You're not listening! Every leaf looks different and has to be graded accordingly. If you'd pay attention, you'd understand. It's not complicated."

This is wrong. Nothing about sorting smelly tobacco leaves could be important. Lord, please get me through this nightmare! I kept chanting.

Two hours into the job, I realized why those women looked the way they did. It was the only way to survive such a menial task. By the end of the day, sticky with a brown foul-smelling tar, my hands rough as a corncob, I headed home wanting nothing more than a shower and warm bed. More than anything, I wanted to erase this painful day from my memory. I vowed to never, which meant the rest of my life, let a friend talk me into anything remotely connected to green leaves or farming. My nails were ruined, my hair kinked from the humidity, and my allergies were killing me.

* * *

When the image faded, I laughed out loud. What memories! They weren't near as painful to bring up as they had been to experience. In fact, thinking back on it, I dare say that day was a turning point for me. I knew beyond a shadow of a doubt, I did not want to end up like one of those female farmhands. I vowed to never fall in love with a boy that bore the mark of the farm, had dirty fingernails or donned denim.

Driving on, I took a right onto Bear Grass Road. Recalling Bob's warnings about how the local cop pulled people over in a heartbeat, I slowed down. Despite its small size, Bear Grass was somewhat of a legend with to-die-for-guys, fast cars, and one angry underpaid cop. Again, I laughed out loud, remembering.

That's when Fred Rogers's face shot up from some obscure corner of my mind. What a joke! He was a seventeen-year-old punk who hung out at the Tastee Freeze, a local teenage haunt. He was known for the "do." His greased black hair was slicked back into a duck's tail, every piece plastered except for one stray curl that hung off center on his forehead. His face was speckled red and pitted from years of acne, popping bumps, and scratching blackheads. But back then, none of that mattered; he was the cat's meow, a teenage idol of sorts. He drove a souped-up '49 Dodge he spent every waking moment repairing, polishing, or sporting around. When he pulled into the Tastee Freeze, everyone stopped what they were doing and ran out to meet him.

Fred knew his status. He would pull into the lot, slow and easy, giving his admirers time to corral around his personal parking spot. Then for drama, he would sit and watch, smoke rings curling out of his half-opened window. Finally he would open the door, unroll his six-foot-four frame, nonchalantly stretch, cock his head to one side and frisk his crotch. At the time, we thought this was like marking his territory, flaming his manhood. Now I see it as too stupid to pour piss out of a boot, but that was then and this is now. Last time I heard about him, he was on his second wife, had six legitimate kids, who knows how many others, and could barely keep a job. Some idol!

Shaking my head, I wondered why I allowed myself to tromp down memory lane—or a nightmare, as it seemed now.

Edging toward Highway 17, I turned north and headed into town. Suddenly, I had a dire need to go to the R&C restaurant for collards, boiled potatoes, and fried chicken. Exactly eight minutes later, I sat at a small booth waiting on my vittles. The aromas wafting from the kitchen were delectable—a coming home to the senses. I closed my eyes, mellowing into the feeling, when someone tapped my shoulder.

"Hey, good looking."

"You haven't changed one bit, have you?" I asked, slowly opening my eyes.

"Nope, and I bet you haven't either. Let me guess, collards, boiled potatoes, and fried chicken? Oh, add sweet tea and fried cornbread. Will you splurge on the chocolate cake or are you on a diet?" Drake asked.

"Diet? You don't know me as well as you think you do; I never diet. And, I can't remember what I ordered, for your information." No sooner had the words left my mouth when Janice plopped a plate of food in front of me. Drake's eyes met mine, an I-told-you-so look spread over his face.

"Everything okay?" Janice asked, my face deepening flush red as I nodded yes. "Go on and eat up while it's nice and hot, you hear?" She ordered, picking up plates and debris off a nearby table.

Janice was an icon at R&C. Her mother and father had worked there until they died, and Janice was destined to do the same. Bets had been taken for years on her age. Her black skin, smooth and unwrinkled, looked timeless. Said to be not-quite-there, she knew everyone's name and their family history. The only people she ever forgot were the ones who left poor tips or gave her a hard time. I, for one, thought she was brighter than most folks I knew.

"Jus like you like 'em, Miss Abby. I done brought you some extra cornbread, too." Janice turned her attention to Drake, but not before taking a rebellious stance, driving one hand into her waist, "And what's do you want, Mista Drake?" She said, curling up her bottom lip.

"Now, now, Janice, have I offended you?" Drake teased.

"I don't cowdown to lowdowns like yuen. I got fresh possum out back. You want some?"

"Ouch! You've dug in too far this time, sweetheart. How's about a cheeseburger, fries . . ."

"A double dill pickle, the ketchup bottle, and a root beer," I continued.

Drake's lip curved up on one side, "You think you know me, don't you?"

Janice's pencil hit her order tablet with a thud, "Ms. Abby, don't you let this scoundrel bother you. I'll get you another table or run his ass off."

"No problem, Janice, I think I can take care of this no-count, but the sooner you feed him the faster we can get rid of him," I laughed.

Janice shuffled off, but not without shooting a smoldering glare at Drake.

"She's always had a thing for you," I teased.

"Yeah, but I'd just as soon have a coon dog love me. That woman is dangerous. If I ran into her in an alley, it would scare the living hell out of me!" Drake said, pretending to shiver.

We both laughed and hit a high five before realizing we had an audience. Two old timers sitting next to us shook their heads. A spinster on the other side had her eyebrows raised in disgust. They all knew us, and I'm sure by nightfall my mother would be asking what I was doing carrying on with Drake at the R&C. Before I had time to get paranoid, Janice slammed a plate down in front of Drake.

"You'd best eat up and get out!" she snarled.

"I love you, too, baby!" Drake's croon sank to a sneer when she walked away, "Did I tell you I hate that woman?"

Delving into our food, we grew quiet, intent. Everything was perfect R&C style—a greasy, wonderful prescriptive to bathe our souls in comfort. I'm sure onlookers, watching us lick our fingers and shovel in one mouthful after another, thought we were a sight for poor eyes.

During breaks in our gluttonous behavior, innocuous talks about the weather, current events, and local politics, we struggled. Between each tidbit of small talk, we downcast our eyes, not knowing what to do with the uncomfortable silence settling in. We ate less passionately and more deliberate, avoiding feelings.

"Abby, I need to talk to you," Drake's voice changed, soft and urgent.

I looked up. He had stopped eating, his hands folded and propped under his chin. I knew the stance. "Huh?" I answered, dreading what would come next, fearing the unwelcome seriousness and soul-searching.

"No you don't, Drake. Let's end our lunch on a light, friendly note, okay?"

"Abby, dammit, listen to me," Drake sighed, shrugging his shoulders, "I have some things I need to say." Seeing my reaction, he lifted up a finger for me to halt, "Don't go get crazy on me, okay? Meet me behind the peanut factory. Fifteen minutes is all I need; I promise." Drake didn't wait for an answer but got up and turned away. He greeted several patrons, paid at the register, pushed open the doors, and exited the building without one backward glance. When he finally got out of my vision, I let my shoulders relax and the import of his words sunk in.

By dammit, I won't meet him anywhere, I thought. *Who does he think he is? I'm not going to go racing after him to some parking hangout and act like two star-struck teenagers. Who in the hell does he think he is?*

I got up defiant and angry. Janice came by checking to see if I needed anything, more likely wanting to get the lowdown on Drake and me. Instead of accommodating her curiosity, I shoved a twenty in her hand and left. Trying to clear my mind proved impossible. I clamped my jaw tight, started the car, and squalled tires out of the gravel parking lot, open and ready for anything to happen. Within one block, I slowed down afraid of an arrest and needing to get control of my raging emotions. At Main Street, I turned left and then a right onto Haughton, heading straight to the peanut factory.

The drive was shorter than my anger. Tears welled at my stupidity; but I kept driving, knowing the exact turn to take to our rendezvous point. When I spotted the old, dilapidated factory, I eased behind it, alongside Drake's car, got out of mine and into his.

Refusing to look at him, I spouted, "I don't know why I'm here. You've got exactly fifteen minutes. Say what's on you're mind, but if this conversation gets out of hand or I get uncomfortable, I leave, understood?"

A lingering silence forced me to look at him. His hands gripped the steering wheel, knuckle white, head bowed, eyes clenched tight. I fought the impulse to reach over and run my fingers across his lightly bearded face to ease the veins rising fitful at his temple.

His voice cracked painful, "Abby, I've always loved you. When you left me and started going out with Bob, my world shattered." He raised his hand to stop me from interrupting, "It took forever to pull myself together and put my life back in order. That's when I met Dora. Abby, she's a real nice lady, and more importantly, she worships the ground I walk on. Hey, you know me, that's like electricity forever keeping me lit up." Drake let out an uncomfortable laugh, "Anyway, when I saw you at the hospital all those old feelings surfaced. There was my Abby, and I wanted you more than I've ever wanted anything in my life. You looked so vulnerable. I just wanted to protect you. Everything I had tried to cover up and destroy came rushing back."

Reaching over, he touched my cheek, pushing the flow of tears into my hairline, causing a fresh rush to cascade. "Abby, I was wrong. I let a moment rocket me back to the past, picking up feelings that should have been buried a long time ago. I asked you to come here to tell you I'm sorry. I'm sorrier because I think I caught you at a weak moment. I didn't mean to take advantage of the situation. I really love my wife and kids. Nothing on earth could cause me to hurt them. God, I'm so sorry, Abby."

Lifting my head, I looked into the eyes of my friend, the man who would always hold a special place in my heart. I took his hand in mind and said, "You have no reason to apologize. You were a part of my life that made me feel very special, needed, and wanted. You made me laugh and push the edges and beyond. It's been a long time since I felt like that, and I wanted to recapture it. I'm the one who needs to apologize." I reached over and hugged him, soaking in his warmth, "There will always be a girl in me who is madly in love with the boy in you, but the woman in me loves Bob."

I gave Drake one last squeeze, got out of his car and into mine. I drove away slowly, waving as I pulled out onto Highway 125, heading toward my parent's house. The heaviness had lifted, and a lightness of spirit permeated through me. Fresh tears rushed forward, but not from regret or loss or what could have been, but as a cleansing. A chapter of my life had finally closed. It had stood open far too long; but the page finally turned, dropped slowly and sealed.

When I arrived at Mom and Dad's, they weren't home, so I used the time to call home. Krista answered, "Where are you? We were worried."

"Hi, Sweetie. I've been riding around the countryside. You know, old stomping grounds . . . the memory stuff."

"Well, Dad's worried, and I'm more worried because he's threatening to cook. Mom, please don't let him. I thought we were going out for pizza. Heck, I'd even take Chelsea's peanut butter and jelly sandwiches," Krista wailed. "Plus, Christmas is almost here. Don't you have to do more shopping?"

Chuckling, I thought, *Welcome home, Abby. This is where you belong.* "Okay, I get it. Now, put your dad on the phone."

"Abby, are you all right?" Bob sounded genuinely concerned.

"Yeah, I'm fine. Just riding around, checking out some of our old haunts."

"As long as you weren't picking up old boyfriends, then that's fine. I was worried about you. After today, I can honestly say I do not envy nor do I want your job. If one thing has been a disaster, there's been forty. You cannot please these kids. I swear they work together to drive you insane," he laughed.

"Sounds like things are normal," I chuckled. "I'll be home in the morning. Let's take the gang out for breakfast. What do you say?"

"Sounds like a plan. Be careful driving home. And hey, I love you."

"Me too, Bob. I love you, too."

Chapter 20

Krista

Christmas was everything I wanted it to be and everything it used to be, all rolled into one. The time of healing continued in our house. I was more grateful about this than any gift I received. I usually didn't get overly emotional about Christmas, unless you consider emotional getting excited over gifts and screaming and yelling with each new item revealed as I tore into the paper. But this year, I could tell there was more to life than material things. So much seemed back to normal. The girls crawled into my bed late on Christmas Eve, and I enjoyed the routine of former Christmases.

Christmas morning, I stayed in bed until I couldn't take it anymore. I went downstairs early and surveyed the gifts, then finished up any last-minute construction before rushing upstairs to get Katie and Chelsea. Our squeals pulled our parents from sleep to join us, and they did so with smiles on their faces. As we plowed through our treasures, our noses eventually led us into the kitchen for some Sausage Fondue and homemade cinnamon rolls. We filled out the few obligatory thank-you notes to what family managed to recall our existence while inhaling Mom's gift of cooking. Even Katie dove in.

I can say I was in heaven. Or at least what I wanted heaven to be.

A few days later, with school still ten days away, we were all gathered in the den and playing with new board games. Thoroughly enjoying my family, I decided to skip the Christmas exchange at Kerry's house. Truth be told, I had forgotten all about it.

"Krista, it's Kerry," Dad called. Entrenched in the game, I hadn't heard the phone ring or noticed Dad getting up to answer it. I'm sure he was annoyed since it was such a novelty for him to have to answer a phone at home.

I trudged over to the phone and gallantly gave up my place for a round. Okay, it helped that I was losing. "Hey, Kerry, what's up?"

"Not too much, what ya' doing?" she asked.

"Mom and Dad are forcing me to play games," I white-lied. I was enjoying all this family time, but I did not need to announce it to the world.

"Oh, that sucks," she said.

Clearly she could not appreciate such things. "What about you?" I asked.

"Just hanging. Cindy stood me up. She said she had something to do with her sister," Kerry said.

A part of me was ecstatic Cindy had blown her off. I bet Cindy figured it wouldn't be much fun without me and decided to bail.

"Bummer," I responded. I was eager to get this chitchat over with and get back to the game. The girls were giggling, Mom was handing out cookies, and Dad was telling jokes. I was missing it for this lame conversation.

There were several more minutes of monotonous chitchat, before Kerry got to the real point. "What do you think of Cindy anyway?"

"Huh?" I asked. She caught me completely off-guard.

"Well, sometimes I think we should buddy up and be done with her," Kerry said.

I had been suspecting as much for months. "I've known Cindy for quite a while, Kerry. She's actually the first person I met when I moved here."

"Well, yeah, but ummmmm, don't you think she can be pushy sometimes?"

"Ha! My sisters would tell you I am," I said, trying to inject some humor. The silence convinced me I hadn't really answered her question. "Sure, I guess she can be. She hears so much from her sister about being popular and all the *must-haves* and *must-dos* of high school. I've never cared all that much for her sister; I think she's mean to Cindy. She's the pushy one actually."

"Really . . . ," Kerry said, urging me to continue. "And her mother . . ."

"Please, don't get me started on her. Even my mom says she just does charity events and ignores her girls." Mom had recently lamented the fact that Cindy's Mom had gone to several charity bashes right after her husband died instead of being there for her daughters. At this point, I saw my mom eyeing me suspiciously, but my teenage ego kept me from seeing the subtle warnings.

"So you don't like her mom or her sister?" Kerry asked.

My brain screamed for a retreat. "I didn't say that exactly. I'm just looking out for Cindy," I replied. "What do you think of them?" *Was there a point to this conversation?* I shrugged my shoulders at Mom who still appeared interested.

"I don't know them as well as you do," Kerry answered and then continued, "Did you know her dad?"

"Her dad? No, I really didn't. He was sick from the time I moved here. I often wondered, though . . . well, nothing really."

"Wondered what?" Kerry pushed. "Don't leave me hanging!"

"Well, I often wondered how close he and Cindy really were."

You could have heard a pin drop, but I was too busy letting my mouth run my world to notice.

"What do you mean?" Kerry asked. "I've wanted to talk to you about this stuff, and I know we're friends and can say anything. So, why don't you think Cindy was close to her dad?"

Wait, how did we get into all this stuff? I wondered. Kerry was a dog on a bone, but I couldn't see it.

"She never really seemed sad when he died. I mean, if I had just lost my dad . . ." I didn't get a chance to finish.

"YOU didn't lose YOUR dad, now did you?" I heard Cindy yell.

Cindy? Wait it couldn't be Cindy. I tried to swallow, to speak, anything; but I couldn't function. I felt the terror I suspected a deer caught in a car's headlights felt. I wasn't sure why the fear gripped me so tightly, but it did. The world stopped spinning for a moment, and it felt like a dream. Actually, it was a nightmare. I felt naked. I had no weapons, no way to defend myself.

"Cindy?" I asked softly. I was telling myself to get a grip. This was my life, not some kind of horror movie.

"Yes, you're damn right it's me!" she yelled again, "why would you say that about me and my dad?"

"I, ummm, I just," I stammered. The world had been standing still, but now the room started spinning. I reached out and put my hand on the desk in front of me to steady myself.

"I told you, Cindy!" Kerry interjected.

"Told her what, Kerry, and what is this?" I asked. Suddenly it dawned on me . . . I shouldn't be scared, I should be royally pissed. These two had ambushed me, so why was I stammering? "I had no clue you were even there, Cindy."

"Well, obviously," Cindy said. "I thought we were friends!"

"We are, and . . ."

My sentence was cut off by the sound of the dial tone. Mom came over at this point.

"Krista?" she asked.

"Mom, I'm not exactly sure what happened, but Kerry called and Cindy was there, and Kerry was asking me all these questions, and then Cindy got mad," I tried to explain. I found myself staring straight ahead, and the tears came like a dam breaking. I wanted to reach out and grab something to hold onto, but I couldn't get my arms to move.

"Honey, slow down," Mom said. She put a hand on either side of my face to try to get me to focus on her, but I couldn't. "Tell me again, what's wrong?"

"I don't understand it! I don't know why they'd do this to me," I said. I knew she was there, but I could not really feel her. I felt sick to my stomach.

"Honey, it's okay, just call her back, and . . . ," Mom said.

Something snapped, and I pushed away from her comfort and said, "No, they did this. They set me up. I don't want to talk to either of them. This is ridiculous." I pushed my way through the fog and headed upstairs. With my head still spinning, I laid down on my bed.

So much raced through my mind. *What was this all about? Why would Kerry do this? Wait, why would Cindy do it?* I could understand Kerry, after a bit of thought, because she always had it in for me. She was jealous of my relationship with Cindy. Heck, she was jealous of everything I had . . . good parents, siblings, money, etc. But Cindy—I was having such a hard time accepting her duplicity. Cindy was my friend. She was the reason I was popular at school, the reason I had done so well in this town.

Insecurities and logic continued their somersaults, but my newfound ego kept pushing back on the old Krista. Maybe I was popular for other reasons. Maybe I hadn't found my niche before. From what I could tell, Cindy had not been popular either until I showed up. And Lord knows Kerry wasn't.

Suddenly, it dawned on me to bounce this all off an unbiased source. It had to be someone who would understand and not parent-it-down. So I called Helen.

"Hey, Krista, hold on," Helen said. I could hear her tell her mother, "No, Mama, un momento por favor." Then she was back to me, "What's up?"

"I just had the strangest thing happen," I explained and then related the whole thing blow-by-blow.

"That's ridiculous, Krista, they shouldn't have done that to you. Those bitches! Excuse me for saying that, but how many times have I told you you shouldn't trust them? How many?" she asked, clearly agitated.

"I know, I know," I said. Various warnings I had ignored were streaming through my mind. "So you don't think I should apologize?"

"Apologize! Krista, you must be kidding? When somebody comes up and punches you in the face, do YOU apologize?"

"Well, no" I replied.

"Don't give them the satisfaction," she said, and then I heard a series of unintelligible Spanish sentences directed at her mother. "Krista, I'm sorry, but I've got to go. Don't you dare apologize to them. Plus, Cindy will cool down eventually anyway. Let's get together and go to the movies before school starts back, okay?"

"Okay" I replied.

Helen was right. Cindy and Kerry had purposefully deceived me. Once Cindy thought about it, she would see how I was set up. Kerry basically held my hand and escorted me down the plank with nowhere else to go but headlong into the sea.

The longer I sat there convincing myself I had no culpability in the matter, the madder I got. I knew Cindy would call. And when she did, I was calling both of their bluffs. They had messed with the wrong girl. I was going to show them!

Abby

I loved the holidays! This year, Bob and I decided to focus on our family and bypass the insane circuit of extended family and friends. We actually sat down, deciding on an itinerary of games and activities to do. For nights, I went through one cookbook after another, deciding on meals, baked goods, and snacks. Bob had Christmas week off, and I wanted everything to be perfect. Never in the history of the Simmons family had so much effort gone into planning.

On Christmas Eve, we bundled up and went to the midnight service. The church was beautiful, lights dimmed and candles glowing. The familiar, traditional music brought our spirits back to the season's meaning and filled our hearts with gratitude. We left the service hand-in-hand, drawing in close to keep out the chilly air while softly humming "Silent Night." Our breaths rose in magical smoke flumes, trailing off into a curtain of blue-black, studded with tiny stars.

Once home, I rushed into the kitchen to fix hot chocolate and serve up gingerbread cookies the girls had made earlier. Finishing their snack in record time, they rushed upstairs to prepare for bed; their voices shrill with excitement and anticipation. I followed to tuck them in but was surprised to see them piled into one bed. I had wondered if Krista would nix the idea now that she was a teenager; but she was right in the middle, thrilled to still be included in the gang.

Within an hour the house was quiet. While finishing the dishes, I heard the back door open. It was Bob, his arms loaded down with gifts. I raced over to hold the screen door open, catching two presents before they hit the floor.

Following him into the family room, I sat down on the edge of the sofa and watched him kneel to strategically place his treasures under the tree.

"I sure hope one of those is mine," I said, teasing.

Bob finished his chore, walked over, and announced, "Here, this one's for you."

He placed a small gift in my hand. I examined the beautiful green foil wrapping and red gossamer bow, turning it over and over. I held it like a first gift—unsure what to do, what to say, excited, nervous, and anxious, all rolled up into one feeling. I carefully untied the bow and peeled the paper away, one edge at a time. I recognized the jeweler's box and trembled, wondering about its contents.

"Oh my," I whispered. "What have you gone and done?"

I held my precious gift, delighting in the euphoric feeling ebbing and flowing through my body. Finally I lifted the lid and found a smaller black velvet box inside. I quickly pulled it from its container, and in one motion, lifted its hinged top. Inside, appearing to be suspended in midair was a dainty pearl, encapsulated in a heart and strung on a beautiful gold chain.

"I've never seen anything so beautiful." Filled with love and gratitude, I stumbled, unsure how to say what was on my heart, "I never expected something like this. I . . ."

Bob put his finger to my lips, "I searched for the perfect gift for the girl of my dreams."

I started to cry, "Bob, I'm sorry," I sniffed, "I only got you underwear."

His face broke into a smile before erupting into waves of laughter. It was contagious, and I followed suit. We laughed so hard I thought the girls would come tearing down the stairs, thinking Santa had come.

"Shhh," I said through hiccups; but every time I looked at him, we started laughing again.

Now it was Bob's turn, "Whew," he said, wiping his face to regain composure. "That's not how I imagined this evening. Do you know how many times, over the last few weeks, I've rehearsed tonight, giving you this gift, and then, taking you in my arms?"

"I guess I threw cold water on a perfect intention, didn't I?" I lowered my shoulders, pursing my lips in a pouty gesture.

"No, no! It was grand! I was so afraid I would somehow blunder the whole thing . . . say something dumb or fall on my face. Never in a million years would I have guessed you would have been caught so off-guard."

"And sound like a damn fool?" Reaching over to kiss him with a promise, I whispered, "I love you, Bob Simmons. Thank you."

"I can think of many ways you can thank me," Bob said, winking.

"Yeah, well let's see how your tune goes after we get the girls' Christmas put together," I teased.

"Yikes! I almost forgot about all the stuff I have to assemble," Bob stood, racing out the backdoor to collect the gifts hidden carefully away under his workbench.

Shaking my head, I thought, *This is a wonderful life! Thank you, God!* I reached down and unwound the gold necklace from its showcase packaging, fingering the dangling heart as it glittered in the light. I slipped it around my neck and clasped it tight. It was home, and I would cherish it forever.

The night proceeded full of laughs and memories. When every gift was placed around the tree, we stood back arm-in-arm in extreme satisfaction. This Christmas would be the best ever. With a last sigh, we both went to bed, totally exhausted, but fully content.

The next few days continued in the full magic of the season. One night, while we were playing a board game, Krista excused herself when Bob called her to the phone. I didn't say a word, knowing how good she had been to put her teen wants aside and join us. Losing one opponent, we decided to start a new game. Katie got very animated, thinking this was her chance to win now that her sister was out of the picture. Naturally, she chose Chutes and Ladders, one step above Chelsea's age level, many steps below hers and totally beyond Bob's patience.

The longer Krista's absence continued, the more concerned I became. At first I berated myself, deciding not to worry, hoping she was catching up with friends; but my mother intuition remained perked. And without deliberation, its radar honed into Krista, especially the voice tones coming from the kitchen.

Feeling I needed to prod her to get off the phone, I yelled, "Hey, are you going to play the next game?" No response. I didn't want to sabotage the game, but I had a sneaky suspicion something was wrong. I hoped my comment would run interference; but when her voice suddenly changed, a sinking feeling filled me.

"Mom, you're losing," Katie said, realizing I wasn't paying attention.

"That's okay. I'll get you on the next round." Something was up. This felt too familiar. Chelsea began her turn, but my mind shifted back to images of my fourteenth birthday . . .

* * *

"Abby, I can't wait for your party. Will there be any boys?" Linda asked, excitedly.

"No! It's a birthday party and sleepover. No boys allowed," I snickered.

"Who's coming?"

"I don't know. I haven't really heard from anyone yet." Until now, I hadn't been concerned, but this sparked some doubt.

Linda turned her head, spotting Sally. Before I could get any more information out of her, she rushed away, trying to catch up with her. Every time this happened, it reminded me how much I hated some girls and their fickleness. They were downright mean. Remembering I had forgotten to pick up my geometry book, I headed back into the school building.

While standing at my locker, Joyce Billings, a twelfth grader, grabbed me, pulling me over to the side, "Who do you think you are?"

"I don't know what you mean," I stuttered, looking around to see who was watching.

"Oh, yeah? Well, let me put it plain. You hurt my little sister because you didn't invite her to your stupid, baby-faced birthday party. So, guess what? No one's coming. Shirley, Peyton, and I made sure of it." Joyce jerked me in closer, "And further off, you are nothing, do you understand? Nothing. No one hurts my baby sister, especially you."

Joyce pushed me hard, causing me to stumble headlong into the half-opened steel front door of the school. I caught myself but hit hard enough to see stars. Jeremy Bailey popped his head around the corner, apologizing, thinking he had hurt me.

"No, no, it was my fault," I said, without looking up.

Walking away, my chin buried deep in my chest, I walked back to my locker. All I wanted to do was get out of there and go home. I fumbled with the locker combination until I finally heard it release. When I opened the door, papers flew across the hallway; and four books slammed onto the floor, barely missing my left foot.

Jeremy showed up again, "Are you having a bad day or something?" He bent down, retrieved the papers and books, and shoved them into my hands.

I didn't respond. Picking out the geometry book from the tangled mess in my locker, I slammed the door closed, twisted the combination, and exited the building.

I didn't get ten feet when Janice called out to me, "Abby? Wait up a minute."

Stopping dead in my tracks, I faced Joyce's baby sister and my best friend. I didn't say a word, fearful of what she would say.

Breathing heavily from scurrying through the crowded hall, she apologized, "Abby, I didn't mean for Joyce to ruin your party. Honest. When I went to her, I was hurt. Why didn't you invite me to your party? I mean, I waited for days after I heard about it, and you never said a word. My mom even baked your favorite rice pudding the other night, and you still didn't say anything. Why?"

I sighed, embarrassed, not fully understanding how or why I could have been so mean, "Janice, I'm sorry. It's like . . . well, it's hard to explain. I didn't mean to hurt you. Looking back on it, I know I was wrong. Deep down I knew you'd be mad . . . hurt, whatever. I just wasn't thinking."

"I thought we were best friends. Guess I was wrong," Janice stated, looking more hurt, less confused. "Now, I'm glad Joyce did what she did. Trust me, you'll be nothing when she's done with you. Everyone listens to her. She's the most popular girl in school."

"You don't have to remind me. I know all about it. Look, how can I make this up to you? I don't care what Joyce does. You're my friend, and I made a mistake."

"You don't know what a friend is!" Janice whipped around and marched away, her nose aimed skyward as if pushing past an offensive odor.

Shame and regret were hard pills to swallow. I walked past staring peers and audible jeers; past the tennis court where someone threw a ball at me; and past Janice's house where Joyce and a group of her friends stood, laughing and carrying on. Yesterday they would have thrown up a hand or yelled "Hi." Today I was nothing. A nobody.

* * *

"Mom, if you aren't going to play, everyone's going to quit the game," Katie said, annoyed. She was close to winning, and I was sabotaging it.

"I'm sorry. I want to play," I stammered. "How about I go pop some popcorn and we can start another game? Just sit tight, and I'll be right back."

I left before any one could protest, my intuition hot wired. I had to find out what was going on with Krista. I arrived just in time to see her hang up the phone. When she faced me, I saw confusion melt into doubt and hurt.

I reached over and grabbed the end of the counter. I couldn't breathe. I felt fourteen again.

You are nothing. Do you understand? Nothing!

Chapter 21

Krista

I sat watching *As the World Turns,* and the irony did not escape me. The joy of the holidays was replaced by a roller coaster of emotions that swooped up toward anger and down into the trenches of melancholy. Cindy had tried calling me back a couple of times and neither conversation had gone well. She would begin spitting nastiness at me, and I was at a loss to fix it. It felt like I had opened our front door to a rainy day and toyed with the idea of running for the car or actually being prudent and opening up the umbrella and walking over to it. The last time Cindy called, I took the phone under Mom's watchful gaze and calmly listened to Cindy's tirade before clicking the life out of the phone with a push of my finger.

"Krista, I don't think that's going to help matters," Mom said.

"I've told her SO many times that Kerry set me up, and I'm just sick of dealing with it now. I don't know what she wants from me," I said and moved to walk away.

"Did you consider apologizing for what you *did* do?" Mom asked.

"And, what was that?" I said with utter frustration, yet knowing I was not the complete innocent. Mom only shot me a telling glance.

"Okay, I thought about that, I really did, until every conversation was her screaming at me," I said defiantly, "Plus, she's never apologized for setting me up."

"Still, Krista, you know that some of the things you said weren't necessarily kind," Mom said, trying her best to get me to see reason.

"I know, but I wouldn't have said that if I wasn't duped. What did I say that was SO bad anyway?"

"Krista, how would you feel if someone said something, anything that could be construed as negative about your dad if he had just died?"

I walked away, but her words penetrated my armor. I knew she was right, but I had not insulted Cindy's father directly. I hadn't said he was a bad man. Helen and I had been talking more and more. She agreed with me. She said it was time I showed the upper hand and had Cindy apologize to me. I didn't give two cents what Kerry thought.

I was thankful for Helen's friendship and support. She told me there were some rumblings about the fight, and everyone was waiting to see who caved. I had a hard time believing anyone was that interested. *Could it have already spread, really?* But at the end of the day, I refused to be seen as the weak one.

I dreaded the next day, the first day back at school since the holiday break. At bedtime, I closed my eyes and drifted in and out, but it was a fitful sleep at best. I kept reminding myself I was worrying about nothing; Cindy and I would squabble for a few more days and then move on. Kerry would ultimately get hurt in this, once Cindy was ready to listen to reason.

When the alarm finally went off, it was a relief to get up and stop trying so hard to lie there. After getting ready and assuring Mom I was going to be okay, I walked to the bus stop. Each step felt like I was trudging through quicksand; but after all the worrying, Cindy was not even on the bus. Maybe she didn't have the nerve to face me. I started to feel more confident with each turn of the wheel and ignored the whispered hum all around me. *Nobody cared*, I thought, *how could anyone know yet, anyway? It's not like there was a public radio address of high school matters.*

I stepped off the bus and sensed people staring, but I convinced myself I was paranoid. After all, I hadn't even seen Cindy or Kerry yet.

I sat down in class, the one I shared with Helen, and she immediately set me straight.

"Krista, the whole school is talking about this, Cindy and Kerry are out to get you," Helen said, looking around the room like we were on some covert mission.

"Come on, Helen, I haven't even seen them this morning," I said.

The words were no sooner out of my mouth than the two of them came in, made some fanfare about looking at me and giggling, and then sat clear across the room. The presence of power they were communicating struck me.

"I told you," Helen whispered, but the teacher beginning the class silenced any reaction I might have had.

I don't think I heard one word from the teacher the whole class. My mind was racing with what it all meant. *Had this gotten bigger than I thought? Had Kerry managed to turn Cindy her way? Should I have just apologized? And, in some way, did Helen seem to be enjoying the whole thing, too?*

As class ended, I started to get up and leave when Cindy walked over. Nobody in class had moved very far.

"Krista, don't you have something to say to me?" she asked very confidently. Kerry was right behind her with what my dad would have called a shit-eating grin.

I didn't want a public confrontation. Trying to sound a lot more confident than I felt, I said, "Come on, Cindy, we'll talk about this later."

"Oh, we'll talk about it now," she said and pushed my shoulder.

The assault drew Ms. Pension's attention, and she reminded us to move on to our next classes.

"This isn't over yet, Krista," Cindy hissed and moved on to the sound of a unanimous ooohhhh from those who had stayed behind to watch.

Luckily, my next class was without the fiends, so I had a small reprieve. But nothing could have prepared me for health class. The teacher was late showing up and another borderline friend, Ashley, leaned in.

"Have you heard what Jeff's been telling people?" she asked.

"Jeff?" I said, confused. *How in the world would Jeff be involved?* The thought unsettled me, and I was barely able to whisper, "No, what?"

"He's telling everyone you like him, but he could never like you because you are disgusting."

"Disgusting?" I asked. I was shaky and could barely put two syllables together. I could feel tears forming.

"He said you guys both went to a pool party this summer and your pad floated up in the pool," she said. She looked mortified to even be saying it, but nobody could have been as mortified as I was.

I sat through the rest of class and heard with sudden clarity every mumbled conversation. I knew now it was about me. I was numb. My stomach was trying to heave up every moment of how I could have handled this differently, and my mind was trying to find a solution to the mathematical equation of how long this could possibly last.

Finally, it was lunchtime. I was lost. Helen wasn't around, and I dared not sit in my normal spot. I entered the lunchroom and tried to ignore the people staring. I had my lunch with me, so I picked an inconspicuous table and sat down. No sooner had I sat everyone, and I do mean everyone, picked up their trays and left. I looked up in time to see Kerry high-five Cindy and Jeff laughing with some buddies.

I got up quietly. I was NOT going to let anyone see me cry, but I knew the tears were coming. I had held them off too long. Suddenly, I realized what was coming sooner, and I quickly made my way to the bathroom and puked up the last shreds of hope that this would go away quietly or thinking I would be the victor. I officially waved the white flag on my day and went to the office.

"Hi, Nurse Perry, I just threw up and I'm not feeling well. I need to call my mom," I said. Mom would save me, and I needed saving right now. I could not face another minute of this insanity.

The nurse called her, and I heard some whispering before she came over and told me I could wait in the office. She then put her hand on my shoulder and told me everything was going to be okay.

I am not sure how long I sat there, but I was glad to be protected by the glass walls of the principal's office. When Mom walked in, I ran into her arms before saying, "Thank God you're here."

Seated safely in the car, Mom tried to get me to talk, but I didn't know where to begin. More than that, I was terrified about how it all was going to end.

Abby

The Christmas holidays ended on a sour note. Krista became Cindy and Kerry's conscripted teenage target. I did not have to fully understand what transpired to see the issue was headed to explosive, dangerous levels.

Krista maundered around the house, dragging from one room to the next. I felt totally helpless. She refused to talk or discuss strategies to handle the situation once she went back to school. The fracas escalated with each phone call and hang up, causing her to grow more sorely quiet.

The first day back to school, the air was brisk and blustery. I watched Krista leave the house, noticing the weather matched her mood. Anvil-shaped, cirriform-capped clouds framed the southwest corridor of the sky, harbinger of an approaching front. I said a silent prayer for her, begging God for a break, anything to bring her out of this slump.

I drove the younger girls to school, arriving back to the house around nine o'clock, rain-soaked and chilled. I put on a pot of coffee and faced the arduous task of packing and putting away Christmas decorations. After my third cup of coffee and eight boxes sealed, I took a much-needed break. My bottom had barely aimed toward the chair when Mama called.

"Abby?" she squealed.

My mother just squealed. My girls did it all the time, but my mother? It was rattling. Something had to be wrong. "Is everything okay?" I asked, frantic.

"Honey, life's a charm! I have a surprise for you," she announced, sounding like a love-struck teenager. "I've got a girlfriend!"

"A girlfriend? Uh? Wow. That's novel," I said, dumbfounded.

"No, Abby, I haven't read any books lately. I said, I have a girlfriend, a true soul mate."

"Mama, soul mates are between a man and woman, not two women. But that's nice," I continued, still confused over her outlandish behavior.

"It was clandestine at first. You know I've never had a girlfriend. I kinda wanted to see how it would work out. But, we've been inseparable. It's magic," Mama squealed again. "She can read my mind before I even say anything. I haven't laughed so much in years."

Looking at the clock, I felt impatience settling in. I had so much to get done. The last thing I needed was Krista and Mama falling by the wayside over girlfriends.

"Well?" I asked. "Who's this girlfriend of yours? Do I know her?"

Truthfully, I would have preferred to lay a piece of fatback on a wound and let it draw out poison than continue a conversation about some obtuse girlfriend.

"Yes, you do. But, you couldn't guess in a million years who she is." She squealed again, this time an octave higher, "It's Quillie Mutherspaw!"

"Quillie Mutherspaw?" I snipped, unbelieving, "Mama, the woman is crazy."

"Abigail Bower, I cannot believe you just said that!" Mom sounded totally horrified.

Now it was my turn to be aghast. "Wasn't that the woman who stripped on the corner of Main Street when she was a teenager? And, later gave up her child for adoption? Then, ended up marrying Billy Skidderpool because she liked his last name?"

Now the gossip came rushing back, "Isn't she the nut who stood on poor Billy's coffin holding a stolen thurible from the Catholic Church, fanning everyone with the burning fumes of his shit she got out of the bed pan?" Horror filled my voice, "Come on, Mama, the woman is a raving lunatic! And, you're hanging out with her? She's not coming over to the house, is she? Beware, she'll be dragging out the Ouija board or performing some kind of voodoo. Dad's going to flip! Are you so hard up . . ."

I stopped in midsentence. I could hear air puffing out of Mama's nostrils, filling the receiver with snorts of displeasure. I also heard her fingernails rapping an unpleasant tune as she tried to compose herself. Years of conditioning had taught me to fear those sounds. I could see her face without even being there; she was about to blow. I had fouled up. Falling on the sword, at this point, might be advisable.

"I don't remember ever choosing your friends, Abby, and I would appreciate it if you would offer me the same courtesy. Quillie has changed. She's been here for me even," Mama sniffed and continued, "in my darkest hour." She gave a sudden heave before blowing out an expulsion of air.

Abigail, you're a royal jerk, I thought, chastising myself.

"Mama, I'm sorry. You're right. I'm so concerned. You've been through so much with Daddy. I hate to see you get hurt or into a friendship that'll cost you more heartache. I promise the next time I see Quillie, I'll be very nice and well behaved. Does that make you feel better?"

Mama didn't answer right away. From conditioning, I knew the silence was to make me squirm. It worked. "Why yes, I do, but mind you, I expect no less from you. I didn't raise a daughter to be inconsiderate of other people's feelings. And, while we are on the subject, Quillie has offered to bring me down to Wilmington for a visit. We want to do some shopping. She's never been there to your home, and she's anxious to meet your family. Clara Wilkinson's coming too. That's the reason I'm calling. I want to know when it'll be convenient for us to come?"

"Clara Wilkinson? Have you lost your ever-loving mind? The woman has a police record. Didn't she murder her nephew for God's sake? What's going on here? Have you decided to take up residence at the loony bin? Please pinch me and wake me up from this nightmare. Do you actually think I want these women around my children? It'll be a freaking three ring circus," I shouted, slamming my hand onto the kitchen counter.

In unison, the phone hung up with a thump, and the back door opened. "What's all the yelling about?" Bob asked, poking his head inside the door, trying to decide if the coast was clear enough to come into the house.

"My mother just hung up on me."

"Why would Nadine do that? What did you say to her?" To accuse never seemed humbling when it spilled from Bob's mouth.

"I guess I didn't approve of her new friends . . . Qillie Mutherspaw and Clara Wilkinson."

"What? You mean the woman who stripped . . ."

"Yes."

"The one who stole . . ."

"Yes."

"And put Billy Skidderpool's . . ."

"Yes."

"Oh sweet Jesus. She's the loon that stood . . ."

"Yes, yes, and yes. That's exactly what I said and mother went . . ."

Bob's eyes grew large, "And did you say Clara Wilkinson?" Feeling safe, he entered the kitchen.

"The one and only."

"Didn't she . . ."

"Yes, I think it was her nephew," I said, my forehead in a scowl.

"Has your mother gone completely mad? You don't befriend women like that, you run like hell!" Bob exclaimed.

All of a sudden, berating my own mother sounded justified, but Bob doing it sounded way too judgmental. "Now wait a minute, let's don't get carried away. I don't know what's gotten into her, but she's not crazy. Mama's lonely."

"Hell, it takes more than lonely to attract those kinda broads. Thank God she isn't bringing 'em here."

"Well . . ." I stammered.

"Abby, don't you dare tell me you've invited those women to my house! There's no telling what they would say to the children. By God, one's a convict! I mean it, Abby, if those women show up on my doorsteps; I'm taking the girls and running."

Indignant that Bob thought he could determine my guest list, I blared, "Are you telling me that my mother and her friends are not welcome in my house?" To further make my point, I shot my hand to my waist in open confrontation.

Bob looked at me like I had two horns. His stance spoke volumes. At this point, I'm sure he thought all women were partially insane. No doubt, if he could, he would round them all up, securely wrap yellow caution tape around the entire lot and tag a notice to say: Proceed at Your Own Risk. By the time my imagination defused, he was walking out of the room, shaking his head.

"Where do you think you're going, Bob Simmons?" I snapped, marching right down behind him. "And, what are you doing home so early?"

He lowered himself into his recliner, picked up the newspaper, and snapped it open. With a moan, filled with echoes of *Go away, Abby*, he threw his nose dead center into black and white. To further silence any more comments, he rolled his shoulders toward the edges of the paper.

I stood, my lips tight and my right leg tapping up and down in open defiance. "Well, answer me. Are you trying to tell me my mother *and* her friends are not welcome in this home?"

Bob turned, giving a disgruntled look. "First of all, I came home early because I had a meeting downtown at one o'clock. I slipped out in hopes of having lunch with you," he stated. Getting up, the paper still in his hand, he drew close, pushing his finger toward my face, "And, lastly, invite whoever you want. The embarrassment will be on your head, not mine." When he had finished, he threw the paper on the floor and walked out the back door. I stood speechless until I heard his car's engine start.

Instantly, I went from anger to fretfulness. Rather than berate myself, I scurried back to work, figuring I would make it up to him later. For now, I concentrated on my conversation with Mama. For the life of me, I could not envision my mother befriending Quillie or Clara. I recalled how people moved to the other side of the sidewalk to avoid them, often spewing sharp-tongued, profanities at them.

Deep in thought, I almost missed another incoming call. When the ring registered, I ran, hoping it might be Bob. I wanted an opportunity to apologize to him for ranting, and more important, for missing a special lunch time with him.

"Hello?"

"Mrs. Simmons?" a young female voice asked.

"Speaking."

"I'm Nancy Bell from Williston. Krista's sick. She's here in the office. I need for you to come and pick her up."

The panic I felt earlier came storming back. A sepia-toned image of Krista, walking naked into a lion's lair, filled my mind.

When I got to school, Krista threw herself in my arms, beyond upset. I tried talking to her; but she was inconsolable, not wanting to discuss what had happened, which was not her usual savoir faire. All she kept saying was, "I want to go home." I was torn. A part of me wanted to shake her, tell her to pull the buckle in, solve the problem, and stay at school. Another part needed to take her home, shelter her from pain, and enroll her in another school. Both ideas sounded plausible, but only one felt right. I knew how vicious teenage girls were. It only took one walk down the school halls to encounter biting comments and roughshod behavior from small packs, facing off on opposing groups. The scenes were chilling.

When we got home, Krista ran upstairs, and I returned to the job of packing up Christmas decorations. Wrapping an ornament Mama had given me brought to mind our earlier conversation. "Quillie and Clara? What in the world could Mother be thinking? And then . . . bringing them to our house?"

Not realizing I had been talking out loud, Krista startled me when she asked, "Is Grandmother coming?"

I shot her an if-looks-could-kill glance, but she did not seem to catch it. She rolled over the back of the couch, and in one fatal swoop, grabbed the afghan and pillow. Before her body made its final landing, she had herself positioned squarely in the center. Her next move was to grab the television remote, turn it on and up as loud as possible.

Lowering the volume, she lifted herself up on one elbow and repeated, "Is your mother coming? Hey, do you have to pick up Katie and Chelsea?"

"Katie and Chelsea are riding home with Ann Lewis. We're going to start carpooling. And yes, my mother's coming."

I was poised, ready to hear her flack, but she answered offhandedly, "Great!" She flopped back down and continued channel flipping.

Within a few minutes, the girls arrived and went straight to the snack drawer. I shouted a hello, taping shut the last Christmas box of the day. Tired and achy from bending over boxes, I decided to go upstairs and take a shower. My foot hit the last stair when Katie called out, "Come here."

She stood by her bedroom door and waited until I entered to slip it shut. Sensing something askew, I went over, sat on the bed, and engaged her with a quizzical look, "So, what's up?"

"Krista's in trouble," she said, flatly.

"What do you mean by trouble?" The hair stood in salute on my arms.

Katie walked over to her record player and started to spin the turntable with her finger. I watched her until I could stand it no longer, "Katie, why do you think Krista's in trouble?" I repeated.

"People are talking . . . saying bad things about her." She strained the delivery before bursting into tears.

Her words sliced the air. In the distance, the clock chimed. I sat, staring at her, seeing each word hang on every note time expelled. I tried steaming them in my brain for purification, but her tears brought the endeavor to a standstill.

I went over and took her in my arms, "Honey, I don't think anyone would hurt Krista on purpose, especially Cindy or Kerry. You know how girls are, one day they're mad and the other best friends. Trust me, by tomorrow this will all be over."

I blew false promises into Katie's hair as I hugged her. And all the while, I felt jammed into a cage with no means to move or escape. Yesterday, today, and tomorrow blurred for a minute. I feared something evil coming our way, and I didn't think it was Quillie Mutherspaw or Clara Wilkinson.

CHAPTER 22

Krista

"Krista, I don't like you missing school, and I don't like hearing from your mom that Cindy is getting to you like this. I didn't raise you to take that kind of shit from anyone," Dad said and lifted a bite full of mashed potatoes to his mouth.

"Bob . . . ," Mom started.

"No, Abby, seriously, the child has to stand up for herself," he said.

"But, Dad, I don't know what to do," I interjected.

"What you are NOT going to do is let this affect your schoolwork," he said and laid down his fork. For Dad to actually stop eating, this was going to get serious. "There are times, Krista, when you have to stand up for yourself. Admit what you've done, but then don't take anymore crap. I know you are stronger than this. And smarter."

"Bob, I'm sure this will blow over soon," Mom said.

"It won't if she keeps calling you to pick her up," he said. Mom threw him a glance. I'm sure she was upset that he would say that in front of me. Mom wanted us to know we could call her anytime.

"Dad, seriously, I'll try to talk to Cindy, okay?" I said, not really knowing what my options were to get out of this nightmare. "Can I be excused?" After Mom and Dad both nodded their heads, I disappeared to my room.

A couple of hours of homework and a kiss and hug from Mom, and I was ready to attempt sleep. I had turned my radio to the sleep setting, hoping the noise might help me drift off. The song "Eye of the Tiger" flooded the room.

> *It was the eye of the tiger;*
> *it was the thrill of the fight.*
> *Rising up to the challenge of our rivals*
> *And the last known survivor stalks his prey in the night*
> *And he's watching us all with the eye . . . of the tiger.*

I stood in the middle of a wrestling ring struggling to keep up with my opponent. Punches were flying. Our quick moving feet seemed like a dance rather than a fight for our lives. I ducked my head but not quickly enough. A punch landed. I stumbled but managed to stay on the mat. This was no longer a friendly match—this was personal. I charged my opponent headfirst. My head hit its target, but the jolt did more damage to me. Landing on my butt, I looked up in time to take a jab to the face, then one to the back. I felt disoriented. *How many people was I fighting?* The crowd had faces, and the stadium suddenly became a small room at my high school. More punches landed. No longer knowing who I was fighting, I knew the importance of the event. Fear began to choke me. My only option was clear. I had to run. Had to get away.

"I must run, must . . ."

"Honey, it's okay. Wake up. It's morning. Time to get ready for school," Mom said, leaning over my bed. She wiped my forehead with the back of her hand. "You're all sweaty, are you feeling okay?"

"Mom, I'm sick. I don't think I can go to school," I said. I didn't have the strength, especially not after that nightmare. I needed to think, to come up with a plan. Maybe it wasn't too late to apologize, but I was afraid it would only make me seem weaker. The last thing I needed was Cindy or Kerry or both going in for the kill.

"Honey," Mom said as she gently stroked my head. "I know this situation with Cindy is hard, but you have to face it."

That's all it took. The words just started pouring out of me like an overfull tub. I knew the possible damage the flood could cause, but I didn't know how to stop it.

"They all got up, Mom, ALL of them. They just left me sitting there," I said through heaving sobs, "and then Jeff said I put pads in the pool. Pads, Mom. They cornered me in class about needing to apologize. The whole school knows! Heck, probably everyone in the county knows."

"That's it, Krista," Mom said as she turned my face to look her in the eyes. "This has gotten out of control. I'm calling their moms, and we are going to get to the bottom of this." She stood up and started walking away.

I jumped out of the bed and ran to her, throwing my body in her way to block her intentions.

"Mom, oh my God, you can't do that," I pleaded. "I have enough problems without everyone at school thinking I'm a baby whose mommy had to save her. I'll go to school; I will. Just don't call them."

"Krista, I'm not making any promises; this can't go on. Now, go get ready for school," she said.

Struggling through my morning routine, I felt like someone had poured liquid lead over me. I was in pain, and my limbs were getting stiffer and less and less under my control.

"Mom, do you think you could take me to school this morning?" I asked at the breakfast table. All I had managed to do was let my spoon swim in the egg flow.

"Okay, but we really need to go," she said and motioned to my sisters, "Come on, girls, let's get a move on."

Having now avoided a possible bus scene, hope began to creep back into my day. I was down in the first set tiebreaker, but I was not done yet. My serve.

Mom dropped me off at the front of the courtyard. A crowd, as always, had gathered. I saw Helen sitting to the side on a bench, so I started toward her. The glance she shot me would have made Spiderman's spidey sense go wild. My senses surely were.

"So, it's not bad enough that you talk about my dad, now you've talked about me behind my back," Cindy said as she rushed up from behind me. The last word landed as her hand pushed my shoulder. My right foot flew back to steady myself from the blow.

"What are you talking about?" I asked. I honestly did not have a clue what was going on. There wasn't much time to reflect because a crowd was forming—most everyone seemed to be behind her. Then again, I did not really have the courage to turn around and see if anyone stood in my corner.

"You've got to be kidding," she said and then threw her arms flamboyantly around before adding, "You are going to stand there and act all innocent."

Good Lord, the girl was putting on a performance. She probably wrote the scenes the night before and would submit her tape for consideration later.

"I told you she wouldn't own up to it," Kerry said.

"Wouldn't own up to what? This is getting ridiculous, and I'm going to class," I said. This public display was certainly not going to solve anything, and it could easily get way out of hand.

I turned to leave and noticed there were only three people standing behind me. One was Helen, and I didn't recognize the other two. One step into walking away, Cindy yanked me by the shoulder to turn me to face her. In an instant, my mind registered her fist coming toward my face. With an instinct I didn't even know I possessed and with Dad's words echoing in my head, I threw up my arm to block the blow. Cindy looked stunned, and the crowd took a collective gasp.

"Oh my God," Cindy yelled, grasping her arm. She was bent over her left arm and holding her wrist like it was broken. "She broke my diamond bracelet!" She took a dramatic pause to allow for greater effect. "The one my daddy gave me." Tears began pouring, and Kerry rushed up to put an arm around her. Then, two teachers were suddenly leading us to the principal's office.

"What's going on here, young ladies?" Principal Laughton asked, closing her office door behind us. "I understand you girls almost had a fight in the courtyard." She waited for an explanation.

Cindy started crying and laying all of her sob story on the line. It felt like eons ago sitting with Susie from Tennessee. The drama of the tears was just too much.

"I'm sick of defending myself, Mrs. Laughton, but I DID not talk about her daddy. And, IF I broke her bracelet, it's only because she tried to punch me in the face!"

"Settle down, Ms. Simmons," the principal said and began pacing her office. "I think both of you need to think about this. You are both good students and smart girls. This feud is not going to do either one of you any good. Krista, your mom is on her way to get you, and we will have a little talk before she takes you home. Cindy, I've got a call into your mom, but we haven't heard back, so you can just sit here and wait."

Just as Mrs. Laughton finished speaking, her phone buzzed, and she pushed the button. "Yes, Ms. Alton."

"Mrs. Simmons is here for Krista," the secretary said.

"Send her in," Mrs. Laughton said, and then she looked at Cindy. "Cindy, you can go wait in the outer office for your mom."

Cindy bent down to get her stuff together, then walked over and opened the door. Mom and Cindy stood face to face for a moment. I could see Mom wasn't happy, but I wished I could see Cindy's reaction. She had always thought so much of my mom, my family, or at least she said she did. So I wondered momentarily, before considering the wrath about to befall me, if Cindy had any remorse when she saw Mom.

"Hi, Mrs. Simmons, please take a seat," Mrs. Laughton said and closed the door. "I'm assuming you know what's going on here."

"I do," Mom said.

"Today, it almost escalated into a fistfight in the courtyard," she said. Mom shot a worried glance at me.

"Mrs. Laughton, I can assure you Krista and I will talk about this, and you won't have to see either one of us in your office on this matter again. And, I appreciate you bringing this to my attention," Mom said and started to stand. Mrs. Laughton appeared prepared for a longer session but seemed relieved not to have to go through it.

"Thank you, Mrs. Simmons. And, Krista, I meant what I said. You are a smart girl, and you don't want this getting any worse."

Mom and I walked out without words. The silence continued until we were safely buckled in the car.

Mom let out a loud sigh before she spoke, "Krista, honey, listen, I know you don't want to hear this, but I'm going to call both Cindy and Kerry's mothers when we get home. You just don't seem prepared to handle this."

Great, the team of Cindy and Kerry prepared to serve up an ace, and Mom just handed them the ball and racquet.

Abby

I had been disconnected and distracted most of the day. Too much was going on, and focusing on any one of a myriad of concerns proved unsuccessful. Being unseasonably warm for late January, I hauled a lawn chair out of the garage and sat on the patio. Some of the flowers in the beds were still struggling to bloom, faint yellow and pink peeking out from under pine needle mulch. It reminded me how much I loved the South—short winters, lingering color.

I put my head back against the metal frame of the chair and tried to settle into a comfortable position. Wrong. I twisted from right to left with nothing but one cramp after another. Restless, I bolted upright in time to catch the sun scintillating off the concrete pavers before disappearing behind a braided cloud. The cloud's tautness reminded me of my own; but I let the thought die, not wanting to ruin the moment.

A fly buzzed around my head, stopping within inches of my nose. We studied one another. Its aqua-green body sparkled in the swirling, dust-cloaked air spewing from behind the neighbor's John Deere. Old man Clester seemed oblivious to the air pollution he was causing, cutting through more sand than grass. Tired of me, my visitor promptly left, leaving a dust trail in his exit.

Sun rays permeated my exposed skin, warm and intoxicating, releasing tension to a dreamlike state. My shoulders lowered, falling softly back into their natural alignment. If life would have allowed it, I would have planted myself in this spot to grow and flourish. I sank into a blissful state. It was wonderful, but there were pressing problems I could not ignore. I had to tackle them before more surfaced. I needed a revised plan, having abandoned the other one when life started to turn upward.

One by one, I mentally listed all the issues pecking at my consciousness. Rather than feeling they were just whirling undercurrents of total confusion, I decided to look at them singularly and objectively. I held up my left hand and began counting them off on my fingers.

Number One. Krista was definitely in a foul mess with Cindy and Kerry. Bob worried she wasn't keeping up with her schoolwork. And the stress had manifested itself in more frequent face outbreaks and a nippy temperament.

Two. Katie was not complaining or demanding attention, but had grown sullen and introverted. Guilt strapped itself around my waist and pulled tight. I had to get to the root of her issues.

Three. Chelsea's seizures continued. Her teachers were overreacting, wanting her to remain inside instead of having recess with her classmates. Yesterday, she cried all the way home from school.

Four. Bob's work demanded longer and longer hours. When not at work, he was traveling or catching up on chores at home. When I tried to brief him on the children, budget or house, he had little to no patience. He seemed annoyed or completely tuned me out. He didn't seem to understand how important it was for me to have a vehicle, a live body, to vent my needs and concerns to. Asking myself questions and answering back was not working.

Five. Dad had not fully recovered from heart surgery, and symptoms of depression were surfacing. I wanted to help more, but didn't know where to find the time. Guilt pulled in two more notches.

I shifted to my right hand to continue counting.

Six. Mama had slid past the point of sanguine behavior. Seeing her stand up as a rubric of her recently defined New Age, whatever that was, or join the vulpine group of Mutherspaw and Wilkinson was more than I could deal with. I wanted to tell her to grow up. I also worried Dad's clinging attitude would push her further into more erratic behavior.

Seven. Me. I was caught in the middle of everyone's life. I feared each step could be on a land mine; every flight could go down in ruins. Everything was complicated.

Seemed like only yesterday, Mama stood pointing her finger, standing like a sentinel, dictating where I could go and what I could do. She'd hammer relentlessly until she extracted a promise of compliance. I remember how I fought with a vengeance to gain independence, even at the expense of a brush across my face. In my mind, she ruined everything, standing a solid roadblock to my perfectly planned life.

The irony of it was, after the fight to get here, to win my independence, I'm floundering. How do you meld the dichotomy of going from being a young girl, thinking you had all the answers, pushing limits and defying reactions, to changing diapers and making decisions for an entire family? The spiky confidence I had as a youth was gone. Now, one bad choice and my confidence dropped like mercury in a thermometer, chilling and stalling future decisions.

The stakes were high. My decisions, the demonstration of strength and confidence, would impact and influence my daughters forever. Growing up, I'd seen a lot of women, like me, like Mama, standing at the same crossroad, falling apart without a clue. Not wanting to accuse or allow myself excuses, I realized change had to begin with me. I did not want to replicate history; I wanted to begin it.

It was obvious life had come full circle. Krista was beginning to resist authority, wanting to establish her own sense of self. I knew her confidence was under attack. How could I empower her? I didn't want her to fall through the cracks while I muddled through my own questions. Mothering had to be the hardest job in the world and the loneliest.

"Mom," Chelsea yelled, "someone's at the door."

I turned toward the sound and breathed a heavy sigh. Please God, no more problems to deal with. I was running out of fingers to count on. Then my cynical mind dove into comic relief. Maybe my visitor would be Michael Anthony delivering a cool million dollars with his soft English accent. The money meant nothing; I wanted the anticipation of magic, excitement, if only for a moment.

I walked slowly and deliberately, savoring each silly thought, wishing something wonderful would be waiting just around the corner. The anvil fell. I looked past the opened door and straight into the eyes of an aging, squattie, gray-haired man with sideburns bushing off his slacking chin.

"Ma'am, sorry to bother you, but your gas meter ain't working right," he announced.

"I beg your pardon?" I squinted my eyes to a slit, trying to erase the picture in front of me. Then, rubbed them hard. Looking up, I realized he was still there, an offense to mankind in his greasy, nasty, blue-striped jumpsuit with New Hanover Gas Works embossed on the pocket. He held a clipboard in one hand and a Camel in the other.

"That meter of your'n is corroded or something. I can't get it to register a reading to save my life on how much gas you people used last month." He cocked his head to one side, scratching his crotch with the edge of the clipboard.

No Michael Anthony. No million dollars. No magic. Not even the Welcome Wagon. Instead, I had been sent a wart on the ass of progress.

"Well, what do you want me to do about it?" I asked, blistering with contempt.

"Listen, lady, I just come by to read the dang meter. It ain't my bisness to fix 'em. You best call the office and get them to send someone else over here. And, I ain't comin' back 'til you cut down them vines. I done been bit by spiders, wasps, and bumblebees whilst I've been on this job. Near 'bout killed me. Gotta keep this here tablet in my pocket or else I'd die."

He slid the clipboard between his knees, fumbling in his pocket until he extracted a little pink pill encased in dark blue pocket-flug. "See? This here pill is all that stands between me and my Maker. My wife done told me . . ."

This was unbelievable! There wasn't enough Calgon in the world for this misadventure. "Look, I'm sorry about the bees and all, but I really have to go. I promise to call and take care of everything." I slammed the door, hoping that thing, whatever he was, would not melt, slide under the door frame or otherwise manifest itself, still talking and all, in front of me.

"He was scary," Chelsea proclaimed, shuttering.

"He was more than scary," I laughed out loud. *Whew! I really don't know what's scarier, my imagination or that character,* I thought.

"Hey, let's go check on the girls. Okay?" I grabbed Chelsea's hand and proceeded up the staircase, counting each step as our foot plopped on the next one. At the top, Chelsea disengaged her hand from mine and ran into Katie's room.

"What ya doin'?" she asked.

"You didn't knock."

"I didn't have to 'cause Mama's here. Huh!"

With such a large introduction, I didn't want Katie to regress into another mood, so I walked in quietly and sat down on the edge of her bed. Katie was drawing bears on a poster board. Bears had become her new insignia. She drew them on everything.

Katie finally handed Chelsea a piece of paper, scooting markers toward her, saying, "Draw something."

I watched attentively. They never looked up from their artwork. It was peaceful to see their heads bent over, almost touching, intently designing their masterpieces.

Slipping out quietly, I headed toward Krista's room. Music blared, spilling down the hallway. I was surprised to see her door open. I leaned into the door frame, studying her. Propped up on pillows, she was reading and listening to her radio at the same time, her foot keeping a steady beat with the music. *How in the world could she be engrossed in a book and music simultaneously?* The thought intrigued me. *Could you merge the two, creating a new sensory perception, more powerful and colorful than one provided?* Maybe this was the antidote for all the misery she had been going through with Cindy and Kerry. On second thought, I doubted it would cure the friction. The boil was coming to a head. I hated it, but knew it was time to get involved.

Krista spied me leaving and asked, "Hey, Mom?" She yelled over the music.

I signaled to her to lower the music volume and listened as she continued, "I'm rereading *Are You There God, It's Me, Margaret*. Sometimes things suck. I actually started to wonder what Margaret would write about today, especially if she had Cindy and Kerry messing up her life. Maybe I should read *Karate Kid*. I need a Miyagi on my side," Krista said, smiling.

Her smile quickened my heartbeat. It had been weeks since I had seen her mouth even bend upward. "I think you're right. There are a lot of good lessons in books. Sometimes I wish I had a Miyagi on my side. It's hard to know what to do when life gets heavy. Hey, you wanna help me create supper out of a million leftovers?"

"Yuk! No. Surprise us," Krista said with an upturned nose. Continuing, she added, "Hey, Mom? Do you think Grandmother's two friends could help me out with Cindy and Kerry? They seem to know how to do it."

"Krista! You don't know anything about those women."

"Yeah, uh-huh. I heard you guys screaming," she said, but when my eyes admonished her, she corrected herself, "I mean talking. I can't wait to meet them. Which one did the shooting?"

"Never you mind, and forget what you think you heard. Don't you have some homework or something?" We drew in the silence, but I knew more needed to be said, "Listen, things have been horrible for you. I don't know how you've managed. I kept hoping this would all blow over, but it hasn't. In fact, it's worse. I think we need to talk to Dad. Let's see what his advice is. Maybe he has a strategy to handle Cindy and Kerry. We might need to get their mothers involved."

"Mom. No! You can't call them. Please! If you do, we'll have to move."

Her panic was real and fresh. It was the same wave of desperation she had displayed earlier at school. I wish I didn't understand it as well as I did. Lessons at this age were excruciating.

"Let's talk to Dad tonight, okay?" Her gulp of apprehension was obvious. I smiled and nodded, trying to feign courage, but my heart was skipping beats. The thought of approaching Cindy or Kerry's mom put my nerves on edge. I hadn't met Kerry's mom, but I had met Cindy's. She was a big woman, the kind of woman that took a stance, narrowed her eyes, snorted and charged, leaving nothing standing after her onslaught. The thought made me want to dry heave.

I closed my eyes and gripped the door frame. *Please, God, don't let me fail Krista. If I do, I'll fail us both.*

Chapter 23

Krista

My life was bound to be the stuff John Hughes needed for his next teen-angst film. What possible good could come from all of the moms getting together? Maybe I could have Mom and Dad send me to a private school for a year or two. After all, we didn't live anywhere long, so I'd only have to hide out for a little while. Heck, Dad could be negotiating a move right now. We had already lived here a year and a half, and we usually didn't make it past three. Still, at this rate, another year and a half here would be excruciating.

That night at dinner, I tried one last time to reason with Mom and maybe even get Dad to help talk her out of her mission.

"Really, Mom, what do you think is going to happen with Cindy and Kerry's moms there anyway?" I started. "Don't you remember all those get-togethers with Susie's mom? They NEVER did any good."

"Krista, we have to do something," Mom said. "Your father and I can't just stand by and watch this happen. You are obviously *not* in control of this situation."

Dad almost choked. He had looked uncomfortable enough sitting through the conversation, and now he was becoming part of it.

"Hey, I did what Dad said. I stood up for myself, and I ended up in the principal's office. Now I have the whole school thinking I broke the freaking bracelet that her precious daddy gave her. And this is also the man everyone thinks I insulted, by the way!" I said, yelling.

"Young lady, it's talk like this that gets you in these messes. You HAVE to learn to control your mouth," Mom said.

"But, Mom . . ."

"I've had a long day at the office, let's keep dinner a little less heated," Dad said. Katie and Chelsea were clearly on his side because they started chatting about their frivolous little days.

"Can I just be excused then?" I asked. With nods from both parents and sisters, I took my food and dumped it in the trash and then put my dish in the sink. I knew Mom would be up soon to discuss the meeting we were going to have the next afternoon. A Saturday afternoon no less. Even with my complete lack of a social life lately, I had better things to do than spend Saturday afternoon in hell.

Later, while lying on my bed, my mind had a million scenarios attacking it. There had to be a way out of this mess. I wished I could go back in time and redo that one silly, stupid phone call. One of my favorite movies was *Back to the Future*. All of a sudden, I had an image of taking my Dad's 300ZX and zipping back to the day before Christmas. I would just keep my fool-mouth shut. Better yet, I would know Kerry was trying to play me and dupe her instead. Even as I imagined it, I realized I needed to come up with a real, executable plan.

I turned over and hit the radio. I wanted noise. But it was not working, so I decided to get up and pull out my diary. I had not written in it for so long. I immediately noticed the last entry was touting my friendship with Cindy and how well things were going. Man, life can be a slap in the face. I even had my distrust of Kerry right there in black and white. I guess this is what my English teacher truly meant in her last lesson by the term *foreshadowing*. Now I understood the saying, "the pen is mightier than the sword". So instead of writing a diary entry, I decided to try something else.

Dear Abby,

I have done an awful thing to a friend. I set her up and I've convinced the whole school to turn against her. I actually told the whole school she hated my dead father and broke the bracelet he gave me. Worse yet, I told this guy she likes that her pad came loose and floated in my pool. How can I make amends? What can I do to let everyone know I'm the jerk and not her?

Also, I think I like girls and not boys. So some help with that would be appreciated, too.

Signed,
Regretful Girl
Wilmington, NC

Seriously, how would the paper know who really sent it? Too bad not many junior high students read "Dear Abby" or it might work. I read it, but I happened to have a fondness for people's troubles lately.

Giving into the inevitable conclusion tomorrow was going to happen whether I liked it or not, I started thinking. Maybe, just maybe, Mom would be able to clear this mess up. She'd saved me many times before; so if Cindy wanted a way out, this might give her the keys and some gas. Clutching my diary, I finally drifted off to sleep.

The next day flew by. Before I knew it, Mom and I were heading over to Kerry's house. I was not sure how the meeting ended up there since the real issue was between Cindy and me. It seemed one of our houses would have been the site of the debacle. I could care less to ever mend fences with Kerry anyway.

"Mom, why are we going to Kerry's? I really don't even care about Kerry, especially now," I said.

"Kerry's mom offered to have it at her house," she said. "And, I thought we'd be less interrupted there than at ours."

"What about Cindy's?" I continued.

"Her mom didn't offer," she said simply.

We pulled up to the brown structure and headed to the door. I imagined myself a prisoner walking the long corridor down death row. The long black car parked in the driveway was an awaiting hearse. We marched slowly toward my doom, toward the preparation of the end. Nothing felt hopeful anymore, and I closed my eyes. I was afraid the angel of death would be there if I opened them.

As we entered the house, I felt I was walking into the execution chamber. In slow motion, Mom and I were ushered to a chair to await the proceedings. I stayed silent, expecting Kerry and Cindy to come over and tighten the straps on my chair.

Some simple pleasantries were exchanged between the moms, but none of us teenagers said anything. Mom started it off with some explaining and what felt like a lot of concessions. I still had not uttered a word. To be honest, I wasn't focusing on the words being bandied about. I could see the characters interacting like a black-and-white film. Someone had forgotten to edit in the dialogue and soundtrack because I was in a fog. I watched the whole thing unfold through distorted glasses. My mom sat forward in her chair and was the one currently speaking. Cindy and Kerry did a lot of eye rolling, Cindy's Mom was looking at her freshly manicured hands, and Kerry's Mom kept folding and unfolding her legs. I saw a cat slink across the hallway and a bird fly by the window. I was jolted back to the land of sound by a tea kettle exploding and then the mention of my name.

"Krista realizes she made some inappropriate comments, but this has escalated into something affecting all the girls at school," Mom said.

"Tea?" Mrs. Treble asked. I could tell she wanted to be anywhere but here. At one point, I vaguely remembered her giving Kerry a bit of a scolding.

Still, I could not help but wonder if this was the best these supposedly concerned adults could do. *How did this end up teatime?* Knoxville and Susie came to mind again. Mom was too nice. I wanted her to go in there, guns blazing and fix things. Instead, she was meeting with her cabinet members to see how to best mend a political mess. She had invited the opposing parties to get their take on things, and I was the screwup who had to be handled.

"I said I was sorry," I started.

"No you didn't!" Cindy screamed and then clutched her mother. It dawned on me she was not only getting attention for this whole thing at school, she was getting attention at home. Something she had been dying for (okay, bad use of words) for as long as I had known her.

"Girls, really," Mom said. "We need to work together on this. You were such good friends before, and I know . . ."

After that it really didn't matter. Cindy and Kerry had gone back to their eye rolling. They would look at each other and time it in a show of solidarity. Kerry's Mom kept moving in and out of the room, trying to avoid any real confrontation. Cindy's Mom sat like a queen on her thrown and offered very little if anything. When she did grace us with her wisdom, she acted like a mama bear fighting a predator for her cub. All this from a woman who did not have the time of day for her children unless one was appearing at a beauty pageant.

The car ride home was barren. I was resigned to a junior high hell, and Mom looked defeated and tired. When we pulled into the driveway, Mom turned off the engine and faced me. I thought she was going to say something, something positive and hopeful. Instead, she did what she knew best.

"Let's go inside and make some cookies. I'll let you lick the bowl," she said and patted my back.

I love you, too, Mom. But cookies can't cure my life this time.

Abby

After dinner, I followed Bob upstairs to the bedroom. I was panicked over meeting the mothers tomorrow. I needed some guidance. Sitting on the edge of the bed, I watched Bob change out of his work clothes and into a pair of pistol-barreled jeans and a North Carolina State T-shirt. He loved those old jeans. He thought they made him look thinner, clinging to his body like cellophane. We might have forgiven the suction cup appearance, but the high-water look had to go.

Bob refused to look at me, adopting a total avoidance strategy. After waiting a customary time, I decided to break the maybe-she-will-go-away mood settling over the room. "Bob, I need your help to sort through the calamity Krista's in. You might think this is a silly teenage thing, but it's serious. I truly don't know how to help her, and tomorrow I face the den of mothers."

Bob rubbed his eye and let out a sigh. He paced the room, back and forth, staring at the floor. One hand was clutching and unclutching, the other uplifted, his index finger extended, tapping against his lip in a rhythm set to a silent melody.

I sat patiently, knowing he had an answer, even though earlier requests for his help had been blown off. One last ditch effort to plead, and he'd come through; I knew it. Bob was an expert. He ran a business and had to deal with personnel issues daily. I continued to watch until I thought I would explode, "Bob, please. How're we going to handle this situation? Krista's sinking."

Bob raised his head slowly and met me eye-level, "I'm trying very hard to understand why you are continuing to insert yourself in Krista's problems. I have to admit I'm really struggling."

This was not what I wanted to hear; his words were totally unexpected. Krista and I were on the verge of a breakdown. I had explained the circumstances—what had happened and where I saw the situation going, but it was clear Bob had no clue how desperate things were.

"What are you struggling over?" I asked, an impatient air to my stance.

"The whole teenage thing."

"Teenage thing?"

Bob shrugged his shoulders, drew his eyebrows together, and ruffled up his upper lip, "Should we get involved? Maybe you should back off . . . let Krista handle her own problems. You know kids; it'll blow over. You could make it worse. Why don't you let it go?"

"Let it go?" I repeated, feeling he had just slapped my face. With a sour note of sarcasm I retorted, "Yes, I probably should let it go, but I can't. My daughter's heart is breaking, and those vultures are attacking her openly. Have you not taken into account that I've already had to pick her up from school because those girls are making it so incredibly intolerable for her? This is affecting her life, her confidence. Don't you understand?" My cool was gone. This was not the support I had anticipated. I needed Bob to listen; instead he walked into the bathroom, dismissing me as if I did not exist.

I followed, watching him stand in front of the mirror inspecting his face, rubbing at a faint growth of beard. This was not happening. Why was he being so cavalier? Had he actually shut me down and up?

I felt a horror movie begin. Maybe Bob was not really Bob, but a clone—an insane, irascible clone of a man I did not know. I tried hard, looking for a semblance of my loving, caring husband, but his face held no clue. I wanted to scream, to warn him, someone or something had invaded his body.

I watched warily. Bob continued, completely unaware of his change. He turned the hot water on; opened the medicine cabinet; and brought out his razor, shaving cream, and English Leather aftershave. Steam threaded upward from the sink, delivering a spiraling acrid, metallic odor that filled the room.

"Damn, I hate having a well," the man flippantly announced. "As soon as the town finishes putting in city water, I'm hooking up. This is disgusting."

The horror movie continued. Suddenly, the whole bathroom warped into a slow-motion scene. I watched him slather his face with shaving cream, pick up the razor, dip it into the steaming hot water, lift it dripping to his face, and glide it in slow vertical strokes. He stopped every other sweep to clean the razor and start anew. Toward the end, he began to whistle a tune. I shook my head thinking, *this is, by-dammit, not happening!* But it was, and then it registered. He was whistling "Dixie." Never in our adult life had I known Bob to whistle any recognizable tune, and never "Dixie." Before I knew it, my mind erupted into the chorus, *Oh I wish I were in the land of cotton. Old times there are not forgotten, look away. Look away. Look away, Dixieland.*

In that terrible minute, I realized I was not in a horror flick, but a Monte Python spoof. Tears sprung to my eyes, not from disillusionment, but from the insanity of the moment. The whole thing was surreal. This could not be happening.

Bob faced me, "You got any objection to my little concert over here?"

"What's so funny, Mama?" Chelsea announced, coming up behind me.

Her voice startled me. I whirled around, first trying to compose myself, and then frantically trying to shake off the last few minutes of an out-of-body experience. "Oh nothing, baby," I managed to say. I stepped around her small frame and edged out of the room. Coaching myself back to reality, I realized I was becoming a performing, superficial shell of a human being, talking in my head one minute and out loud the next. On one hand it relieved stress; but on the other, it made me feel torn, literally torn in half. The dilemma? Back to issue number four, Bob. I had no one to discuss life with; and my mind, left unchanneled, mystically went back where it did as a child, seeing the world as a giant cartoon, a movie or a musical finale. It's how I managed to escape ugliness, turning myself inside out, from Abby to Looney Tunes, from color to black and white. It was the only way I knew how to manage loneliness and hurt.

What I needed was someone who would listen, letting me explore ideas, strategies, and alternatives. If given that opportunity, life might not seem so daunting.

Later that night, after everyone had gone to bed, I tossed and turned, trying to figure out what to do and say the next day at the dreaded meeting. I imagined how I would look, the clothes I would wear for a perfect impression, and how I would sit with a composed, confident air. I would listen to everyone and give a graceful nod of acknowledgement to shared commentary. I would articulate and select the right words, pausing for a punch line performance, the way actors do when they're trying to get a point across. I rehearsed opening remarks, points and counterpoints. Feeling my argument memorized and firm, I finally drifted off to sleep, proud to have a plan.

Saturday morning came and went quickly. By the time breakfast was over, and the household chores were done, it was time to collect a sulking Krista and head toward the Treble's house. I felt confident and steady about my mission. When Krista started questioning me, I instantly put on a big smile, "Honey, we have to do this. Trust me."

The ride over was quick. Mrs. Treble met us at the door and ushered us into the formal living room. I knew from Krista's painful sigh, the Sheppards were already there. A large black Lincoln Continental spread itself across the driveway. I figured it belonged to Cindy's mom. Somehow it didn't fit the humble Treble dwelling.

"Mrs. Simmons, why don't you sit here," Mrs. Treble said. "Krista, you sit next to your mom." Once we were settled, she exited through an arched doorway.

Alone, Krista reached over and clung to my arm. I sat stiff-backed, both feet firmly planted on the floor to anchor myself. Distant chatter announced the arrival of our assailants. Mrs. Treble led the pack in, ushering them to their prospective positions, right across from us. The jury was in and court open.

We silently observed one another. Mrs. Sheppard, a large overpowering woman, stared at me, one eyebrow pasted to her hairline. There was no denying the look of you-haven't-met-your-match-until-now-baby. Cindy and Kerry huddled together on her left; and Mrs. Treble, a small-boned, mousy little woman, sat to her right, leaning away, keeping her eyes downcast toward the floor. Immediately, I sensed a disconnect. These two mothers were not on the same page. Strike one. I felt power surging. I had one up on Mrs. Sheppard. Maybe this was not going to be the end of the world.

Staring continued. No one spoke. Determined to not give in, I continued the uncomfortable stance. Mrs. Treble's cough provided the needed signal to start. I decided to be the first one out of the gate, "I want to thank you for having us here today. I realize how uncomfortable this is, but I'm sure we can find a solution to the girls' misunderstanding."

Silence. And more silence.

"Tea, anyone?" Mrs. Treble broke through again.

No one answered.

Finally, Cindy tugged on her mother's sleeve. Mrs. Sheppard bent toward her, and Cindy cupped her mouth and bent in to her mother's ear to whisper. Mrs. Sheppard erected herself into a viper and said, "I'm very disappointed in your daughter, Mrs. Simmons. Her conduct has been inappropriate, spiteful, and damaging to my daughter and her dearest friend, Kerry."

She shot her eyes toward Krista and continued, "Krista, I thought you were Cindy's friend? What do you have to say for yourself?"

"I said I'm sorry," she stammered.

"No she didn't!" Cindy screeched, clasping her hands to her tear-streaked face.

Krista reached out for my arm and dug in for support. In response, I slipped my hand over hers, offering encouragement. She was shivering; so I patted her hand, turning to give her a smile before setting my gaze on her accuser.

In a controlled tone, I proceeded, "Girls, really. We need to work this out. You're friends. That's what we need to focus on."

Next, I turned my attention to Cindy's mother, "Mrs. Sheppard, I know we can get to the bottom of this. I agree Krista was out of line with her comments during the phone conversation in December, but I do believe all the girls were equally to blame for the disaster that has followed. Rather than hashing out each incident, isn't there a way we could end this nightmare and leave here on friendly terms?"

In unison, Cindy and Kerry began spouting off things Krista had supposedly said to them, their friends and teachers, with it's-just-not-fair comments sprinkled in between each assault. When Cindy started to cry, Kerry put her arms around her, drawing a stricken look across her face.

Mrs. Sheppard, still stiff, screwed up her mouth and squinted at us, "No, Mrs. Simmons, it will not be over in a quick apology. You can visibly see the stress and anxiety my daughter is under. She trusted Krista. It breaks my heart to think Cindy has to deal with this slander right on the heels of losing her beloved father. It was viscous and maligning. I think . . ."

My resolve, my plan for patience shattered, "Hey, wait a minute. We're talking about teenage girls. Sure, they can be very cruel to one another, but they get over it. I know my daughter. She's not the person you're painting her out to be. I truly feel Cindy and Kerry are getting way too much attention over this, and it's fueling their behavior."

Mrs. Sheppard sprang up as if released from a jack-in-the-box, "How dare you insinuate my child is lying to get attention. I've seen her cry herself to sleep every night since this horrible incident happened. I've overheard her on the phone with friends who called to apologize for your daughter's cruel actions. I've forbidden Cindy to have anything further to do with your daughter. She has not been raised to act like white trash. And, you will be getting a bill from my plumber for damage done to our pool when your daughter stuffed her sanitary pad in the pump." She grabbed Cindy's arm, yanked her up, and stomped out of the room, slamming the door so hard a vase on a nearby curio crashed, shattering on the foyer floor.

I faced Mrs. Treble and Kerry. Both had a panicked look on their face. I stared at them long and hard before saying, "I had hoped for a civil meeting, but it did not happen." I got up and put my arm around Krista's shoulder. She looked totally dazed.

Mrs. Treble rose, "Mrs. Simmons, I apologize for how this turned out. I have instructed Kerry not to say another word about what happened or discuss this meeting. If I hear anything else about this, she will be punished. Krista may have made an insensitive remark, but these two girls have capitalized on it." She turned toward Kerry and delivered a "Do you understand?" statement before walking behind us to the door.

Two dejected females left the Treble household. We parted long enough to get into our respective sides of the car; but once inside, Krista slid over close to me. We drove in silence. I reached over to rub Krista's arm but said nothing. My mind whirled as I reviewed the events of the meeting. What a disaster! My worse scenario did not match what had happened today.

We pulled into the driveway, and I resolved to make things better. At the moment, I could think of nothing better than baking cookies. How could anyone stay sad under the aroma of brown sugar and chocolate?

When I asked Krista to help, she didn't say a word. She exited the car and went into the house. Katie, Chelsea, and Bob were standing in the kitchen, waiting to hear about the outcome. Krista walked by them, silent. Seeing this, Katie took Chelsea's hand and left, too.

Bob came over and put his arm around my shoulder, "Abby, I don't know what to say. I still think you made a mistake going over there, but what's done is done. Maybe I should go upstairs and talk to Krista."

Bob turned to leave, but I grabbed his arm, "No, you won't! I don't need you to rush in and be the hero, leaving me the screwup. I'll go. I'm the one who owes her an apology, not you."

I laid my pocketbook down on the kitchen counter, prepared to go up and do damage control. Not two steps out of the kitchen, Katie arrived at the door, announcing Mama was on the phone.

"Hi, what's going on?" I inquired.

"I wanted you to know we're coming tomorrow. We'll arrive sometime in the afternoon."

I wanted to scream, but I leveled my voice instead, "Tomorrow? Oh, okay."

"Abby, you don't sound well. Are you all right?"

"Just tired. I'll see you tomorrow. Drive safe."

Bob started to tune up, "Abby, don't tell me . . ."

Hell flew in me, "No, I'm not going to tell you shit. But yes, my mother and those crazy broads will be here tomorrow. At this point, you have the right to remain silent, get over yourself or be an ass. It's your call."

I left, trudging upstairs to fall on my sword in apology to Krista.

Chapter 24

Krista

I had tried and failed, Mom had tried and failed, so it was time to get Dad involved. I didn't call on him much, so this would be the perfect time for him to prove what he could do for this family. Every time we had moved or he had missed an important event due to work obligations, we heard the lecture. His work provided for our lives, allowed us to have Mom at home, gave us the clothes on our backs, and would one day give us an excellent college education without us owing a dime or mortgaging our future. Particular attention was always paid to this last part because both Mom and Dad's families couldn't afford to send them to college.

What a relief today would be for him. I was going to ask him to jumpstart his career and go for the next promotion. After all the grief he had gotten about our moves in the past, I bet he would relish the chance to do one by his daughter's request and play the hero.

I was mustering up the courage to talk to Dad when Mom came into my room.

"Hey, I'm sorry today didn't go any better. I really am," she said as she sat down and pulled me into a hug.

I enjoyed the warmth for a minute. I was angry at so many people right now, but I could only be mad at her for so long.

"Mom, do you think Dad could start looking for a promotion?" Surely she would see the win-win in my case.

She pulled away and looked into my eyes, "What do you mean?"

"I can't stay here, Mom. On top of everything else, when I go to school on Monday, the wolves are going to have fresh meat. They are going to know about the parent conference and who asked for it," I said. "If we could move, I could start fresh. Now would be a good time to do it if you think about it. We are getting towards the end of my freshman year, so I could go somewhere new as a sophomore. That gives Dad six months to find something."

She looked at me intently for a while. *Should I continue with my planned points or await some feedback?* Just as I started to gear back up, she countered.

"Krista, first of all, we don't run away from situations. That's not what all this moving has been about. Second, believe what you may, your dad does not orchestrate these moves. It's not something he can just call on when he needs it. Plus, we have to consider your two sisters and my dad's health," she started.

I flung myself off the bed, ran to my door, and shouted for Dad. "Well, I'm talking to my dad. All these moves we've made when I didn't want to, and now I want to. He's got to help me," I said.

"Krista, really," Mom exhaled.

"Christ, child, what is it? What's all this screaming?" Dad called out coming down the hall toward my room.

"I need you to move us," I said. No need to beat around the bush.

"Move you?" he asked. "Abby, help me here." Then the light came on, "Is this about today?"

"Dad, I'm talking to you, not her," I yelled into his face. Dad responded by grabbing my arms and giving me a knowing shake.

"Watch it," he sneered. He loosened his grip when Mom put her hand on his arm.

"Dad, listen, please, I can't stay here, especially after today. I figured you'd be relieved to know that I'd like you to move us," I said. He still looked bewildered, so I continued, "I want you to look for the next job level and go for it. I could be in a new state, new school by the fall."

For a minute, I was not sure if he was going to laugh or turn and walk away. Even Mom offered nothing.

"Krista, it doesn't work like that. And, you know we don't run away from anything in this house."

"I can't do it, Dad. I've tried what I know. I realize I could have done some things differently in the beginning, I really do, but I can't fix that now. And, I can't, I just can't live like this. Why can't you just move us? You've managed it well enough for fourteen years, and I didn't want it. Now I do. I want it NOW!" Tears were streaming down my face and fogging my vision.

"Abby, I'm not doing this. I can't deal with another insanely emotional woman right now. I can't. YOU explain it to her," he said and pointed at Mom. "Krista, I'm done! You need to figure this mess out. Quite frankly, you got yourself into this, now GET OUT!" He turned and left so quickly, I half expected a gush of air to knock Mom and me down.

"I hate him," I grunted to Mom. "I hate him, and I hate myself. I hate everything." She touched her hand to my back, but she didn't offer anything else, neither consoling nor ridicule. I felt like both of them had given up on me—appropriate since I was giving up on myself.

Sunday brought even better news. My grandmother and her insane friends were coming for a visit. Mom waited a day to tell me. I'm sure she figured I couldn't handle anything else. Well, I couldn't. I needed her to tell them to stay away. Unfortunately, it looked like Dad would not move us, and Mom could not stand up to her mother. Wonderful! I spent the rest of the day wrapped in a thick blanket of self-pity.

Time never stands still, and Monday came as it always does. I forced myself to walk past the whispers and looks. Actually, I had gotten used to it the last few weeks. It was the direct challenges and confrontations I couldn't deal with, and avoiding them never helped for long.

"What a sad little baby you are," Cindy yelled down the hall in between classes.

I didn't need to turn around to know who it was or that at least twenty other people were looking my way, too. I kept walking and pretended I didn't hear.

"Hey, I'm talking to you." I could feel her getting closer. Suddenly, she was jerking me toward her, "Don't walk away from me, you little coward."

I grabbed her arm with what strength and self-respect I could still muster and tossed it to the side. "I'm done with you, Cindy. Done. I'll spend the rest of this year staying away from you. I'm done talking about it, and I'm done trying to deal with you. Nothing I say is going to matter or change anything," I said in her face and started walking away. *Please let it be done, please just let her be done,* I thought.

"Then why'd you have to get your mother involved?" she spat at me.

I stopped dead still. I felt like someone had poured lead over me in midwalk. "My mom cares about me, which is something you'd never understand. I couldn't have stopped her if I tried. But, like I said, I'm done fighting with you. Grow up!"

She was enraged, and the kids in the hall had started chanting, "Fight, fight." Two teachers jumped in and stood between us. They escorted us back to the familiar territory of Mrs. Laughton's office.

"Girls, you can't really still be at it, can you? We have talked and talked about this. What do you suggest I do at this point?" she asked, looking drained. "Krista? Cindy?"

"Mrs. Laughton, I told Cindy I'm more than happy to stay away from her if she'll stay away from me," I said.

"Fine with me," Cindy huffed and crossed her arms over her chest.

"Girls, listen, this has got to stop," she said and sealed it with the look of *or-else*.

Cindy pushed past me to be the first out the door. I headed to my choir class, yet another one I shared with Cindy and Kerry. Luckily, it was the last class of the day. I walked in and saw Mrs. Haliburton at the piano while everyone was shuffling toward their seats. She often said I could be a first soprano if I wanted because I had a nice range. I stuck to being a second soprano to be closer to Cindy. I walked right over to the teacher and decided the only way to deal with the rest of this year was to separate myself as much as possible and become invisible. Even if I could not get Dad to move me, and I wasn't done working that angle, I was going to Hoggard High in the fall, and hopefully nobody would remember any of this. I could start fresh. I only had to endure a few more months of this year. *How many people really enjoyed their freshman year anyway?*

"Mrs. Haliburton," I said.

"Yes, dear," she said and started waving her hands to get everyone's attention.

"I think I'd like to give first soprano a try," I said.

I could tell by the way she looked at me, the way a lot of adults seemed to lately, she understood.

"Sure, sit down right here next to Rolanda," she said. She gave me a reassuring pat and led me to the new section.

Rolanda, a nice short-haired and rather round black girl I was ashamed to say I had made fun of in the past, moved some stuff out off the seat next to her.

"Glad to have you, Krista, you can share with me," she said.

My fate of invisibility was sealed, and I put that cloak on willingly.

Abby

> *Parenting... the Final Frontier. These are the voyages of the Starship Motherhood. Its ongoing mission: to seek out new life, defend, and uphold the values of the United Federation of Mothers, to boldly go where no one has dared before.*

It was beyond me why a stupid rendition of Star Trek's theme song kept running through my mind. I started to laugh, but then remembered my mission, damage control. My foot hit the top landing, and I walked down the hall to Krista's room. The door was slightly ajar, which meant we didn't have to knock, but we did need to announce ourselves. Funny, in the last year, new rules for privacy had been established. Katie would be next; Chelsea would follow suit. That's how it went with these three.

My feet plodded against the pine flooring in sync with troubling thoughts. What was I thinking when I headed over to meet those women? What could I possibly have hoped to accomplish? Bob was right. I should have never gone. I'd made a mess of everything. How in the world would I ever make this right with Krista? Had I, in fact, closed the deal on any reconciliation between the girls?

I rapped lightly on the door frame and walked into her room, "Hey, I'm really sorry about today. It didn't go well, did it?" Krista sat, clutching a book with both hands. She did not look up or answer. It didn't take much instinct to see how broken and vulnerable she was.

I edged toward the bed, sat down, and put my arm around her, pulling her in close and protective. At first there was a slight resistance, but slowly her shoulders softened, and she folded into my arms. I wanted to hold her like this forever, wishing away all the ugly parts of life.

Krista's defeated voice echoed in my chest, "Mom, do you think Dad could start looking for another job?"

My heart sank. In an endeavor to convince me a move was her only salvation, Krista went into graphic detail about her fears and the ordeals she had been through. I sat, conjuring up horrible images of her facing the jackals at school on Monday. I envisioned the sidling herd, forcing her into a crowd of teenage man-eaters, saliva dripping from their eye teeth, gnarling in pleasure as they tormented her. I was terrified, depicting the pain on her face as she filed by this maniacal engine of pain.

When she stopped, the room hushed around the blackness created in her word pictures. Shocked and stunned, I spoke cautiously, "Krista, I don't know what to say. You've been through so much. I had no idea how bad it's been for you."

She looked down at her lap and began picking at imaginary fuzz. I watched and waited. Tears dropped, huge pellets falling on and popping off the book in her lap. She raised red, rimmed eyes toward me, "I know. It's not your fault. You tried."

"I appreciate that. I certainly learned something about myself today."

"What's that," she asked.

"I'm stronger than I thought. My knees knocked the whole time, but I felt strong, even with that mean old woman hawking me!"

Krista snickered. I lifted her chin with the cup of my hand, "Look . . . I'm not sure how this is going to play out. Maybe it'll blow over. Regardless, you have to remain determined not to let these girls ruin your life or stop you from making new friends."

I paused, giving her time to process what I'd said, and continued, "And, about your dad requesting a move, it doesn't work like that. Dad has no control over transfers. But more important, we don't run away from uncomfortable situations. Trust me, I've learned the hard way. If you don't face your problems, they'll face you, weakening your confidence for the rest of your life. Plus, we have to consider your two sisters and my dad's health. Determination . . . that's what you need. You have to be determined to stand up and not let those girls ruin things for you." I stopped but too late. I had lost her.

Her furrowed brows were a clear indicator I had gone from preaching to meddling, not a good sign. "I'm talking to Dad!" Her face left no speculation; she was fired up.

Next thing I knew, Bob was in the room screaming, and Krista was yelling back. Voices slammed, overlapped and swallowed one another. Both refused to listen, and none of it made sense.

Bob spun out of the room red-faced and angry. Krista continued shouting in isolated bursts of fear, discouragement, and isolation. When she stopped, she sunk to the floor. Instinctively, I reached over, touching her. I had no words of consolation. Like her, I was ransacked of emotion.

After I got her to lie down, I eased myself out of the room and walked down to the kitchen. That's when I spied the underlined note on the counter. *Mom and crew due in on Sunday for dinner!* One responsibility rolled off my shoulder, and another took residence. I grabbed three cookbooks, my grocery list and began planning.

On Sunday, we pretended nothing had happened, the true measure of a Southern family. I decided to put Krista's problems on hold until Mama and her cronies left. We went to church and out to lunch. After a quick grocery store run, I prepared dinner, dreading my company. There was no telling where this mélange of characters would lead. The house was sure to be a freaking circus.

At five o'clock, I heard a clunk and a chang followed by a sputter. Rushing to the kitchen window, I saw six women pile out of a car. Mama slipped out of the passenger side, but I did not know the purple-haired lady she pulled out behind her. Quillie emerged from the driver's seat, howling with laughter. Barreling out of the back in cue were three frumpy women dressed in varying colors of tight polyester pants and mismatched blouses. Painted faces blotched with harsh pinks and reds crinkled over Quillie's continued antics. Two of the women were actually doing some type of jig right there in the driveway. Fearful Bob was looking out of the family room window, I hurried out to greet and contain my guests.

As soon as I opened the screen door, Mama pushed through. "Move, Abby, I'm about to pee in my pants," she squealed, shoving me aside. In a steady flow, five more women rushed by with no introductions except, "Hi, Abby. Sorry, we've had too much tea."

"What the hell just came by me?" Bob said, entering the kitchen, his hands upraised as if to pray he hadn't witnessed what he just saw.

"My mother's here."

"Mama?" Chelsea began tattletaling, "Grandma's friends are upstairs. They're all jumping around and holding themselves. You said it wasn't nice to do that." Chelsea continued her announcement all the way into the den.

"Jesus Christ!" Bob yelled up to the heavens.

Half whispering, I chided, "Not one peep, do you hear me? We'll get through this. I don't want to hear one word out of any of you!"

I turned toward Chelsea, "Go tell Krista and Katie I want them on their best behavior."

Chelsea reached up and whispered, "Which one killed . . ."

I shot her a bulldog glare, sending her running up the stairs. I rushed back into the kitchen to finish dinner and compose my shaking limbs.

"You're sure perty," a voice behind me stated. "You need any help thar?"

Startled, I whirled around, facing a very circular woman, equal measurements both vertically and horizontally. Her eyes were a brilliant blue, highlighted more so from rosacea that spread from one cheek over the bridge of her nose to the other. Her earbobs, as Mama called them, were large pink and lime green balls dangling from earlobes stretched almost to her turned-in shoulders. A vivid madras shirt blazed in hatches of lime green and yellow, hugging her rotund stomach, pleating and folding onto her periwinkle blue polyester pants. Red tennis shoes completed the outfit.

"Ya catch a fly?" Quillie asked.

"I beg your pardon?"

"Looks like ya caught a fly. Better shut that trap or who knows what else you'll catch."

I did indeed try to pull my bottom lip up, but it still trailed somewhere near my chin when Mama waltzed into the room. "Boy, something smells yummy, Abby. Been cookin' all day, haven't you? Why, I can't wait to introduce you to the girls. Well, I should say sisters cause they're fast becoming my dearest angels in the world, aren't you sweeties?" She grabbed an unidentified oversized woman and hugged her greedily.

A new voice brought me back to my senses, "Hi, you must be Abby. Your mama talks a lot about you. She thinks you're the cream dela cream or ever how that goes. As you can tell, I'm not up-to-date on my Frenchie words." Another peal of laughter erupted before she continued, "We haven't had the privilege; I'm Annias Alexandra Atwood. My mama always said she named me like a double Dutch hopscotch—single *A* to double *A* to triple *A* asshole." Everyone in the room, except me, howled with laughter, and more laughter erupted when the other women crowded into the kitchen.

"Yore mama kept a sack full of silliness around, didn't she, Annias? Yore brother, Titty, I mean Titwell, never did have much of her sense of humor. He was like a plastered pancake most of his life. Yore mama gave him the only charm she could, and he just shrugged it off like he did with everything he teched."

"Salem, let's not spoil the party," Quillie announced, stepping closer to me. "I haven't even hugged my little girl, now have I?"

Quillie wrapped her large fleshy arms around my neck and squeezed me into her breasts, forcing my head to push her ample bosoms to the side, while she cradled me between mounds of Cotillion cologne.

"Let me tell you, this was the prettiest baby girl on Doodle Hill. And when she got to be a teenager, it was like a dog in heat. You've never seen such a flock of sorry-tail boys what merged on our little neighborhood. I was as proud as a prostitute in church on revival night!" Quillie expelled a cough, then a gag, before joining the revelry.

"Mama," Chelsea said, tugging on my shirt, "are these Grandma's friends?"

Quillie didn't miss a trick. She swooped Chelsea up in one giant sweep, saying, "We're your Aunties, child!" She pulled her in, placing a slobbery kiss on her cheek, and then whirled her around to face her audience.

"This is Chelsea Diane Simmons, Abby's youngest. There's two more somewhere. You'll meet them later. Now, Chelsea, let me introduce you to everyone. I'm your Auntie Quillie Mutherspaw. I'm the ringleader of this motley crew." She went around the room with vim and vigor, explaining who was who, always prefacing their names with *Auntie*. There were glees and claps from each as their names were called. Finally, Mama took Chelsea and gave her a big kiss and hug.

"This is wonderful!" she exclaimed, "One big happy family!"

"Well, don't be so quick to judge," Bob exclaimed.

Everyone swung around in a giant wave of color, moving in one fluid motion. "Oh, this is my husband, Bob," I quickly announced.

Quillie pushed through the wall of women and grabbed Bob in a clinching bear hug. "What a little man." She squeezed him hard. Below her shoulder I could see the top of his head glowing brilliant red. "Nadine said you were cute, but sweetie, she underestimated your damned good looks." Quillie spun Bob around to face the *Aunties*.

"Girls, this is Bob. He doesn't realize the treat he's in for, now does he? Okay, Bob, let me go through the intros. This is my sidekick, Clara Wilkinson. Remember her, sweetie? She accidentally shot her nephew's peenis off. It was a sordid little affair, but we look at her like a kind of celebrity, don't we, girls?"

"Quillie?" Bob and I yelped at once, pointing toward Chelsea.

"Oh, hush. The child knows what a peenis is, don't she?" Quillie turned to Chelsea who hadn't stopped smiling and said, "Honey, you do know what a peenis is, don't you?"

Chelsea didn't say a word. She could have cared less what they were saying; the inclusion into their conversation was enough.

Quillie didn't miss a beat, "Clara, I hope you don't mind that little intro about your nephew, but you know as well as I do that little incident made your family famous." Clara smiled, nodding in approval.

Reaching over, Quillie pulled the smallest of the large women out front. She dipped her head, tilting it from side to side. "This is LeRoux. We had the most wonderful opportunity to meet this special sister today. She was just sitting there at Joe's Diner lookin' all lonesome and such when your precious mama started up a conversation, and before we knew it, she was one of us." Quillie hugged LeRoux like a long-lost friend.

Very theatrically, she dabbed at her eyes and continued, "And these two sisters are Salem Nadona Spruill, and I think you've already met Annias. We were introduced many years ago when I was down on my luck. These two women helped me through the worst possible times of my life, and I'll never be able to repay them for their generous love."

The *Aunties* and Mama fell into a group hug. Bob and I looked over their heaving, crying bodies in total dismay and disgust. Out of the corner of my eye, I caught Krista and Katie staring wide-eyed from the dining room. Bob edged his way out of the kitchen trauma-unit and headed out to waylay comments from the older girls. Chelsea was smack in the middle of a flesh tent, loving every minute of it, while the *Aunties* caressed her back and hair, whispering sweet things.

"Okay, well, since we've all had our grand introductions, can we proceed with dinner?" I projected, around streams of unintelligible blubbering.

I walked over to the sink, swearing under my breath, clanging dishes, and slamming drawers in open defiance of their intrusion into my already-tense-filled weekend. How dare my mother bring unexpected guests! Thank goodness I had cooked extra food, hoping to have leftovers for a weeknight. Well, forget that! I glanced back at the ham hocks waddling to the dining room table for affirmation.

"Abby," Mama said, laying her hand on my shoulder, "You're so sweet having us here for supper. I wanted you to know that we're leaving right after we eat. I hope you aren't disappointed. We want to drive straight to Myrtle Beach. We've got shows booked and shopping waiting!"

A lukewarm rush of relief spilled over me. I reached around and hugged Mama tightly. "I'm glad you're here. That group better tone down their dress or the cops will be following you for suspicion of circus fraud!"

"Darling, you haven't seen my new git-ups. I'm going to be blazing in glory!"

Supper went by in a fanfare of laughter. The *Aunties* recalled, embellished, and rehashed a multitude of stories. Most were self-deprecating accounts of their youth, lost loves, and near misses with tragedy. My girls laughed so hard I felt God had poured medicine over our household. Even Bob . . . yes, Bob, got into the lovable riot of six women expounding the virtues and misfortunes of life, sometimes at once, but all with the gusto and joy of lives well lived. When they left, we followed them out to their car like stray dogs, hugging them and begging for more scraps of their intoxicating view on life.

When we crawled into bed that night, Bob apologized for his temper with Krista and for not trusting me to talk with Cindy and Kerry's mothers.

"You know, after listening to those old broads tonight, I realize I take life way too seriously. Those women laughed a lot at the injustice in their lives. They laughed at themselves; that takes a lot of courage. I hope you'll never remind me about what I'm fixing to say, but tonight was one of the liveliest meals we've ever had. Makes you stop and think, doesn't it?"

I snuggled in deeper, "I know."

Bob slipped into a snore. So much for talking, but I realized those ladies had eased some grave tones in our household. They gave us a reprieve; laughter had been in short supply lately. Those silly old women infused us with it, at least for tonight. Maybe that's what drew in Mama—the offer of a lighter life perspective. Maybe we got a dose of it, too.

Chapter 25

Krista

I was surviving, nothing more and nothing less. This in itself was a small victory. There were many times I felt like nothing special. Have you ever realized it was not the same old robin or morning dove that caught your eye while you were staring out the window? Usually, it took the brilliance of a blue jay or the sound of a red-headed woodpecker to draw your attention. The other birds flew in and out of our lives with no effect, not even a conscious awareness of their existence. I was starting to view my life this way . . . a small ripple in one heck of an ocean. If nobody remembers you when you've gone, did you ever really exist?

My thoughts stayed in those melancholy places most days. There would be moments of humor followed by deep despair. I was past crying and past anger, just numb. Even the insanity of the crazy women my grandmother brought gave only a moment of lightness. It quickly evaporated when Grandmother called the day after they left to inform Mom Grandfather was not doing well. Even my sisters were giving up on me. Nobody wanted to constantly be under a dark cloud; they were sick of wasting their rays of sunshine on me.

Helen and I were getting together more often, and I had even eeked out a group of people who allowed me to associate with them. But I did not have much left for effort. I hoped Dad was secretly working on a move and, any day now, would come home and unwrap the bow on the gift of life for me.

I woke up one Saturday and pulled myself out of bed. I had a babysitting job later that night; I was socking the money away at this point. Unpopularity and social exclusion were good for the pocketbook. There was a knock on the front door, but I was in no rush. Lord knows it was never for me.

"Krista, it's for you," Mom hollered.

I walked out of my room and peered over the landing to see my next door neighbor, Annie.

"Hey, Annie, what's up?" I asked. She rarely popped over. She was a high schooler after all. Sometimes she would have me over; although if I thought about it, she had done more so lately.

"I want to talk to you, got a minute?" Annie asked.

"Sure," I said and escorted her to our living room.

She plopped down on the sofa. I sat down beside her. "I wanted to tell you about this group at Hoggard. It's called Tea Ladies. I think you know Brittany down the street," she said and continued when I nodded, "Well, she and I are both members. This is an invitation-only club." She must have thought I'd be instantly hooked because she took a dramatic pause. "We want to invite you to our summer initiation party in late June. It's actually at Brittany's house. We all get together and the upperclassmen ask the incoming sophomores questions and either invite you to join or not. Everyone who's anyone is part of it. I was lucky someone told me about it, so I wanted to tell you." Apparently, she thought I would already be sold at this point.

"What do you do?" I asked.

"We have tea parties during the year and get all dressed up. Some of our moms hold them, and sometimes we go downtown and go fancy. But, only some girls are going to make it. It's kind of like Civenettes. Oh, and by the way, they start their interviewing in the fall."

It was starting to make sense. I had heard about the importance of making Civenettes. It was a high school sorority really. Sure, the popularity measure used to interest me, but now it was more important that it was a good thing to have on my record for college admittance. Getting into a good college was becoming a lot more important to me than the hope of ever being popular again. If Dad would not move us, then college would be my next chance to escape.

"Think about it, Krista. Here's the invitation," she said and handed me a frilly pink and white polka dotted piece of paper, "You have to dress up, nothing like a prom dress or anything, but a tea dress. Ask your mom; I bet she'll know what we're looking for. Brittany and I really like you, and that's good for you, but I have to tell you . . . Cindy has also been invited."

"And, Kerry?" I asked.

"Kerry who?" she responded.

Sweeter words were never spoken. I relished the thought that she had never heard of her. Of course, it probably helped that Kerry lived on the poor side of town. Well, not in black town, but it was poorer than most of the neighborhoods in the school district.

"So what kinds of questions will they ask me?" I asked her. I was becoming more and more interested in reentering the ring of life.

"I can go over some of the questions with you later. Hey, I'll even come and critique your dress for you, if you'd like."

"Perfect," I said. She stayed for a bit and drew some designs. She was always drawing different dresses; and in her room she had all kinds of Paris pictures, including pictures of high-heeled shoes and mannequins with various outfits. She swore she was going to be a designer some day. And for the first time lately, I wanted to stick around long enough to find out if she actually would be.

"Mom, Mom, guess what?" I yelled after closing the door behind Annie. Obviously, nobody had heard me this excited for some time because Mom and my sisters came barreling around the corner.

"What's wrong?" Mom asked. Katie and Chelsea were behind her wide-eyed, wondering.

"Annie just invited me to be part of the Tea Ladies. It's this group in high school; an important group that could help me get into Civenettes later. They have tea parties, and you can only be part of it if you are invited. You have to go through these interviews and get a dress, too."

"Whoa, Krista, whoa," Mom said, but she was beaming from ear to ear. "That sounds great. Why don't we go and get you a dress? Does that sound like fun, girls? We can do a fashion show and tell her what looks best." Excited, Katie and Chelsea ran to get on their shoes.

Dad walked in, and we filled him in on the news. He just handed Mom his wallet in a show of humor, "I guess this is all you guys need from me, unless there's some benches that need warming at the mall."

"Oh, honey," Mom said, leaning into him, "come with us. You might even enjoy it."

This was the first day in a long time I touched on happy. I must have tried on six dresses before we found the right one. I knew it as soon as I put it on. It was a midlength apricot-colored dress with subtle white stripes going down the bodice. It had a wide white collar that went into a v-shape, and to set it off, apricot polka-dot piping edged around an eyelet ruffle. The final touch was a pretty ribbon tied into a bow in the back. Perfect.

I stood admiring myself in the dressing room mirror, Katie standing with me. She had seemed bubbly when we first came out tonight but quickly became quiet. I noticed she was staring at herself in the mirror. She looked to be in a trance.

"You know, Mom will buy you a dress when you have something like this," I said, thinking she must have been jealous I was getting all the attention.

"I don't want a stupid dress," she said.

Trying to ignore her, I pulled at the sides of the dress where there was a bit of a gap due to my boyish shape, "What I wouldn't give for some hips to fill this out."

"You mean like me," she said and stood up. She was angry. I was confused.

"What hips?" I said trying to defuse the situation.

"These!" she said and put both hands on her hips and stuck out her stomach. "It's so gross."

"Please, you've gotta be kidding me," I said. There really was no other way to respond to her lunacy. If she wanted attention, she would have to come up with something more plausible than this.

"You just don't get it because you are so skinny. You eat whatever you want and you just stay skinny. Mom had to buy me a size four the other day, a FOUR, in the adult section no less." I looked down to see the dress I had was a zero.

"Do you want to look like a boy?" I asked her. "I'll trade you."

"Whatever," she said and pushed past me into the waiting area. I could see Mom's face as she shot me a look. I couldn't tell if she was mad or wanted to ask me something. I decided to enjoy my bit of a moment and move my thoughts onward.

After the dress and dinner in the food court, we went to Payless Shoe Store and picked out some white open-toed sandals with a bit of a heel.

Never one for wearing dresses, at home I proudly hung it up in my closet. It was my beacon of hope. If I could get through the next two months and finish up this junior high existence, I might have a shot. There would be a chance to surface again in high school since only about twenty percent of Williston students went there. That would leave a whole lot of people who didn't know me. Granted, Cindy already had a leg up on me since her sister had been so popular. And from what I understood, she was as popular with the boys. But maybe Cindy would have let all of this go by then. I had no hope, and quite frankly, no real desire for a friendship with her. I just wanted a mutual understanding we would go our separate ways.

There were even moments of great maturity. I told myself this would be insignificant in years to come. I would look back and laugh at the silliness of it. Something as simple as a teenage argument could never really affect how one's life turns out. *Right?* But for now, it was my whole life.

As my dad always said, aging sucks, but it was better than the alternative. So I had not lost all perspective. Annie's visit had given me some much-needed hope that I could come out on the other end with a life again. It made surviving these last two months at Williston almost bearable. I was going to wear this experience like my red badge of courage. All the times I had to listen to how my parents survived poverty with no pot to piss in; well, now I could say I survived months of ostracism—assuming I could actually survive the next two months.

The next day, Annie came by again, but this time she was with a boy. One of the best looking guys I had ever seen. He was taller than my dad, which was no great feat—Dad being only five foot eight. He had dark brown hair and the most piercing blue eyes.

"Krista, this is Jack. He and his family just moved here. They are four houses down on the other side."

I think Annie wanted to save me from my awkward gawking. "Hi, Jack," I managed.

Obviously curious, Mom and Dad entered and made introductions. Jack went on to tell us that his father, a colonel in the marines stationed in nearby Jacksonville, had just moved them here. Then he turned to introduce his sister, Alyssa.

All this time, Alyssa had been standing there looking like a lost puppy. Yet until Jack introduced her, I hadn't even noticed her.

"Alyssa, why don't you and Krista hang out in the kitchen for a bit? I made some cookies, and Krista can tell you all about Williston. I'm sure she has nothing but pleasant things to say," Mom warned as she motioned us into the kitchen. Jack and Annie were heading off to her house to do the same. I would say she drew the luckier straw.

Alyssa had the same dark brown hair as her brother, so long it went halfway down her back, and the same piercing effect with her eyes, though hers were green. She was quite gaunt in the face, but very pretty. I kept thinking it was too warm for the sweatpants she was wearing, but just chalked it up to having moved here from the Midwest and not understanding our climate. Weird to think of this town as "ours," but I was not the new girl anymore.

"Would you like some cookies?" Mom asked.

"No thank you, just some water," Alyssa said. Mom went to get water for her, and I was hoping cookies for me. "So what do you think of Williston?" Alyssa turned to ask me as we sat down.

I gave her warnings about the area, honest ones about the race situations at school, and then a watered-down version of my own issues. I felt like God had placed her in my lap, and I wanted her to get my take on things first. We were really getting along when the doorbell rang again.

"Alyssa, your mother's here," Mom said, walking into the kitchen. Mrs. McConnell followed behind her. She was so slight she seemed breakable, like a minor wind would take her away. And she was tiny. Obviously, the colonel had some height because not only was Jack tall, but Alyssa was several inches taller than any female in my family.

I stood to shake her outstretched hand, "Hello, Mrs. McConnell."

"I'm so glad you girls got to meet," she said as much to me as Mom. "My husband has been gone for six weeks and I'm still trying to get everyone situated."

"I totally understand," Mom said, and the two women walked into the foyer.

Alyssa and I talked for a few more minutes and bonded over the number of moves we had both made. Then they left.

"Alyssa seems nice," Mom said and put an arm around my shoulder.

Exactly what I was thinking.

Abby

The *Aunties* had been on my mind since they left. Sitting at the table with them, surrounded by the weaving of entwined stories, I felt a magical canopy had been drawn over us. Watching my mother created new images and memories for me. I'd never seen her laugh to tears, or jib and jab jokes in such a playful manner, or be at ease with anyone. I was heart-warmed amused by her.

Memories of Mama were always wrapped around my thoughts of her hiding from us, sinking her nose in books, books, and more books. Even when money was scarce, somehow our book collection grew. I can't recall how many times I would go looking for her, finding her sequestered in a corner, puffing on a cigarette, reading. Sometimes she would find something clever and motion for me, "Abby, come here. Listen to this!" What I listened to was her voice. A voice I seldom heard and craved so much.

Watching her made me wonder what life would have been like had Mother taken a different path and where the *Aunties* would be if they had veered even slightly from the road they chose. My mind stuck in this fold of thought. Intrigued, I decided to see what Bob's views were on the subject.

"Bob, do you ever wonder how things would have been if your decisions had been different?" I asked, settling into bed.

There was a long pause. He was obviously pondering my question. Selfishly, I had wanted him to say, "No! You know all my dreams have come true." But I knew the sentiment was unrealistic. We all wondered, at times, what our lives would be like if our choices had been different.

Bob punched his pillow and repositioned his head, "Yeah, I guess I do, but I don't dwell on it."

Once he broke the pause, I jumped right in, "You know, even before the *Aunties* came, I've been thinking about this. I know we can't change the past, but it does make you wonder." I tried a half-hearted laugh. "I guess it's nonproductive to dwell on what-if's and could-have-beens." What I desired more than anything was to talk to him, share our day, our dreams, our disappointments. Bob and I had been so wrapped up in life we'd forgotten to take time for one another.

Bob let out a loud moan, a reaction I knew only too well—a shut-up-go-to-sleep-Abby moan. But the hunger to be with him refused to be put off. I would have listened to a discussion on how to shuck oysters just to have him talk to me.

I continued rambling, "Oh, did I tell you Mama called? They're leaving Myrtle Beach in the morning. Daddy's not feeling well. I'm worried about everyone right now, me included. I haven't found my niche lately, and it's beginning to bother me. I usually acclimate better than this. I'm lonely."

Pouring forth like an opened facet, I continued, "You don't think there's something wrong with me, do you? I worry that the apple doesn't fall far from the tree. Take a look at some of my relatives. There's no doubt my genetic sewing kit came lacking. Too many of them are on public assistance or disabled or high as a kite on liquor or drugs. Thank God, my parents are sane. At least Daddy's always held down a job, and Mama's sharp as a tack. Well, until recently. I'm not sure what's happening with her, but she appears to be happier, certainly more out-going. But, I can't ignore I have a strong family history of nut cases. Throw me in the middle of that mess, and you have a crazy quilt, for sure," I made a desperate chuckle, hoping to pull him into the conversation.

Bob refrained. He sat looking at his cuticles, one finger at a time.

My thoughts took a spin, landing on an entirely different subject, "I've been thinking about going back to school. The part-time job made me feel useful and productive. I often wonder how different life would have been if I had chosen to work rather than stay home. Not that the girls aren't satisfying, but I know I need to begin planning for the time when they'll be leaving home, pursuing their own lives. It makes me question why I dropped out of college. It's been so long ago, I've forgotten. Do you remember?"

Bob huffed and puffed, finally sitting straight up. He ran his fingers through his hair, deliberate and slow, and crossed his arms, not in open defiance, but to calculate every word. "Abby, I don't know what you want me to say. I believe what my dad used to say, when it comes down to where the rubber meets the road, the consequences are set. It's all in motion. We don't look back. When you quit school, you must have had your reasons. That is that. If you want to go back to school, I'm behind you. If you decide to stay home, I'm in your court."

Bob stopped, exhaling a burst of air, "But, I have to be honest, I think your plate is full. Trust me, I understand this job might not be as satisfying as others, but look at how many women want to stay home and can't. A lot of women would give their eyeteeth to be in your position. You can do anything you like. You're not the breadwinner stuck in a job with no options; unless, like me, it is to move and learn a new job every two years. I wake up at four-thirty every morning, and most nights don't get home until after seven. Frankly, I don't have the time, energy or indulgence to wonder about *what-ifs*. I'm sure this sounds rather callous, but, Abby, I don't get it. Look at those women your mother brought here. Several of them would take the pennies off a dead man's eyes, and not shudder once. 'Take all you can get before someone robs you' is most people's motto. The rest of us, who try to do it honestly, don't have as exciting a life as those who push all the buttons. For me, I like to sleep soundly. An honest day's work, and being able to face the man in the mirror, helps me rest at night. Beyond that, I must be a dull boy 'cause I don't rehash every corner of my life, and how I rounded it."

Bob's eyes never left the wall where his glare had settled during the whole course of his dissertation. Cold permeated the room, chilling me to the core. My chest ached, each exhalation painful and tight. I truly did not know what to say. What happened? How in the world did I go from a lonely woman wanting a little attention to the scalding bath I had just received?

Slowly, I edged under the sheet, pulling the comforter up to my nose to stop my teeth from chattering. Bob followed suit. For hours, I watched the clock. Sleep evaded me, replaying our conversation over and over. I knew I was guilty of exorcising some kind of demon, and from the silence figured it was sitting between us laughing. Guilt bled my consciousness until exhaustion lulled me to sleep around four o'clock.

When the alarm went off thirty minutes later, Bob got up straightaway and headed for the bathroom. I got up, groggy and unsteady, stumbling downstairs to make coffee. Bob passed me on his way to work, not one good-bye or kiss my ass passed his lips. Nothing. Silence strapped the mental straight jacket tighter.

By seven o'clock, I had the girls up, breakfast ready, two loads of laundry completed and the downstairs vacuumed—any task to ensure I did not have time to think. After everyone was off, I put on my tennis shoes and headed out the door. Maybe pounding the pavement would clear my mind and help me figure out what to do next. At first I walked, then I ran. The harder I ran, the harder it was to breathe, but I didn't care. I kept pushing and pushing until my foot hit a pothole, and I fell headfirst into a mud-filled culvert.

When I stopped rolling, I lifted my head and saw my body sprawled, covered in muck. All of a sudden, a picture of me in this position, in this predicament, hit me square on, and I began to laugh. I sat up and grabbed a hand full of thick mud, rolled it into a ball, and threw it as hard as I could. Images of being a little girl making mud pies in hammered tin pans filled me with tenderness. So for the next few minutes, I played in the mud, feeling its texture, molding and shaping it, and finally picking myself up out of it.

I walked home with no idea what I was going to do about last night's debacle, but resolved to do something to move us forward. After showering and lunch, I looked up the telephone number for the community college.

By the time the girls got home, I was steeped in planning, the sheet of paper in front of me filled with options. I knew I didn't want to work full-time, but my brainstorming activity proved there were viable opportunities for me. I had to find out which one fit best.

I vaguely heard the doorbell but clearly heard Krista say she would answer it. I sat staring at my handiwork, crossing through and adding new ideas.

Chelsea walked in with a pad of paper and pencil, "Mama, I don't know how to write my name. I want to write it with curlicues just like Krista."

Chelsea and I were working on writing letters and curlicues when Krista burst into the room. "Mom, guess what?" Her face bloomed with joy.

She talked about her excitement over Annie's visit and the prospect of being in the Tea Ladies, which required a new dress. Chelsea clapped in delight, fueled by Krista's energetic pace. When I mentioned shopping, Chelsea ran for Katie.

Coming through the back door, Bob must have overheard. With his hand stretched out, wallet perched in his palm, he surrendered, "Shopping? There goes the paycheck!"

Bob came over, touched my shoulder, and knelt to kiss Chelsea. In that minute, in only one touch, I realized last night had been about two people who stood away from love to indulge in a spat that generated from selfishness and misunderstood feelings. We weren't willing to put our own agendas on hold to listen to each other's needs. Touching Bob's hand, I implored, "Oh, honey, come and go with us. You might enjoy it."

Bob didn't hesitate, yelling, "Supper's my treat! Let's hurry before the stores close down!"

Krista found the perfect dress and looked adorable. Katie kept quiet, too quiet. It was another nudge. Somehow I had to find the energy to seek her out or maybe wait her out, but I had to find answers. Every time I approached Bob about my concerns, he thought I was creating a problem. "Leave well enough alone," he always said. But leaving it alone didn't feel right.

The next day, I made several calls about classes. Finally, I connected with Dorothy Elwood at Cape Fear Community College. Within minutes, she scheduled a time for me to talk to an advisor. I hung up the phone, exhilarated and scared. Before I could indulge in another minute of second guessing, the doorbell rang.

I was surprised to see Annie again. "Hey, you put us in a tailspin yesterday. Krista had us rush out and buy a new dress. It was perfect timing since the new summer styles are out. I realize she won't need it until August, but her excitement got the best of us."

Behind her, I noticed two young people, "Oh, I'm sorry. I didn't see your friends." I reached a hand toward a teenage boy, "Hi, I'm Abby Simpson. It's nice to meet you."

Annie interjected, "This is Jack and his sister, Alyssa. They're the new kids on the block."

I made the same overture to the young girl, but she ducked behind Jack. It felt odd, but she was a teenager.

"Come on in! I'll go get Krista," I announced, stopping halfway up the stairs, shouting, "Krista, you've got company."

Katie poked her head around the doorframe, took a glance downstairs, and slammed the door shut.

Bob came through the door as I got to the kitchen, "Hey, how was your day?" His brow was furrowed so deep, I stopped, waiting for a response.

"Shitty!" he said, laying his briefcase on the table.

"Do I dare ask how shitty it was?"

Bob did not answer. Instead, he kept walking. Hearing the kids talking in the foyer, I decided to rejoin them. Bob got there first and had the young man cornered in a conversation about a Bronco's win.

I intervened, "Bob, do you remember Annie from next door? And, this is Jack and Alyssa, our new neighbors." Bob glanced over, gave a half-nod, and rushed back into his conversation with Jack.

While Bob talked, I corralled Krista and Alyssa into the kitchen for snacks and to talk. Once they were settled, I stepped back into the foyer to rescue Annie and Jack, "Bob, if you don't mind, I need your help."

I sent Bob off to change his clothes and ushered the teens out. Picking up a basket full of clothes, I was halfway to the kitchen when the doorbell rang again. I ran back, figuring Annie or Jack had forgotten something, but instead was startled to see a very short, petite woman standing on the stoop. Dressed to the hilt, she had on panty hose and heels in the middle of the day. My first thought, before saying hello, was to wonder where she went to buy such tiny clothes, and then quickly reprimanded myself for being so shallow.

"Hi, I'm Mrs. McConnell, Alyssa's mother." Her voice rolled off her lips like soft, creamy butter.

"Oh, it's a pleasure to meet you; please come in." I welcomed her in, motioning her toward the kitchen where the girls were still chatting.

"Krista, this is Alyssa's mother," I said.

"Hello, Mrs. McConnell."

Mrs. McConnell tenderly took Krista's hand in hers. I couldn't help notice how frail her hands were; a map of blue bulging veins lined their back.

"I'm glad the girls got to meet," she said, "My husband has been gone for six weeks, and I'm trying to get everyone situated."

"I totally understand."

After everyone left, Krista and I sat on the bottom stairstep. "Well . . . Mrs. McConnell looked nice."

"Yeah, I guess," Krista said.

Chapter 26

Krista

The last few months came and went like a typical Nor'easter. It slammed in, causing a lot of damage in a short period of time. But I survived. A rainbow full of hope was on the horizon; I could feel it.

Summer was finally here. I certainly couldn't say I was upset at the idea of this chapter at Williston Junior High coming to an end. There was so much to look forward to in high school. Thankfully, I was missing my true freshman year, although in my wildest imagination it could not have been any worse than this past year had been. I was looking forward to joining Tea Ladies, and Alyssa and I had fast become friends.

It only took a half day sitting in my surgeon's office to have most of those hopeful thoughts disappear.

"So, explain this to me again," Mom said to the doctor. "My daughter will be scheduled for her surgery in August and will need to be wired for a minimum of six weeks."

I started adding days in my head and came to a sudden suffocating conclusion. "Wait a minute. That means I'll have to start school with my mouth wired shut, and the Tea Ladies induction is mid-August," I said, panicked.

"It is customary to be wired for six weeks to give the bone time to heal, but if she requires a splint, and mind you not all patients do, it would be more like eight weeks. Also, if she has a splint, it would be a bit more difficult for her to talk and eat," he said, not at all concerned about my teenage problems. He kept talking, directing it all toward my mother.

"Can't we do it sooner than August?" my mom asked. At least she had heard me.

"No, not with my schedule the way it looks now. It would either be this," he paused to look at his schedule, "or, I could do the surgery later in the fall, but she would need some recuperation time."

"No, I wouldn't want her to miss any school," Mom said.

Miss school. *No, please good Lord above, please do not let me miss one day of school. Oh, and please be sure I enter high school with my mouth wired shut. I look SO forward to sitting in a lunchroom with a bunch of high schoolers sucking my sustenance through a straw.* No need to bother this retard with such menial concerns, but I was mentally making a list of everything I could use to ambush Mom as soon as we left his office.

The doctor mumbled some more technical jargon, and then Mom committed me to an August 5 surgery date. On top of everything else, this was the day AFTER my fifteenth birthday.

I kept quiet for the rest of the appointment. We left his office, stopped by the receptionist desk to get some paperwork for insurance, collected my sisters who had been tormenting each other in the waiting area, and headed toward our car. Somehow, I managed to make it there and actually close the door behind me before the tirade came rushing out.

"Mom, you've got to be kidding! I can't start high school and do this Tea Ladies thing with my mouth wired shut. How am I supposed to do that? Plus, I'm going to be drinking my lunch from a straw for every kid in the world to see. Oh, and my birthday will be ruined, thanks very much," I could not stop the ferociousness of the outpouring. "Oh wait, did I mention the fact that he said I'd probably lose fifteen to twenty pounds and I look like a boy as it is!"

I heard Katie mutter in the backseat, "Oh, the problems you have!"

"Listen, Krista, we don't have a choice here. These headaches are getting progressively worse and are a direct result of the TMJ problem. I have to look at the bigger picture and do what's best for you. We'll just have to explain to your teachers what's going on. And, if you have to go the six weeks, then it will only be a couple of weeks into school before you'll be fine again."

"Mom, seriously, did you ever attend high school?" At this point, I just wanted to scream. Her life as a teenager always seemed so long ago and obviously on a different planet.

"Please, Krista," she said exhausted.

"I'm scared. I have to be under and stuff. And, isn't that like being dead?" I asked. I was scared. I heard all the stuff the doctor was saying even though he appeared oblivious to the fact I was in the room.

"Krista, of course not. You will not be dead, just kind of asleep. Let's go get ice cream or something, okay?" she said. This got Chelsea's attention. It shocked me Mom was not three hundred pounds, because for her, food was the answer to all life's problems.

When we got home, I called Alyssa to check about some plans for the weekend. I told her all about my surgery and the unfairness of life.

"You want to talk unfair," she said. There was an elongated pause, and then she asked, "Can you come over for a few minutes?"

"Hold on," I said. I yelled down to Mom to see if I could run over. "Sure. Mom says it's okay, but I need to be back in an hour or so."

I walked the five houses to her house. I did not tell her I also had to be back before it got dark because I was a big, fat fraidy cat that would have conjured up flying vampires trying to catch me in the path between her house and mine. We were getting along great, and I certainly wasn't ready to freak her out.

She thrust a pile of papers into my hands as soon as we walked into her room and said, "You have to see this."

There was bold print with the word *contract* at the top of the page. It went on to detail how much food Alyssa would have to eat at every meal, and there would be periodic weigh-ins with required weight gains and strict punishments for any weight loss. I was stunned. I was not sure how to process all of it. I knew Alyssa was having some issues, and I knew she was way too skinny. I had recently met her colonel dad—seriously, he scared the fool out of me. I felt the need to salute, say, "*Yes, Sir,*" and run when I saw him.

"I don't understand, Alyssa. What is this?" I asked as I handed her the sheets back.

"The doctors say I don't weigh enough, and that it's a problem. Dad just freaked out."

"How do you feel about it?" I asked, concerned.

She kept ranting about her father as if she didn't hear me. "He told me he's not putting up with this. All I had to do was eat and be done with it. He watched me like a hawk at dinner tonight," she said so vehemently spit was flying out of her mouth. I thought the best thing to do was sit and listen. "He can make me do a lot of things, Krista, but he CANNOT make me eat."

"Don't you want to eat?" I asked. It seemed a stupid question but an obvious one. "In my house, food is the answer to everything. If you are sad, it's cookies. If there's a need for celebration, Mom makes a huge dinner with all of our favorites."

She cut me off and flopped on her bed, "I don't want to eat if HE wants me to."

"Maybe you should eat just enough to avoid confrontations with him. Why don't you come eat at my house? As I was saying, we are never in short supply around there," I said. It sounded like a fine solution to me.

"I'm sorry to bother you with this, it just makes me so mad," she said, throwing the wadded-up paper into her trash can.

"No problem, that's what friends are for, right?" I said. It felt good to call her my friend.

"Right," she said. "Hey, do you want to go bother my brother?"

Boy, do I! I thought to myself, but just responded with a simple, "Sure."

We goofed off for a little while, and then I headed home. The vampires were kept at bay, the sun still too high in the sky.

With all the talk about food, I managed to clean my plate and then some. While I was helping Mom with the dishes, I told her about what Alyssa's dad was doing.

"Why is he making it such a battle, Mom? Why can't she eat what she wants?" I asked.

"Krista, I didn't know if I should tell you this or not," she said and laid down the dishrag and turned to me. "Alyssa's Mom told me that Alyssa was diagnosed with anorexia right before they moved here. Do you know what that is?"

"No, not really. I mean I think so. Didn't Karen Carpenter die from that?" I asked. I had been such a fan of her music. I loved her soulful sound, and Mom had bought and played the *Carpenter's Greatest Hits* album over and over when she died.

"Yes, she did," Mom replied.

Abby

Outside the kitchen window, hummingbirds fluttered and scurried around. They always amazed me. Their panic in flight coupled with their serenity at feasting seemed quite a dichotomy for such a small creature. For the next few minutes, I stopped to indulge in their antics. Then it struck me. My life was a lot like theirs, rushing from one thing to another, settling in for a second, before flitting off to the next thing on my list.

I wondered if God chuckled over my correlation. If so, then He was doing a lot of chuckling lately. Surely in God's world order there was a template for behaviors and designs that he often reused with a minor holy tweak. In that vein of thinking, I surmised He was in the middle of creating me when a curious little hummingbird came by, annoying Him. As payback, He decided I would be a human draft of that creation. The possibility intrigued me, spiked my curiosity, and filled me with joy. Me, a human hummingbird? It all made sense.

But the joyful daydreaming stopped, bringing reality back to the forefront. The last two months had been hectic with end-of-the-school-year activities. Between the insaneness, I scouted around, checking out future options from my list. A lot of them got crossed off, still others generated new ideas. More than anything, the discoveries bolstered my confidence. Several opportunities seemed promising, but the most appealing option was to complete my teaching degree. Gloria had been cheering me on enthusiastically. In the last three days, I had received two cards from her, chastising me for not getting off my duff and getting on with life. Even her move hadn't stopped her infectious persuading personality from touching me.

But by the time summer unleashed itself, my excitement about going back to college or anything else dwindled to a fizzle. With the girls home and more on my plate, I found myself tripping over things I used to whiz through. Mundane chores became demanding. I went to bed exhausted, tossing and turning. Sleep became a precious commodity.

When I did fall asleep, questions floated around me like spirits I could not harness or suppress. None had answers. All I heard were whirling bits and pieces of nagging words, strung with no rhyme or reason, leaving a sad cloak wrapped around my days.

One night after tossing and turning for hours, I decided to forget sleep and head downstairs. Rather than ponder unsolvable questions, I turned on the television, stuck in a video tape, *The World of Suzie Wong*, and curled up on the couch. Halfway through the movie, I heard someone coming down the stairs. It was Bob.

"What's wrong?" he said in a sleepy voice, one eye closed, the other half open. His hair stood up comically in a point on the top of his head.

I turned around and continued to watch the movie. Bob slid next to me on the couch, grabbed the afghan, and snuggled close.

Without looking away from the television, I fledged into Suzie's dialect and euphemisms, "I feel sorry for you, Robert. You are a little man, and little man have little heart."

"Huh? What are you talking about?"

"Everybody talk about this, for goodness sake," I said, agitated.

"Okay, whatever you're on, I don't want any. In fact, this damn thing might be contagious. So, I'm outta here." Bob threw the afghan on the floor and stood up to leave.

Grabbing his arm, I pulled him back down beside me, "I'm just having fun. I've been watching Suzie Wong. Don't you remember the movie?"

"No, I don't. What's going on here?"

Bob's sullen face did not deter me. "I can't sleep. I feel totally run-down. All these things, totally disconnected, keep swirling around in my head . . . words, faces, a feeling like I've forgotten something important. I feel out of sorts. I can't decide what I'm supposed to worry about first—Krista, Katie, Chelsea, you, me, Mama, Dad, the neighbor."

"You're worried about the neighbor? Maybe we ought to talk about this tomorrow." Bob got up again, poised to run out of the room.

"It's now tomorrow, Robert, and we need to talk," I whispered, defused of energy. Suzie Wong had said the same thing when Robert Lomax would not listen, would not see truth hiding behind a falsehood.

Bob got up and walked away. As he left, I opened up, shouting louder and louder, "To Whom It May Concern: I can't do this by myself! I don't know who to turn to or what to do. Hey, Buddy, I'm sinking over here, and you'd better take a few minutes and check out who's drowning!"

When I stopped my tirade, the only one who remained was the dog. Sugar was wagging her tail, hoping her allegiance would earn her a treat. My life had ended up with the dog. The pit opened wide, beckoning me.

How do women get to the point where they feel tacked to a wall by the very people who are supposed to love them? I felt like a painting—created to be silent, then hung on a wall to admire. If I tipped, I prayed someone would notice, come over, and carefully upright me. But more often, I remained askew. I feared falling or worse cracking. At that point, I knew I'd be hauled up to the attic, replaced or discarded. Was this nagging fear part of those ghostly visions? I envisioned dropping the hummingbird into a pitcher of water and watching it grow into a turkey buzzard, ready to feast on me at any moment.

I finished watching the movie, got up, and went into the kitchen. Turning on the faucet, I began washing dishes, relaxing into the sound, the repetitive, circular motion of plate washing, and the slippery feel of bubbles. Reality snapped me to attention when I realized the sudden warmth I felt on my pant leg was water trickling down from an overfilled basin.

"Shit!" I yelled, pushing hard on the faucet handle and at the same time grabbing several dish towels. "This is not what I need . . . one more mess to clean up."

I was on my hands and knees when Katie walked into the room. I looked up, offering no comment about how early it was and continued. Finally, her silence and staring grated on me, "What?"

Katie leaned her head into her left shoulder, indignantly put her right hand on her hip and announced, "I'm not going to school!"

Right then, being tossed into the attic, alone, seemed very appealing. Maybe being cast off wasn't such a bad idea. *But then*, I thought, *who would replace me? Who would take care of this family?*

Slowly, I got up, wrung out the wet dish towel and addressed her, letting each word unfold as a sentence, "So, you're not going to school."

I paused, before continuing, "Is there some part of this picture I'm missing? Has something happened I'm not aware of? Have you declared yourself an emancipated youth, able to make life decisions on your own? Please, pray tell me, what do you mean by the statement that you're not going to school?" I sputtered, stewing and sarcastic.

"I'm just not going."

Slowly I started to rock, back and forth, snorting through my nose like a bull ready to gore their opponent. I lowered my head in an attempt to stave my mounting anger. What I wanted to do was beat the living tar out of her, for being here, at this moment, with these insane words.

Katie sank to the floor, crying. I wanted to shake some sense into her; but internally I chanted, *keep your cool, there's more here than meets the eye.* I cradled my arms rather than cradle her, not able to think or react. By the time I did, she was gone, and I was alone. I felt like branding a *G* on my chest, broadcasting guilt to the public. I had failed again.

In the middle of my world imploding, Mama called. "I'm sorry to call so early, but I wanted to update you on your dad. We went to the doctor to see why he's dragging around. The doctor put him on a new medication. Your dad thinks he's already getting better. I'm not so sure though. His coloring's pasty, and he's not eating. All he really wants to do is sleep. Guess we have to give the pills time to work. But, I've got some really exciting news. Hold onto your britches . . ." Mama paused for effect, before adding, "Quillie's engaged to Dickie Washburn. Can you believe it? We were eating at Griffin's Quick Lunch yesterday, and oh, by the way, their lunch specials are really nice, collards and backbone every Monday! I bet they'll even give R&C a run for their money!"

"Mama!" I shouted, totally unnerved.

"Now, Abby, don't get so testy with me. Anyway, Quillie just up and told us right in the middle of the lemon meringue pie. We squealed so loud the lady next to us almost fell out of her chair. You would have thought we guessed the right door on *The Price Is Right*. Quillie beamed from ear to ear. Who would have guessed? Dickie is fifteen years younger than her. So, you know where that conversation went. We teased and taunted her about Dickie's . . ."

"Mama, please! How could you go there? And how could one midwife's observation about a private area on a baby still be a topic of discussion today? Please, let's change the subject! When you see Quillie again, tell her congratulations for me, and remind Daddy I'm praying he'll soon get better." I hated to rush her but knew I had to check on Katie. "Listen, Mama, I've gotta go. Katie isn't feeling well. I promise to call back when I'm not so rushed. Heck, I think I'll just drive home in a few days, okay?"

"That'd be wonderful. You sound like you need a break. We'll take you to Griffin's for collards. A good colon cleansing will relieve all that pent-up anxiety you have. Give the girls my love. Bob, too. Call me!"

I hung up the phone and headed upstairs to present a peace offering. When I opened Katie's bedroom door, she was sitting on the bed, holding her bear tightly.

"Hey, I'm sorry. You really took me off-guard earlier. I apologize; I didn't know what to say. Can we start over?" I edged in the doorway but stood far enough away to give us both some room. "Tell me why you don't want to go to school?"

"People laugh at me," Katie announced with no preface, no explanation.

"What do you mean?"

"People laugh because I'm fat."

Katie was serious. Right now, in this very moment, I could make or break the situation. It didn't matter if I was looking at a dwarf of a child. In her mind, she was fat. In her mind, people were laughing at her. In her mind, she was miserable, vulnerable, and ready to explode. And unfortunately, I understood.

"Honey, I don't know what to say," I stalled, trying to feel my way through her pain. "I can tell you're very hurt and sad. That makes me sad, too. Why do you think this is happening?" I felt I was sinking in quagmire with no lifeline.

"Well, when I look at them, they're whispering and staring at me."

"Do you mind if I ask who 'them' is?" Treading only sank me further. I didn't know how far to take the conversation, fearing the narrow line between receptivity and total shutdown.

"My friends," Katie blurted, hugging bear tighter.

Moving closer to the bed, I finally sat on the opposite end. Katie made no move to acknowledge my presence. "Honey, I'm really trying hard to understand. By your friends do you mean Madison and Allison? Are they whispering behind your back, saying you're fat? Have they come right out and made those ugly comments? I mean, right out of the blue with no care about hurting your feelings?" As soon as it spilled out of my mouth, I knew I had slipped over the line with one foot already sinking.

"You don't believe me! You think I'm crazy. Is that why you want to send me to a doctor? Krista told me you were talking to Dad about it," she yelled, throwing her bear to the floor. "Go away! I don't need you!"

Krista bounded into the room, sparked, "Hey, Mama, I forgot to tell you Dr. Cherry's office called yesterday to finalize the surgery date. Guess what, I AM NOT GOING!" she screeched, before turning around and running back down the hallway.

"Why don't you just go after Krista? I don't need you," Katie yelled.

My feet felt paralyzed. I frankly did not know what to do. The sad part was I had to know what to do. "Katie, I will not be talked to in this manner. I'm here because I love you. We need to get to the bottom of this, okay? You're going to have to open up with me and be honest. I'm sorry Krista told you about a private conversation. But I do think we need to get some help. Do you understand?"

Open defiance pasted itself over her face. No words were needed to express her full understanding, and that was to not understand.

My gut reaction, which flamed, won over common sense. "I'm not through with this conversation. I'm going to give you fifteen minutes to get your story together. I want a full-blown version of 'what Katie is really thinking,' not the shortened one you offered me earlier. The time clock is ticking. I'll be back!" I left the room without one backward glance.

Chelsea caught up with me, halfway down the hallway, spouting, "Mama, I've been thinking . . ."

Without slowing my pace, I asked, "Thinking what?"

"Thinking this house is going crazy! Everybody's acting loco!" she said, laughing. "Katie thinks she's fat. Krista says you're ruining her life. And this morning, Dad kissed me and said, 'Mama's on the warpath, beware!' Am I the only happy person in this family, Mama?"

Pausing, I stopped to study her. Her smile was contagious, and she was right. We were all loco. "Hey, come here." I bent down to whisper, "I need you to come with me to Krista's room, and afterward to Katie's. Your job is to put a smile on their face. Do you think you can handle it?"

"Well, I'll try. We need some smiles, don't we, Mama?"

Trying hard to lighten my own spirit, I tiptoed to Krista's room, glad to have Chelsea as my partner in crime. She followed suit, exaggerating the tiptoeing with high-stepping movement. With her index finger poised on her lip, she looked like a child sleuth in full imaginative play.

Reaching Krista's room, I said, "We have to talk. We've been through all the options for the surgery. I'm very concerned about the headaches you're having, and the only way to stop them, per the doctor, is to have the surgery."

"Oh, so you think Dr. Cherry is God; he knows all the answers," she rankled.

"Well, no, but who do we believe? This is our second opinion. I didn't rashly take one doctor's advice. We have to stop the teeth grinding, too. You could have permanent damage, problems that will go on into adulthood if we don't get this fixed," I reminded her, then continued, "Why don't you think about it? I'll be back in ten minutes for your answer. If you are dead-set not to do the surgery, we won't do it. But, young lady, I will write down your request and have you sign an affidavit stating your wishes. I don't want you to come to me when you're grown, saying I ruined your life and now you're facing even more severe measures to correct your problem. You'd better think hard about your choice."

Exiting behind me, Chelsea exclaimed, "And, Mama means it!"

"Mama," she said, "you did good."

"Well, for a seven-year-old, you're quite intuitive, aren't you?"

"I don't know what intu . . . , well, what that word means, but I'm glad you're taking me along." Chelsea walked faster to keep up. Somehow she was getting a kick out of being in the middle of her sister's turmoil. And I'm ashamed to admit it, but having her with me bolstered my courage.

We marched into Katie's room. For months, I had been standing outside her door, seeking permission to enter, but not now. This was my house, and I was tired of placating children.

I faced Katie and demanded, "I'm back, and I'm early. Start talking."

"You don't understand. You are all thin and beautiful. You don't know what it's like to be fat and ugly."

"No, and you don't either," I stated. "Why are you doing this? I know how you girls pick on one another. You say things you don't mean; we all do. But you also know it's not true. Look in the mirror." I stopped to watch her, giving her time to process.

Then I continued, "How am I supposed to react when my family and friends comment on how much you look like me?" I threw my hand up over my forehead, acting the true form of a thirties actress, and dramatically crumbled to edge of the bed.

When I lifted my head, I whined, "Oh no, you've got to be kidding. You think I look like Katie, the fat, ugly one? Shoot me, now! Put me out of my misery!"

Chelsea laughed hysterically. I caught myself from doing the same and watched for Katie's reaction. At first she looked stunned, even tearful. Then a small smile crept around the corner of her mouth.

"You're gonna smile!" Chelsea squealed.

"What's so funny?" Krista smirked, standing in the doorway, a stern stance settling over her face.

"You," Chelsea responded. "Wise up, wise-cracker!"

Krista shut her eyes in tight disgust at Chelsea's screwed-up version of an old saying. "You never get anything right!" she stated.

I looked around the room. Everyone cast their eyes on Katie. The hurt and madness had only slightly faded. The upturned mouth remained a mere suggestion of a smile. I thought we had scored, at least for the moment, but I was wrong.

Chelsea broke the tension, "Let's play Old Maid," she announced, running out of the room to retrieve her card game. Krista went over to Katie's nightstand, picked up a book, and sat down to wait.

Then my eyes slid over to Katie.

The faint rise of her mouth had disappeared. And in a split second, the change happened. Like the parietal eye of a side-blotched lizard, her soft brown eyes changed to an onyx black before instantaneously returning to their natural state. I blinked and refocused; but nothing seemed out of the ordinary, except the strange weight on my chest. Fear laughed, and a cold shiver ran down my spine.

Unsettled, I left the room, reminding them they had thirty minutes to play the game, finish dressing and get ready to leave for school. Entering the kitchen, I went straight toward the coffee pot, hoping a strong black brew would ease my shaking hands and calm my fear. The phone rang as I was reaching for a coffee cup.

"Mrs. Simmons, this is Frank Stanley. I'm the admissions counselor at Cape Fear Community College. Dorothy Elwood asked me to give you a call and set up an appointment to discuss our fall session. I apologize for calling so early, but thought this may be the best chance to catch you."

"No problem. I feel I've been up for hours. What's your schedule like? I'm ready."

I penciled in the appointment date on the calendar and gave my final warning, "We have to leave in ten minutes. Breakfast will be Pop Tarts on the run."

I dropped everyone off at school and headed to the grocery store. An air of excitement sent a refreshing coolness up the back of my neck. This just might be a turn of events for me, a new beginning.

Walking into the store, I kept mentally reciting the list of five things I needed. When I reached for a bag of sugar, a tap on my shoulder caused me to jerk and lose my grip. In one swift movement, I secured the sugar as it slipped from my hand and turned to face Mrs. McConnell. "Oh, hi!" I said, placing my first item in the grocery cart. "How're you doing?"

"Abby, do you have a minute?" she asked.

Hells bells! I thought, and then chided myself. "I've got an appointment . . ." I started explaining; but when I saw her face, retracted, "But, I've got time. What's wrong?"

"I'm worried about Alyssa. Her health is very fragile. You do know she's anorexic."

I let the words wash over me.

I kept looking at her, wondering why her tone was so flat and unemotional. My ineptness to answer must have propelled her to continue.

"We've been dealing with this for some time. I wanted to thank you for Krista's kindness. She has been a true friend to Alyssa. I think that, more than anything, will help her manage her disease." Mrs. McConnell touched my arm, smiled, and walked on saying, "Continue your shopping, my dear. We'll talk later."

When I got back to the car, I realized I had only gotten four of the five items on my list. Oh well. What I could not forget was Mrs. McConnell's sunken, ghostlike face and edgy behavior. More than anything, I could not believe I was still calling her Mrs. McConnell. She couldn't be that much older than me. And every time I allowed myself to think about Alyssa or anorexia, a tightness wrapped around my throat and a hollowness widened in my gut.

After dinner, Krista offered to help with the dishes, which left me questioning, what's next. Before long, she started talking about her visit with Alyssa and how her dad insisted Alyssa eat more.

"Why is he making it such a battle, Mom? Why can't she eat what she wants?"

I didn't want to talk about anorexia, but knowing it had been a recurrent theme on the news lately, decided to go into a clinical observation, hoping she would be bored and brush me off. I tried to explain about the devastating effects of the disease. It wasn't long before a light bulb went off in Krista's head, and she remembered her favorite vocalist, Karen Carpenter.

"Didn't she die?"

It was hard to answer her. Another light signaled. An awful image filled my head—Katie with eyes sunken, cheeks hollow . . .

Chapter 27

Krista

Alyssa and I hung out quite a bit—when she wasn't being grounded for missing a meal or losing a pound. Her father's newest threat was putting her on the sixth floor of New Hanover Hospital. And as Alyssa so delicately put it, that was the floor for the crazies.

My surgery was two weeks away, and school would start three weeks after that. Supposedly, this was enough time for me to recover for the most part, but I would still be starting my sophomore year in high school with my mouth wired shut for another three to five weeks. The thought was horrifying. I loved food, unlike Alyssa, and was not looking forward to getting sustenance through a straw.

Still, with all these worries swirling in my head, the more pressing concern was the social test of the Tea Ladies tryout. It was two days away. The closer it got, the more nervous I became. I hadn't seen Cindy all summer. I wanted to believe I was like a squashed ant to her, totally irrelevant now that the deed was done. *Would high school girls even care about some tiff from junior high?* I would have felt more comfortable heading into a room full of boys, their attention span would have been advantageous. I didn't know if I was up for female scrutiny with Cindy standing there waiting to pounce if it suited her. *Was I just a pathetic soul for still fretting over all of this?*

I stood in front of my dresser mirror with my beautiful peach dress on trying to find the answers when Mom walked in. "That dress really is pretty on you, hon," she said.

"I love it, but I'm worried," I started. I needed some guidance, but I hated the thought of her knowing fear was still such a reality for me.

"About what?" she said and started picking up my room. She was on automatic pilot, not aware she was opening drawers and moving things around.

"I don't think I can do this, Mom. I really don't," I said and took the dress off. "I'm sorry."

"Honey, are you worried about Cindy? Is that what this is about?" she said and came over to help with some buttons.

"That's a big part of it, Mom. I wish I wasn't still worried about what she might do, but I am. And, I just found out that once you make it, if I made it, there are a lot of parties before school, and I'm going to be wired shut or even miss some while recuperating. That will give Cindy time to get into the group and make everyone hate me or get them all to like her." There, it had all flowed out. Some of it, I wasn't even aware of until Mom's concern opened up the floodgate.

"Oh, sweetie. You don't really believe Cindy would still be such a child, do you? Scratch that, because who knows with some people. But, honey, you've got to stand up to her sooner or later, you know that," she said.

Mom's constant need to be forthright was not always an endearing quality. *How could one woman use so many honeys and sweeties in a dialogue and still manage to remind you that women are cruel?*

"The ceremony is not for a few more days. Take some time and think about it. Your dad and I will support whatever you do here, because when it comes right down to it, your academics are the most important thing," she said. She leaned in to give me a quick peck on the cheek then rushed out to handle other issues. I was shocked she didn't squeeze in one more sweetie before she left.

After several fitful days of tossing and turning (somehow even when I was awake), I decided I couldn't do it. Annie understood and told me she would still help me with the Civenette tryouts. She even attempted to make me feel better by reminding me it was the better group for social reasons, and also to have on my college applications. I was thankful for her support. After the ceremony, she was quick to tell me Cindy had made it and came up to her and made a snide remark about my absence.

"But, Krista, you've got to stop worrying about her. Trust me, my friends are not going to want to hear about some year-old feud coming out of junior high."

"That's good to know," I said, truly thankful for her friendship. "Now, give me some pointers on Civenettes."

"Every person wanting to join has to go through individual interviews with current members. They just ask questions, try to see your personality. It's kind of fun, and they only take like forty or fifty sophomores. This year is weird with the freshman coming in, so we decided not to take any," she said.

"Wait, when are the interviews?" I asked, my brain racing ahead.

"About two weeks after school starts," she said. "Hey, I gotta go. See you soon."

I sat there, stunned, realizing I would have to try out for Civenettes with my mouth wired shut. Wonderful! I had given up an opportunity to have a normal tryout with the Tea Ladies for that. But something snapped, and I decided I wouldn't let Cindy's presence preclude me from trying for something I wanted ever again.

Another week went by and summer was fast coming to a close. Even though tomorrow was my surgery, and I was getting scared, I decided to focus on today, my fifteenth birthday. We never did big birthday parties, but Mom always planned a special day. And it didn't hurt that today was also my last day to eat without wires.

Mom let me pick my favorite meal for breakfast, lunch, and dinner. She also brought Alyssa with us to do some shopping at the mall. Such a nice, simple day. There was cake and ice cream after dinner, too. I couldn't eat after midnight, so I savored every bite. I stuffed myself all day long. Alyssa commented about how excited one person could be over food. It was her crucifix, while it was my salvation.

The next morning, Mom got me up very early to head to the hospital. Dad was staying behind until the sitter came, and then he would join us. I was not concerned with who was taking care of my sisters or where Dad was. I was scared. I went into the bathroom to brush my teeth, and I felt it. The cruel fate of being a teenage girl was even crueler today. I immediately sat down on the toilet and dreadfully pulled down my underwear.

I tore out of the bathroom and yelled, "Mom, you won't believe this!"

"Honey, keep your voice down, the girls are still asleep. What is it?" she asked.

"I started my period! Did I do something in a former life, Mom, really?" I said with as much drama as I felt. She seemed to hide a grin, playing Rhett to my Scarlett.

"I'm sorry, honey, but we really do have to go," she said and ushered me out of the house and into the car.

We parked in the patient lot and walked in. I felt miniscule. We continued into a narrow sterile room, and the nurse handed me a giant paper towel (okay, that's what the darned thing looked like) and told me to take off all my clothes.

"Oh, I'm on my period, so I can't," I said.

The nurse gave me a look lacking patience or support. "Honey, you have to take off everything for surgery. It's procedure," she said without even looking up from her clipboard. She turned to Mom, "Once she changes into the gown, we will be setting up the IV and having Dr. Abrams, the anesthesiologist, come speak with you about what he'll be doing."

"I won't take my underwear off, I won't," I said to the nurse over Mom's pleading eyes. "You guys can't make me!" *How the hell was I supposed to have on a pad and no underwear, and for what reason did they need me to be stark naked?*

The nurse gently laid a hand on my mother's arm before telling me, "Okay, honey, just leave them on."

I wanted to pump my fists and jump up and down. It felt like so long since I'd won a battle, and I was overwhelmingly happy with myself. Then it took the same nurse three attempts to get the IV in, so she found a way to get even. Mom held my hand the whole time, and I was glad. I focused on the warm feeling her hand spread to mine rather than the intense pain of the needle punctures.

As the nurse wheeled me toward the surgery room, she told me to start counting backward from 100.

"100, 99, 98 . . ." I started.

"Oh, as soon as you wheel her in there," I heard the nurse say to my surgeon, "you'll need to take her underwear off. She refused."

A brief moment of wanting to fight and scream was drowned by the blackness that took over.

Abby

"Hey, about last night," Bob started, "I'm sorry. I should have stayed and listened."

I heard but refused to acknowledge his words. I kept scrubbing the spills and crumbs off the countertop, my hands keeping a steady pace with the beating of my heart. My discomfort? Embarrassment? The truth of last night's outrage seeped in slow, reddening my face. I knew he couldn't grasp the pressure I felt. He had no clue. But instead of apologizing for my outburst, I bound a silent, suffering sign of virtue to my stance.

It was not a conscious attempt, but a learned behavior, a protective mechanism. How many times had I seen my mother and grandmother do the very same thing when caught red-handed in an uncomfortable situation? Or use it to camouflage imperfections rather than expose real feelings. The reason? Most people run from hurt. They don't want to deal with or understand it, so they ignore it or label it as troublesome. In actuality, it's a simple fix. It's the heartthrob of people's lives, T-I-M-E. Taking the time to listen, without counting the cost, even if it means doing it over and over. But they can't. People fear if they give their time, they'll lose it forever. And I bought into it . . . hook, line, and sinker.

Bob pushed his chair away from under the table. My heart skipped a beat, hoping he would come over and put his arms around me. At that moment, I needed him badly. I almost ran to meet him halfway, but halted when I saw him lean back, stretch his legs out, and yawn.

"Man, I've had a long day," he exclaimed in the exhalation of a yawn. "If I never saw another manufacturing plant in my life, it would be too soon. You can't imagine how much crap gets thrown at me in a day." He paused, but my hope for a quick ending was dashed, "Oh, by the way, did I tell you that John Angler's wife had a baby today? I heard she worked at her law office till late last night, couldn't stand to think of missing a beat."

Bob shifted, sighing, "John said she'd probably take off a couple of weeks, and head right back. Sorry, but I don't get it; I told him as much. Women need to take care of their kids and home. Trust me, there's no way I could do my job, the way it needs to be done, if I had to constantly worry about what's happening here. He had the nerve to say he was thinking of taking a leave of absence to help out. That's crossing the line. I won't put up with that nonsense for a minute! How can I run a business if my employees start signing up for diaper duty?"

Throw me off a cliff, now. Let me tumble and bounce on the rocks below to take off the splintered edges. Cold threaded my extremities where heat had just resided. The Ice Queen erupted.

"What is that woman thinking! And her husband? For Christ's sake, he must be hormonal!" Flatness moved my speech. My sentences shrank to periods, slamming the air with frost needles. I turned to face Bob. He sat rigid, carved into the kitchen landscape.

"I wonder . . . did they consider the benefits of her staying home? Let's see, stay home with the baby or work? Huh?" Word darts landed with perfect accuracy, "Work? Get paid. Home? A nonpaying, thankless job. Work? Elegant dinners out . . . with adults. Home? Cook it and clean it up. Work? Accolades. Home? Screaming kids. Mundane, boring, repetitive work." I paused, "Oh, in my excitement, I forgot sex. Guess that one isn't offered at work, or at least not that I'm aware of."

Bob slowly slid his legs into a ninety-degree angle and stood in one swift movement. "Abby, I don't know what the hell is going on with you. I was trying to have a friendly conversation, and you openly attacked me."

"That was anything but friendly. You just made me feel like a hired hand instead of a partner. Did you actually say she needs to be home to take care of her baby and keep her house clean? I'm insulted for John's wife and myself. She chose to work. Maybe she realized she wouldn't be a good stay-home mom. Or maybe they're so far in debt she has to work, giving up her time to raise her child to partner financing. Who knows? As for me, I thought we chose for me to stay home. I didn't necessarily want to put my life on hold, but I did it for our family. But, if I choose to do something else in the future, I expect you to encourage me, and to pitch in and help whether you work or not. Let me remind you, this is a partnership, Mr. Simmons."

Bob looked stunned, "Abby, I didn't mean to insult you. I guess I didn't say it right. Let me back up. What you're doing here is imperative. I couldn't get the promotions I need for the financial success of this family if I had to manage your job and mine, too. And, I can't afford to hire someone to do what you do. I wouldn't know where to begin to pull it all together. The girls would have to go to foster care and well . . . they'd have to put me in a home for men who can't even balance their checkbook."

Bob lowered his head, shaking it from side-to-side, laughing. Not a snicker, but a rip-roaring belly laugh. He threw his head back and slapped his knee simultaneously. I didn't know what to do. I'd felt I'd been teleported from a trial scene to the Barnum and Bailey Circus.

As he continued, his whews and sighs got louder and more demonstrative. Finally, in total exhaustion, he sank back into the chair, breathless.

Katie walked in, looked from her father to me, and left without a word. I heard Krista yell at Chelsea to stay out of her room. I heard Peter Jennings's voice delivering the nightly news. Minutes elongated and stagnated.

My feet felt strapped to the floor. I had to control swallowing to squelch the drumroll in my head. The little girl in me climbed out of her hiding place with two boo-boos, and I couldn't stop her. I tried, but she wouldn't hear of it. She needed a hug. I walked over to Bob, kneeling down in front of his chair. I stared at him for what seemed an eternity before he finally leaned over and took me in his arms. With each squeeze, with each warm breath on my neck, the little girl was able to retreat, doll in tow, to her safe place.

"Are you okay?" he said, flinching.

His squirming made me realize I had his shoulders in a vice grip. "Oh, I'm sorry," I said, disengaging my fingers. "I think I'm overtired and strung out from too little sleep. I've been watching too many old movies; I guess. You know, the melodrama and all."

Blowing butterfly kisses up and down my neck, he asked, "How about a date?"

"A what?" How could I concentrate on words when this felt so good.

"You know, you and me, a nice restaurant, no kids, actually talking."

"Oh, that kind of date. Yeah, it'd be nice. It's been too long, hasn't it? We'll take a look at the calendar," I responded.

"No, I mean Friday night. This Friday night. I have the perfect place in mind. I'll call and make the reservation."

Krista interrupted, "Great, now my parents are dating again. My life is falling apart and you two are going out for a romantic evening. Great! Just great! Can you at least take that pest upstairs with you? I'm fixing to muzzle her!" After her interruption, she abruptly left the room.

Bob put his hand to my face, drawing me back to his request, "I mean it, Abby. We need time away from this chaos. You're buckling under everything going on. I may not be the most observant guy or one with a lot of sympathetic words, but I do love you. I need to be with you. Okay, have we got a date?"

"Yes! Tomorrow I'm going shopping for the perfect date dress, whatever that is. And, I'm going to think of being with you right up until the magic moment gets here." Hope bloomed like a rose offering.

The next day, I woke with a start. I had an appointment with Mr. Stanley at the college. Afterward, I hit the shopping mall. I carefully got dressed, did my hair, and applied makeup. When I walked out to the car, I greeted the day with a quirk of amusement and expectation. College. For once, I did not ponder the hows, except how quickly I could get started. Our date night would be the perfect time to lay out my plan to Bob.

It didn't take long to fill out the paperwork, get a fall semester course outline, and take a quick tour of the facility. It was thrilling! Course sign-ups for new enrollees were in a week, a week to start a new life.

At the mall, I found the perfect outfit. A lightweight navy dress, simple and sexy. The sweetheart neckline made me look taller and thinner, and the wrap around closure cinched my waist to perfection. The necklace Bob gave me last Christmas would set off the attire to a classic finale. Today had been too sweet!

Walking into the kitchen, I dropped my shopping bags on the table and reached to silence the screeching phone. It was Mama.

"You sound out of breath. Are you okay? Did I call at a bad time?" she asked.

"No, no, you're fine. I've been shopping and running errands. In fact, I've had a fabulous day! What's up with you? Is Dad all right?" Concern made my stomach lurch.

"We're fine. He actually tried to play golf today. I didn't ask how his day went; I didn't have to. I heard him whistling. Anyway, Quillie wants to know if you'll be a bridesmaid in her wedding. I told her you might be coming home soon. She'd like to talk to you about what to wear. It's so exciting. We're planning every day," Mama's voice sounded different, younger, lighter.

"Well, I guess I could be in the wedding. Listen, let me get back with you in a day or two. Actually, I'd better make it next week. I've got some things going on here that might change my calendar. Remember, Krista's surgery is coming up."

"Abby, I forgot. You know Dad and I can't come, but I do expect a phone call from you to let me know how she comes through it. Tell her we love her!"

A loud ruckus erupted from the garage. Before I had time to react, the door flew open with Katie in hot pursuit of Chelsea, screaming, "I told you NEVER to say that again!"

"You're not my boss! I don't have to mind you!" Chelsea responded, breathless.

"What's going on?" I yelled, sprinting after them. Krista's voice stopped me.

"Just leave 'em alone. Katie won't kill her." Krista sounded too much like her dad, except her voice projected deadpan and dejected. I started to walk away, but backed up, wondering about her response and hoping to gain some insight into her mood, "Do you know what's going on with Chelsea and Katie?"

"Nope. Don't know; don't care," she said.

Insults tumbled down the stairs, but I relaxed, realizing there was no blood curdling yells or objects being thrown. Maybe Krista was right. Let them figure it out for themselves. I switched my attention, "Now can I ask, what's going on with you?"

"You can ask, but I'm not sure I can tell. Mom, I'm scared. I'll be fifteen tomorrow. I have surgery the next day. Then, there's the Civinette event. And, I get to start my sophomore year with my mouth wired. Everyone will think I'm a mute!" Krista summarized.

Even I knew it was a lot for her to handle. I sat down beside her, putting my arm around her shoulder, and hugged her tight. A crash and a scream caused us both to jump and run. Arriving at the top of the stairs, I saw Katie and Chelsea's doors closed. Silence hung like a flag over the upstairs foyer.

I took the coward's way out and slipped back downstairs. Krista followed, plopped down in front of the television, flipping channels. I felt like tiptoeing, afraid I'd cause a wake in the moment of peace.

At the kitchen table, I grabbed the course outline brochure, sat down, and started thumbing through it. Right then, I decided on classes and how I would surprise Bob with my initiative.

Krista's birthday proved to be one of her best. Alyssa accompanied us on our shopping trip but refused to stay for supper, cake, or ice cream. I didn't push the issue, nor did Krista. We both realized it might be more enjoyable without her and her rejection of food. I knew we did not need to highlight the situation around Katie. Daily I became more sensitized to what she was eating or not, realizing my little girl was sick. Every day, I promised myself I would get to the bottom of Katie's issues. But I could not get there. I needed help to figure out what to do and where to go. Maybe, Mrs. McConnell would know; I should ask her. Surely she would know the right doctors and resources.

The next morning, the alarm went off at five o'clock. I tiptoed into Krista's room and softly called her name. She got up quickly and went into the bathroom to dress. I headed back to my bedroom to finish putting a small bag together for the hospital.

Before we could get out the door, Krista had already started mumbling and bellyaching. After check-in, we were taken to a small prep room where she was told to remove her clothes and slip into a surgical gown. Krista openly defied the nurse, refusing to remove her panties. The nurse didn't say a word. I knew I didn't need to get involved; she had the situation firmly under control. The nurse's expression guaranteed she wouldn't waste another word on a smart-mouthed teen. There would be no doubt, Krista would come back pantyless.

The doctor let us peak at Krista in the recovery room. Bless her heart, after all she'd been through, and would go through over the next few weeks, she'd have to wake up and realize she lost the underwear battle.

The post-op nurse signaled our time was up. Bob excused himself to make a few calls to work, and I headed down the hallway to the surgical waiting area. Two quarters had just clinked into the soda machine when I heard, "Abby? Abby Bower? Is that you, girl?"

I didn't bother to turn around. I knew who she was; the devil herself. I shuddered.

"Jalessa Reddick." Hate, hidden years ago, surfaced and stood like sharp bristles across the back of my neck. In one insane moment, I wanted to thrust my hands around her cocoa brown neck and strangle her to death. And it wasn't the first time in my life I had felt this way.

Chapter 28

Krista

"That's so gross. Can I eat in my room?" Katie asked Mom at the dinner table.

"No, we will eat as a family," she said. Then looking toward me, she added, "Krista, I know it's difficult for you to get the food in, but use this napkin and be considerate of others."

"Please!" my dad added. Mom shot him an evil glance. Lacking any sort of real intelligence, he continued, "Come on, Abby, it is gross watching her syringe mashed potatoes into her mouth. I think I'm losing my appetite, too." He put his fork down and pushed his chair back.

"Cmm on. M chrying," I said. This was frustrating. I could not talk. On top of being wired shut for eight weeks, of which I had only completed two, the doctors had put in a splint. So unless I wanted to live on straight liquids, I had to use a syringe with a tube attached to push food into an opening at the back of my mouth. Mom was kind enough to puree whatever we were having for dinner. And although I admit pot roast and mashed potatoes put through a blender looked gross, it sure did taste good. Of course, having to follow it with my required serving of prune juice killed some of the pleasure.

"We know, honey," Mom said, and Dad got another shot of evil his way.

Knowing I could not muster too many intelligible words in a row, I pulled out the trusty pad and pen—faithful companion the last couple of weeks. *I'll just go eat in the kitchen*, I wrote, *I'm almost done anyway.*

Nobody really argued anymore; they were all relieved to go back to eating. The recovery from this surgery had been nothing short of hell, making the situation with Cindy look like a field day. First, I had complications from the anesthesia and ended up in the ICU. What an experience that was. All I remembered was waking up and seeing a nurse and Mom and Dad next to my bed and realizing with great terror I couldn't breathe. It was a nightmare. I tried to open my mouth to take a deep breath, but it was wired shut. My gut told me to try and breathe through my nose, but it was blocked, too.

I could hear all kinds of strange beeps. Funny, those beeping noises were so clear and yet I couldn't make out any other sounds. There was a rush of nurses into the room, and I looked to my left just in time to see Mom faint and Dad carry her out of the room. Then the doctor pulled what I guess were tubes out of my nose. I felt like the monster from *Aliens* had just popped out of my chest. I swore they were pulling out my intestines through my nose. But then, miraculously, I could breathe.

"Mom, how is Mom?" I tried to say. This is when I first realized how limited my speech was.

"Honey, I can't understand you," the nurse said. She turned to a side table and then handed me a yellow pad of paper and a marker. "Use this."

Where is Mom? I wrote simply, but my eyes were pleading. I wanted to write more and beg someone to go help her, but I kept it short. *I think she fainted.*

She took the pad and then gently stroked my hair. "Sweetie, she'll be all right. I'll go check on her in just a few minutes. Okay?" I nodded, so she continued, "We need to change your pad. We have you on a catheter and IV right now, but we hope to get you out of this room and into recovery tomorrow."

Finally, Mom came in the room. She was pale, and Dad was holding her up. I was so relieved to see her.

I spent the next two days in recovery and received the most flowers ever. It was great. Grandparents, aunts, and uncles, all the people who didn't know there was a day every year celebrating my birth. They kept me longer because of how sick I was. Sadly, with my mouth wired shut, it was rather problematic to throw up. I was given wire cutters in case but was told what a mess I could be in if I ever actually used them.

I was now a mere eighty-five pounds, having lost about ten pounds in a week. I was home, and horror of all horrors, on a steady diet of pureed meals, baby food, and Ensure. To be fair, Mom was doing a sweet job of mixing in a lot of milkshakes. My sisters were being nice, coming in and playing games in my room. It was a big thrill for me, albeit not for anyone else, when I finally got to come downstairs and have dinner with everyone.

School was only two days away, and I was getting more and more nervous. Normally, this would have been an exciting time, shopping for clothes and enjoying the last days of summer. Not this year.

Alyssa, who was obsessed with my weight and how much I was losing and exactly what I was eating, had gotten her brother to agree (with some help from her parents I'm sure) to take us to school every day. He was going to be a senior, so we were only going to have to walk from the senior parking lot, which was much closer than where juniors and below had to park. There was also the perk of being seen with a hot guy, brother or not. Having the transportation issue handled was a relief, and it was comforting that Alyssa and Annie had been by several times to check on me since my surgery.

Two days went by in the blink of an eye. *Why does the world work the way it does?* When I was excited about something, the days turn into weeks; but the minute I was anxious and worried about an event, it arrived like the days were mere minutes. The first day back was nothing short of interesting. Thankfully, my mom had talked to almost all my teachers, and they were aware of my issue and wanted to help. I must say I was blessed to have no classes with Cindy and several with Alyssa. I did share homeroom with Kerry, but it was only an hour a day.

Unfortunately, the only teacher Mom was not able to talk to was my typing teacher, Mr. Broomfield. It was sixth period, and my day was almost over. Thanks to Mom, I made it through lunch, which I was extremely worried about. She packed tomato soup and a straw; it was a cinch! I promised myself I would kiss her as soon as I got home. After surviving lunch, I thought I was home free until Mr. Broomfield's class started.

Each desk in the class had its own typewriter. For some reason, mine was humming loudly when I sat down. The teacher barely started class when he noticed it, too, "Ms. Hmmm," he glanced down at his attendance sheet, "Simmons is it? Is your typewriter working?"

How the heck would I know, I've never used one before, I thought, but I shrugged my shoulders. He seemed to get miffed rather quickly for the first day. "Ms. Simmons, an answer please."

"Mmmm nut shrr," I said, feeling the heat of the whole class staring at me.

Mr. Broomfield walked toward me. I was writing down my situation on a piece of paper to try to explain everything when a boy sitting next to me said, "Mr. Broomfield, I think she's retarded or something; she can't talk."

There was a low murmur of giggling. The students did not seem to know whether to laugh or be somewhat protective of a poor retarded girl. I stood and thrust the notebook paper at Mr. Broomfield while giving an evil stare to the doofus on my left.

"Oh, Ms. Simmons, I'm sorry. We can chat about this after class," he said. I could tell he was caught off-guard and somewhat embarrassed. Still, he didn't attempt to clarify to the class I wasn't retarded, so the fellow flanking me probably thought I had just explained my mental disorder in writing.

Not the best first day of high school. Certainly not how I imagined it a year ago while planning life with Cindy and Kerry. Alyssa thought it was funny when I relayed the story on the drive home.

After Jack dropped me off, I walked into the kitchen and directly into Mom's waiting arms. Enveloped by the smell of cookies, I could see that she had prepared me a cookie shake already.

"Hey, darling, how was that first day?" she asked, handing me the milkshake. Famished and exhausted, I took a big gulp and then got my pad of paper from her desk.

Other than the fact that I'm the resident retard, it was great! I wrote.

Abby

"Lordy me, you look like you could kill someone. Could that someone be me?" Jalessa asked, a smile rounding her large lips into a perfect C."

I continued to stare at her, not trusting my thoughts or my hands as they clinched and unclinched. I hated her. For me that was a very strong statement. I had hoped to never see her again; and here she stood, my worst nightmare.

"I'm not an apparition, Abby," she crooned.

"I happen to know exactly what you are."

"Ouch! You still have the touch or should I say, bite. Seems like nothing has changed since the last time I saw you. How's Drake?"

"You hussy! How dare you mention his name to me."

"I bet a lost summer day and images of old man Griffin's tobacco barn are swirling through your pretty little head right now!" Her nostrils flared, but not in anger, in triumph.

"Jalessa, is that you?" Bob exclaimed, way too cheerful.

Jalessa's face softened, "How could I get so lucky in one day? First Abby, and now the best looking boy from Williamston High shows up. You still as hot as you once were, baby?" Jalessa reached out for Bob's arm, gently caressing it.

It didn't take a quarter of a second for me to pounce, slapping her hard across the face, "I've wanted to do that for a long time." Swirling around, I left in a fury.

Behind me, I heard Bob exhaling and stuttering an apology. I could have cared less what he said to that bitch. I ran until I saw the nearest exit sign, pushed open the door, and stumbled down three flights of stairs to the lobby. By now, I could barely manage a full breath; but I continued out the front door of the hospital, past the side parking lot, and into a gardened area near the back. Collapsing on the first open bench I found, I sat panting, trying to calm my breathing and my anger. I had to regain composure and get back into the hospital; my daughter needed me. Why did that bitch have to show up? Here? Now?

There was no doubt seeing Jalessa had been a frigid shock—a startled voice out of the darkness, a flattened toned memory I thought had faded. An onslaught of forgotten feelings, coupled with concern over my daughter, brought me to my knees, vomiting, dry heaving, and vomiting again.

Drained, I pushed myself up and onto the bench, draped my arm across my eyes and tried to still my wired nerves and throbbing head. What I dreaded, what I hoped was dead and gone . . . returned. Images flared like July 4 fireworks and refused to retreat.

* * *

It was late summer, 1967. Drake had called the house and asked me to meet him at Griffin's barn, our old hangout, saying he wanted to talk. I hesitated on the phone not knowing if I wanted to rehash the breakup. I also did not want to get into a conversation about Bob, my newest boyfriend. But resisting Drake had always been hard for me. I knew he loved me, but I needed more than he could offer, more than he could freely give.

Trying to act coy, I arrived at the barn an hour later than scheduled. Drake's car was parked on the east side of the barn, like always. When I pulled up beside his car, I noticed an old battered black truck parked around the back. I'd seen the truck before, but for the life of me, could not remember where.

I checked the mirror to make sure my makeup was perfect. Usually Drake would have been propped up against his car waiting for me, his left leg posed over his right, arms folded loosely across his chest, dark glasses tilted low on the bridge of his nose, and those brooding, smoldering eyes, watching every movement I made. Then, he would let out a long slow whistle, saying, "Good God Almighty! That's what I call a woman!" Thinking about this made my legs go gimpy.

I walked slowly to Drake's car and peered inside. It was empty. Edging toward the barn, I heard laughter. My alert antenna went up, but I kept moving toward the sounds, feeling my stomach cramp and ache.

When I reached the barn door, I saw them. Drake had his back to me, bare and glistening with sweat. I heard a high pitched laugh, and then her face appeared; it was Jalessa. Every girl in school hated her. Every guy challenged her. She had been caught in the janitor's supply closet, the boy's bathroom, and the agriculture department's storage building, buck naked, having sex with half the football team and Mr. Hayes, the agriculture teacher. Nobody believed it about Mr. Hayes until he unexpectedly left without even a good-bye. After that, we girls called her the black widow spider.

When she saw me, she looked straight on and smiled, caressing Drake's neck, running her long tapered fingers through his thick curly hair. Then she danced her hands down his back, nibbling on his ear. Her eyes begged me to observe her conquest, witness her seduction, and feel the power of her sexuality.

I felt strapped to the earthen floor. I stood and stared as she wound her naked legs around his waist. His groans; his pumping buttocks; and her eyes, intent on mine, made me want to kill them both. I longed for a sawed-off shotgun so I could level it right at the back of his skull, knowing the bullet would go straight through her forehead.

Jalessa's laugh, mocking, swelling into the overhead, wooden rafters, broke the tension. I ran, seething in anger. Worse, the knowledge of what I saw and my lingering gaze made me feel repulsed. I was a filthy Peeping Tom, destined to go to hell.

I stumbled to my car, shaking so furiously it was hard to open the door or find the ignition key. As soon as I started the car, I spun away, leaving a trail of dust so thick it obliterated everything behind it. I never looked back. For weeks, I refused Drake's calls. I tore up his letters. I turned away his friends and walked away from anyone wanting to defend him. I ached but clung to pride like a fragile treasure.

At night, alone in my room, bitter tears left black mascara tracks on my pillow, black as my heart. The experience saturated my alone time and fractured my belief system. Mama asked if I was sick. My friends avoided me. I understood; I wanted to avoid me, too. I felt I had lost an essential part of myself, innocence. I stopped praying my nightly prayer. Every time I got to the part, *I pray the Lord my soul to keep*, I panicked, so terrified God had turned His back on me.

* * *

"Abby?" Bob called out from a distance.

I didn't bother to change my position or get up from the bench and face him. His voice only added to my splintering headache.

As he drew near me, I heard the raspy tone of his breath. With an effort he asked, "Abby? Are you okay?" He sat down beside me, picking up my hand, "Please . . . talk to me. What's wrong?"

I wrapped my fingers tightly through his and squeezed. Looking up, I took a deep breath and said, "I'm fine."

"You aren't being convincing in the least. Hey, I've known for a long time Jalessa was not your favorite person, but I'm confused. What was all that about? That's not like you, Abby."

"Bob, you wouldn't understand. It's something personal between Jalessa and me. It has nothing to do with you. Plus, I'm worried sick about Krista. The combination proved lethal to my stomach and nerves."

Bob studied me, and I didn't flinch under his scrutiny. When the change occurred, I was waiting for its arrival. His face went from concern to hurt. Hoping he would not make the connection was foolish on my part.

"Oh, I see. There's more to this drama than Jalessa, isn't there? For some damn reason, I see Drake McGregor written all over this scenario. Are you ever going to get over him? Every time I think he's a dead issue with us, I get reminded that he still lives." Bob grew quite, withdrawing. I knew it would only take a little reassurance to still his doubts, but I didn't reach out or offer one consoling word. I refused to open up the scene again or acknowledge my part in it. What would Bob think of me? Would it destroy our relationship?

Standing, stiff and stern, he stated, "Okay. Do what you want, Abby. I'm going back into the hospital and see about our daughter. I suggest you do the same." I watched him walk away and ached, but I couldn't tell him the truth. I couldn't risk having him look at me in a different light.

A loud blare from a car horn jostled me to my feet. I glanced at my watch and winced. Too much precious time had been lost. I sprinted toward the hospital door, not wanting to miss another moment with Krista. I also needed to see Bob and know he was all right. How would I ever repair the damage I'd just caused? Why did that wench come back into my life? I could kill her for this. If I ever saw her again, I'd go straight for the jugular.

By the time I got back to the waiting room, a woman motioned for me to go to the nurse's station. Bob was talking to the doctor when I arrived. I stood far enough away not to unsettle him, but close enough to listen.

"Krista has a small mouth," the doctor said. "Her bone structure is very delicate. We had to do a lot of internal wiring. The procedure was more complicated than we expected. We'll monitor her closely for the next few days. My goal is to make sure she's coping with the feeding tube. I've instructed the nurse to show you how to use the wire clippers in case Krista chokes or has breathing problems. Do you have any questions?" the doctor asked.

I was in shock; and from the look Bob shot me, he was, too. No one had prepared us for this conversation. The doctors we had consulted thought the surgery would be routine, and Krista would have a quick and uneventful recovery.

"Can I see my daughter?" I inquired.

"Yes, but she's still in the recovery room. Please don't stay long. Once we get her moved to a regular room, you can stay as long as you like." The doctor motioned for a nurse and instructed her to take us to see Krista.

My knees buckled when I saw her. Bob caught my arm and steadied me, "Are you all right?" he whispered.

I nodded a yes but held his arm tightly.

Krista's face blended in with the stark white bed linen. Tubes spiraled from her nose, her arm, and from beneath the sheets. She looked frail and tiny. Bob stayed close, tightening in further as we approached her bed.

"Krista? Can you hear me? It's Mama."

With an effort, she began opening her eyes. Relief only lasted a moment. When she realized she couldn't talk, panic scribbled itself on her face. I glanced up at a large machine at the head of her bed just in time to see the red, green, and yellow wavy lines flatten, slicing alarming blasts through the icy chill of the room.

White coats blurred around us, as a team of nurses ran in to attend to her. I reached for Bob to tell him something was wrong, but the room turned upside down.

When I opened my eyes, Bob was sitting next to me, holding my hand. Momentarily I panicked, not recognizing where I was. He must have sensed my dilemma because he reached over and stroked my forehead, reassuring me, "You're okay. We're in the hospital."

"Krista?" I asked, suddenly remembering.

"She's fine. I think she needs to see you. You scared us," Bob looked tearful. I reached out and touched his face. A face so familiar, one I had loved over half my life. Realizing how fragile relationships were, and how shallow harboring old memories could be, I reached up and pulled him close, "I'm sorry. You're my only love . . . now and forever. Will you forgive me? It was an incident that happened a long time ago . . . before we dated."

His hug, tightly given, was answer enough. "Let's go check on our daughter," he said.

Bob and I never mentioned Jalessa again. When we got home, we were so intent on making Krista feel special and helping her get back to normal it didn't seem necessary. The black hole Jalessa left, healed. Blood pulsed, renewing and reviving. Bob and I had several more dates. At first I hesitated, not feeling confident about Krista taking care of the girls, but soon realized we needed our time to renew, and she proved capable.

Our biggest hurdles were at mealtimes. Poor Krista. Her feeding tube had to be inserted between two of her back teeth and attached to a one-and-a-half-inch diameter syringe that was filled with pureed food.

Determined my ninety-pound daughter would not waste away, I decided everything could be blended into a delectable concoction. But Krista and I seemed to be the only ones who thought so. Bob, Katie, and Chelsea gawked as I filled the syringe, slid it into place, and gagged watching the overage spill out from between Krista's teeth.

Krista's main concern was school. Trying to pave the way for her, I went to the high school and talked to the principal and each of her teachers, only missing one. They were wonderful, assuring me Krista's needs would be met.

From there, I went to the community college and signed up for the fall session. Bob was a little reticent but finally caved when he saw how much it meant to me. He even stayed home from work to watch the girls for me while I visited my parents and met with Quillie to find out more about her wedding plans.

During my trip home, Mom insisted we celebrate by going to the *Aunties* favorite diner for collards and backbone. Quillie beamed like a young girl. She waited until she had everyone's attention before announcing she was having a Doo-wop wedding, using Neal Sedaka's music. We laughed, tears coursing into brown chunks of backbone, spilling over, making oil lakes on top of the collards.

Quillie had her plans in place. As bridesmaid, my job was to find something to wear in pink and lilac and get pink gingham dresses for the girls. Bob needed to wear a suit, any color would do. At the end, Quillie announced, "Girls, this isn't my only surprise. I'm saving the rest for my wedding day. At my age you have to design surprises!"

The days before the wedding were hectic. Krista's first days of school went well, except for a disaster with the absent teacher I failed to talk with. Chelsea loved her teacher the first day. Her seizures were under control, and we were breathing a sigh of relief. Then there was Katie who rarely joined family time, kept to herself, becoming a shrinking shadow, flitting from room to room, hoping not to be seen.

Thankfully, Bob was home more, talked about work less, and was more attentive, even offering to help in the kitchen. I had my school books, had met with my professors, and was familiar with each class syllabus.

But a storm was brewing. The winds were whipping up, the clouds were darkening. And when it began, it was relentless.

Chapter 29

Krista

Slipping the straw into yet another milkshake sent a shiver down my spine, and not because the liquid was cold. I was stuck in a weird sort of punishment where milkshakes had become my albatross. Like chocolate, I never thought milkshakes would lose their delectability. Yet here I was, slowly moving toward the straw with a lump of vile coming up in my throat.

Sensing my struggle, Mom said, "Only two more weeks at worst."

"Two more weeks is an eternity," I said. My mouth had become accustomed to the wiring, so most times I could speak understandably. People at school finally figured out I was indeed NOT a special education student.

"Aren't the Civenettes's tryouts tomorrow?" Mom asked.

I managed a few more gulps before pushing the milkshake away and giving up. Mom went to get me the cup of prune juice I was to drink next.

"It is. I'm also trying out for Key Club just in case. I need one of them," I said.

"Go for it, Krista. It's time for you to blossom," she said.

"I know, and I'm ready," I said, "Well, I think I'm ready." To reward my good spirits, she took away the quarter cup left of my prune juice. Sweet!

The next day, like all important and nerve-wracking days, came quickly. More to the point, getting to school was suddenly upon me, but the lunchtime tryouts for Civenettes took forever. Alyssa was scheduled to go before me, so I eagerly awaited her take on what went on behind closed doors.

They made the whole process intimidating. Each girl was interviewed by a select group of current members. Challenged, I could feel the stirring of that old spirited girl and wondered why I would sit there so long worried about what a couple of girls thought. Knowing she still existed, even as only a small blip, greatly encouraged me.

"Hey, how did it go?" I asked when I saw Alyssa.

"Not bad at all. They asked gay questions like, 'What type of activities do you want to do in high school?' and 'If you were a tree, what tree would you be?'" she said.

"Tree, huh. What did you say?"

"I said I would be an oak, because they stood strong. So gay!" she said and giggled.

Alyssa, for a girl with such emotional baggage, could be self-confident and attractive when she tried. So many people liked her, and her brother was instantly popular (his insane good looks certainly helped). Yet her achievements were not enough. She had yet to gain weight, and her dad was coming down on her worse than ever.

"Looks like you're up," she said, popping her signature hard candy lunch in her mouth.

"I'd have you give me one of those butterscotches for good luck, but where the hell would I put it," I said with a wink and a smile. Some of her confidence had rubbed off on me.

I walked into the room and immediately noticed Annie sitting over in the corner. The sense of relief almost buckled my knees.

"Please introduce yourself," a perky blonde said and motioned me to the center of the room.

The teacher sponsor, Ms. Rounder, then decided to cue in the girls to my current situation. I heard one girl whisper to her friend how brave I was to interview with my mouth wired shut.

"If you were on a deserted island, what would you like to have with you?" asked one of the girls. I looked in her direction and realized it was Brittany from my neighborhood.

I started to answer, but my voice did not project very well, so I turned to the blackboard and wrote my simple answer. Although humor is very dear to my heart, I really was just answering in the most honest way I could. The fact that everyone in the room thought it was endearing and cute was a wonderful side effect. I wrote simply, *my wire-cutters*.

After several more questions and getting into a rare groove, I left the room. I ran to the Key Club tryouts, or Keyettes as they affectionately called themselves, and barely got there in time to hear my name called. The interview was very similar to the one I just completed, so I stuck with humor and used my situation to my advantage.

The next day in homeroom, we would know if we made the group one of two ways: we would get a rose (Civenettes) or a carnation (Keyettes). I often wondered if the cruelty of the teenage years was used in ancient times as some rite of passage. I guess not everyone in homeroom would know you tried out, but most everyone was going for one or the other. Few put themselves through two interviews, but I was a true pragmatist at this point in my life and wanted to double my chances for a successful high school run.

When I got home, I found Mom in an unusually frantic tizzy. She was running around collecting things, mumbling under her breath about something she needed for class. She started school about a week ago and had been running on fumes ever since. We noticed the rushed dinners and cleaning up, but nobody dared say anything. At night, I helped Chelsea with her reading and often helped Katie with her math homework.

To avoid the steamroller insanity of cleaning occurring downstairs, I grabbed Chelsea and went looking for Katie. We found her in her room. She had her earphones on but noticed us.

"Dinner should be ready soon, but I'm so done with milkshakes, so I think I'll skip. Plus, Mom is tearing up the downstairs, so I don't think it's safe to go down yet anyway," I said.

"I'm not hungry either. Why don't we just hang and listen to the new music Mom bought me?" she asked and went over to her desk and popped a hard candy into her mouth.

Something dawned on me with such clarity it was scary. A shiver of awareness coursed through me followed by the numbness of not knowing what to do. Not only was it hard candy, it was the same butterscotch Alyssa always ate during lunch and on our way home from school. The mood was good, so I didn't dare say anything. I thought about asking Alyssa some questions, but she was in such complete denial about her issues it would not produce much helpful information. And I knew I could not add this to Mom's plate right now either. The best thing I could do was keep the girls occupied and happy.

"Mama's busy right now," Chelsea said, reading my thoughts.

"I know," I replied, happy for the reprieve. "But we have each other, and we know Mom will always be here for us, right?"

"Always," Mom said from the doorway. Chelsea ran to her and threw her arms around her waist. I wondered how Mom kept herself upright when Chelsea did this.

"Dad says we all deserve a night out to get a burger and a milkshake," she said. I rolled my eyes but was actually thrilled at the prospect.

Sitting in homeroom the next morning, I was more focused on the fact that Mom said we had an appointment in three days to let the doctor check my progress than the flowers being handed out around me. I only noticed when a petite brunette with flowered shorts handed me a carnation. Although a bit disappointed, I knew I made something. I looked over to see Kerry holding a carnation and sporting a pathetic smirk. I laid my carnation on my desk and lifted my head in time to be handed a rose.

I had made both. Both! This was a moment I knew I would always remember. Not only did the smell of acceptance emanate from these two beautiful flowers, but the quick loss of Kerry's smirk made my insides scream with pleasure. As I headed out to go to my next class, Alyssa ran up with her rose, and my good day was truly complete. Well, almost complete. As we walked into English, I saw Cindy cross the hall with her rose pinned to her shirt. Show off!

"I sure hope my dad didn't take me seriously when I told him to look for the next promotion," I joked to Alyssa as we tried to find her brother for our ride home.

She put her hand on my shoulder and stopped me in my tracks. "What do you mean?" she said, suddenly very serious.

"Nothing really, I had asked him to get me out of Dodge. Mom says he doesn't control these things anyway, and with Mom starting school, I doubt he'd even try it right now," I said.

"You can't leave!"

I wanted to, in that moment, beg her to eat. Just eat. I could offer to take out some of my precious and hard-earned babysitting money, which was flowing in with a lot less volume now that I had her and a life, and buy her any food that would entice her hunger. There must be some food temptation. After all, I had been salivating for days at the prospect of getting my jaw freed and chowing down on some good grub.

In the last few months, I became savvy to Alyssa managing to look like she was eating. She would get a hot lunch and push everything around, sucking on hard candies in between. Other days, she would forget to do her homework and skip out on lunch period to make it up. Add this to the baggy clothes hanging on her in the insane humidity. Making matters worse, I saw it all over again at night with Katie.

Unless it was raining, Jack parked in his driveway, and I walked home. It was only a few houses away, and I usually hung with Alyssa and chatted for a while. Today, I could sense Alyssa's tension as we rounded the bend.

"Oh God, Dad's home," she said. "Can't you stay for a while?"

I did not want to, but I knew I had to for her. I hoped when Mom and Dad figured out what was going on with Katie they would take a different approach. Certainly, Colonel McConnell's attempts at eating boot camp were failing, and I could not bear the thought of my house becoming a battleground.

Abby

D. H. Lawrence once said, "I can never decide whether my dreams are the results of my thoughts, or my thoughts the results of my dreams."

Regardless how dreams became reality, mine had. From the moment I attended my first class, it felt right. I bought books and carried them like rare gems. I knew I had a lot of work ahead of me but couldn't wait for the challenge. The best part was the support my family showed.

In the middle of this, we had to get ready for the big event, Quillie's wedding. I met with a seamstress, told her my needs, and she went to work, loving the idea. I met with her twice, once for measurements and again for hem lengths. The final creations exceeded my expectations.

My bridesmaid dress, a soft lilac, displayed a pink tulle overlay and a deep purple taffeta cinch belt. The full crinoline underskirt rose and fell with each breath. To continue the theme, I added black cat-eye glasses and a pair of black-and-white, chunky heel, wingtip saddle oxfords.

Mama and I found a sweet Swiss polka dot fabric to make the girls' dresses. Its playful array of pink and white dots on the bodice complemented the black ribbed satin for the skirt. Patsy, the seamstress, really got into the theme by adding a bright pink poodle toward the bottom of the skirt. For fun, she tied pink wool loops across the poodle's head to resemble a kinky pompadour. For a finishing touch, she adorned it with a sequined collar and leash. I even found saddleback oxfords for the girls to complete the outfit.

On the day of the wedding, it took two hours to get ready. Then we struggled to get out the back door, our crinolines so stiff and full. We pushed and pulled. When it was my time, I had to lift the whole skirt up and over my head, and still had to turn sideways to accommodate the fullness. Bob laughed and catcalled, watching me brazenly exposing my underwear.

Standing in front of the car, we stopped, looked down at our dresses, and shook our heads. It looked hopeless. One at a time, Bob directed the melee, amidst moans and groans, flying elbows and shoving hands. By the time he got to me, he was done. It took three tries to get the car door open. My dress slammed it shut before it could ever fully open; and once inside, another three tries for me to be able to sit down.

"Okay, the fun is over. Everyone quit bellyaching!" Bob threatened, his patience wearing thin.

"Mama, there's no way we're going to make a trip stuffed in here like this," Katie yelped, pushing Chelsea hard to move over.

"Isn't wurkng," Krista stammered, her head slamming against the car window. "Ouch!" She thundered in concert with a slap to Katie's back, who then turned to Chelsea and walloped the back of her head.

"That's enough! You're not the Three Stooges!"

Now it was my turn to whine; I wasn't having any better luck than the girls. "Bob, you've got to come over here and shut the door. I cannot manage holding this monstrosity and closing the door, too. Also, do the same for Chelsea. It might help end the riot going on in the backseat."

Bob's tension broke, looking at the spectacle. He got out of the car to help but was laughing so hard he had to hold onto the car for support. "I need a truck to haul this load! Well, on second thought, if anyone sees those pitiful looking dogs on those dresses . . ." Bob bent double, heaved and almost choked before he finished, "Abby, no joke, they look like skinned Chihuahuas with an Afro."

"Shut up, wise guy, and get me in the car. We're going to be late," I exclaimed. "By the way, those poodles are adorable, regardless of the stupid picture you just planted in my brain."

I cannot describe the harrowing trip from Wilmington to Williamston. My skirt flew up twice, blocking Bob's vision, almost causing us to have a wreck. I finally had to completely pull up the back of the skirt and lean against it while I pushed with both arms to keep the front from slamming my face. In my head, when the wedding was over, I vowed to burn all four costumes in an incinerator.

When we turned the corner onto Spruce Street, the churchyard was filled to overflowing. Bob let out a "Whew!" when he saw the whole left hand side of the church parking lot lined with beautiful antique cars: a 1949 Lincoln Cosmopolitan convertible, a 1950 green-on-green Ford Crestliner, a 1950 crème-on-tan Mercury Club Coupe, a 1951 Lincoln Cosmopolitan, a 1955 Chevrolet Bel Air Convertible, a 1955 Mercury Montclair, and a 1956 Packard Caribbean, to name a few. Each car was draped in purple and pink crepe paper streamers and coordinating paper roses.

Bob almost wrecked the car again, straining his neck to look at the vintage beauties. As soon as the stick shift landed in park, he jumped right out to join the group of men hovering around the parade of cars, totally oblivious to our dire circumstances.

I called out to him, but it didn't bring him back. "Dammit," I said under my breath. Knowing it was up to me, I said, "Chelsea, see if you can stand up, reach over, and lift the handle on this door."

"Mama, I can't move!" she yelled.

"Dis is dumit idea you had," Krista added.

Just then, Titty, I mean Annias's brother Titwell, opened the door, spilling me to the ground.

His monotone voice strung each word on an invisible line, "Thought it was the decorator with lots more stuff to pile in the church. Who would have thought bodies were under this mess."

"You're a lifesaver. Bob ran off to look at those dang cars and left us stranded. Open the back door and let the girls out," I replied, brushing off my dress.

It took a lot of energy to pry the girls from the back seat. Once out, I popped the trunk and retrieved five hat boxes. We decided to keep the theme, placing our gifts in hat boxes, and tying them up in bows of pink, lilac, purple, and magenta.

The grounds were filled with people strolling around, taking items in and out of the church. If I hadn't known better, I would have thought we were late, considering the amount of activity going on, rather than being two hours early.

We struggled to the church. Going up the stairs proved another problem. The girls could not see over their full skirts to gage where the steps were. I was in the process of coaching each one where to place their feet when I heard Daddy yell my name. I leaned around the church column; and there he was, standing in the middle of about twenty men.

"Hey!" I hollered. "What're ya'll doing over there?"

"Been cookin' a pig since three this morning. Yore Mama's in the church decorating. When ya'll enter in all that finery, she'll stop dead in her tracks!" Dad barely finished his sentence, busting out in fits of hysteria, along with several of his buddies.

Huffing, I threw him a glare and continued pushing forward. When I opened the door, several women looked up and stared. Mama stood in the middle, a look of astonishment crossed her face, "Why, Abby, look at you . . . and those girls. Wow! I'm speechless!"

"Mama, please spare me from any dicey talk about these getups. I've just spent two hours in wrenching pain to get here."

"Grandma, Mama's right. It was the most horrible ride of my life," Chelsea said. Flopping down on a pew, she forgot about the crinoline and immediately landed on the floor, "See what I mean!"

Krista and Katie edged down purposefully beside her, not wanting to have the same thing happen to them.

Looking around the sanctuary, I took in a breathtaking panorama of flowers. So many flowers: purple and white asters, purple ironweed, yellow black-eyed Susans, and an abundance of pink and white roses.

The banisters on either side of the altar area were draped in tulle, mounds and mounds of tulle. Every so often, a sprig of greenery, tied up with pink satin ribbon, scalloped the design, adding more color. Every windowsill had a candle arrangement with more greenery and pink ribbons. Pink ribbons adorned the end of each pew. Pink was everywhere.

Walking closer, I noticed the first six pews on the left side had cards printed, "Reserved for Pink Ladies" and the right side, "Reserved for THEIR Men."

Mama motioned for me to help, right as Quillie entered the room. Everyone grew silent as coins. She was beaming!

"Okay, who's got your tongue?" Quillie asked, dropping a box of boutonnières on a pew.

Katie beamed, "Ms. Quillie, you look beautiful!"

She smiled so big her gold-filled back teeth sparkled, "Today's the beginning of a new life for me. I've been waiting a long time for this, and it's going to be perfect."

Salem stopped pinning tulle, got off her knees, and ushered Quillie out of the room, "It won't be perfect if that groom of yours walks in here and sees you. That'll bring on bad luck, for sure. Now get!"

Mama finished the last touches, stood back, and said, "Perfect!" Turning her attention to the girls, she pointed, saying, "Stay put! I need your mama's help in the kitchen.

If I thought the church was brimming over with adornment and frills, the kitchen had met its match. Silver trays and punch bowls decorated pink draped tables. Food was mounded on trays, and more food was arriving.

"Mama, who fixed the food or did Quillie have it catered?" I asked.

"Child, as soon as Quillie mentioned a wedding, the phone rang off the hook. Everyone in the surrounding three counties begged to bring something. We don't know how many folks will be here, but from the looks of things they'll be standing out on the front lawn."

Someone grabbed me from behind, "I'm so excited! I can't stay out of the bathroom. I just know I'll pee in my pants from sheer joy before this here day winds down," LeRoux was gasping, her rosacea a brilliant red.

"LeRoux, are you okay? You seem mighty winded," I asked, leaning over to hug her neck.

"I'm fine, just nervous. Quillie has us up in the front, you know."

Smiling, I added, "That explains the Pink Ladies section."

"We are the Pink Ladies, aren't we, LeRoux?" Annias grabbed me, giving me a big hug, "We've just had our cue. The wedding's about ready to begin. Nadine, we need you and Abby to finish up."

Mama grabbed my hand and pulled me down a corridor toward the back of the church. At the end, Annias threw open the double doors, "Wedding's almost on, gals!"

Walking into a large room, it was evident the party was here! Over in one corner, a bar had been set up. The bartender, a cute little blonde sporting a mile-high beehive, talked and took orders from a long line of colorfully dressed women. She didn't miss a beat: shaking and mixing concoctions, dangling a cigarette out of the corner of her ruby red lips, and squinting around the curled stream of smoke rising around her false eyelashes.

Mama and LeRoux let out a "Hot Damn" and ran to get in line. I stood, wishing for the thousandth time I had remembered my camera. Pink and poodle skirts were everywhere!

Looking around, I saw several women I remembered from my youth: women who poured communion, women who sat in nursing homes holding hands and praying, women who had pushed every button on God's green earth and beyond, and women from uptown who stood on the other side of the counter demanding assistance. All were in this room. They looked and acted the same. They were reaching over and hugging one another. They were pressed in talking and laughing. For a moment, it seemed like the world had become one and differences had collapsed within the framework of an obtuse wedding of chance and circumstance.

I turned to a tap on my shoulder, "Hi, Abby. Quillie sent us over to meet you. My name is Piney Jodelle Mizelle, and this is Tobaccie Manning. We've known Quillie and yore mama for years, but only recently rekindled our old friendship. Quillie and I put in tobacco when we were youngins. Yore mama hated it. Seems like a long time ago, thank God. Our lives sure have taken on different and necessary journeys, but here we are."

"Oh, I'm so pleased to meet both of you. Tobaccie, I remember Mama telling a tale on you and your sister, Inelle. Wasn't it something about a little tête-à-tête at the Tasty Freeze one night?" I joked.

Tobaccie shuffled her feet, her face ran red from her chin to her forehead, "Shaw, your mama's in trouble, for sure. Ain't no need to stir up a pot that's been spilled for years. You can rest assured it ain't gonna be me."

A tall rotund woman pushed open the door and yelled, "Show's on!"

Women tipped the last drop out of their Dixie cups. Most scurried, making sure to get a good seat. The rest got into position, waiting for the cue from the processional music to enter the sanctuary.

"Where's Quillie?" Mama asked, scanning the group.

"She stepped out back a while ago. Guess she didn't hear the roll call," someone answered.

Mama got out of line, headed toward the back door when Quillie walked in.

"Hang on to your britches, Nadine. I was settin' my girdle on right. The dang thing was pure more up my crack, than on my tail. Who in the world invented this dang torture chamber?" Quillie belched loudly, "Whew! I think I finally got up the last gas bubble. Some surprise if Dickie goes to kiss me and I belch in his mouth."

"Gas is natural when you're nervous. My daddy always said to just let it fly so as to control bloating. I can tell you, that man was never bloated," Salem offered.

"Gas or not, we've got the signal to start. Is everyone ready?" Mama asked.

I couldn't believe my ears when the door opened to the music of Neal Sedaka's "Calendar Girl." Mama, Salem, LeRoux, Annias, Tobaccie, and Piney Jodelle started into the church dancing a conga to a variation of the Watusi. I went right down behind them, trying to keep in step, knowing my face was a crimson red. When I finally locked eyes with Bob and the girls, he eased my humiliation by giving me a flirty look and smile. I got into the rhythm, lifting my legs high, pumping my shoulders with the beat of the music, and enjoying the craziness of it all. The music stopped; and I slammed into Piney Jodelle, who caught herself before toppling Tobaccie.

Quillie, who was last in line, halted at the door. Dickie was not standing at the front of the church, which was customary. Heads turned back and forth in anticipation. The suspense grew. When the music began again, the front door of the church blew open, and there stood Dickie. He had on a white T-shirt underneath his tuxedo suit coat, plaid shorts, black socks, tennis shoes, and an apron. The apron brought the house down. Bold black letters across its front read, "Play my tambourine, anytime, Baby."

Paying no attention to the roars of laughter, the men whistling or the women screaming his name, he held up an old wooden spoon and began to sing, "I love, I love my calendar girl. Yeah, sweet calendar girl, I love, I love, I love my calendar girl each and every day of the year."

At his prompt, Quillie started toward Dickie, doing the shuffle, lifting her hand in greeting as she turned from one side of the church to the other. This was indeed their hour of glory.

When Quillie reached Dickie, they held hands and strolled back toward the pulpit, finishing the rest of the song, dancing and carrying on. Bob caught my attention, motioning to his chin, then pointing to mine. My mouth was gaped wide-open.

The minister raised his hand to quiet the mayhem and delivered the opening prayer. I wondered if this would be the only sane, traditional part of the service. When he said, "Amen," Dickie began his vows, reciting his version of Sedaka's "Oh, Carol."

> "Oh, Quillie, I'm but a stupid fool,
> Darling I love you tho' you treat me cruel,
> You hurt me and you make me cry,
> But if you leave me, I'm gonna waste up and die.
> Darling there'll never be another,
> To thrill me like you do.
> Don't ever leave me or ever let me go.
> I'll always want you, no matter what you do.
> Oh, Quillie, I need and love you so."

Seeing Quillie sniffle, Mama slipped her a lace hanky over her shoulder. She dabbed at her eyes and began her vows,

> "Dickie Washburn, you're my man,
> You make me feel like no one can.
> Let's have fun and never look back,
> At all those secret things we packed.
> I'll be true as long as you,
> God forbid you mess up, too.
> Forever true I'll be to you,
> You belong to me, so don't be blue.
> We'll be together, stuck like glue."

My mind reeled, trying to process the whole menagerie. A tissue flapping in front of my face snapped me back to nose blowing and tears. All the bridesmaids were bawling! A swift glance around the church confirmed my suspicion, I must have missed something. There wasn't a dry eye in the house, except mine. But what I didn't miss was the minister's final words, "And now I pronounce you man and wife!" Dickie grabbed Quillie, kissing her from one side of her face to the other, stopping long enough to yell, "I did it!"

The pianist started the recessional, and everyone stood to sing, "Love Will Keep Us Together." Dickie and Quillie walked down the aisle, stopped, waved, kissed, and hollered "Glory Hallelujah" until they exited the church. People pushed and shoved down the aisles behind them.

Bob and the girls slipped out the side aisle and met me up front. "Now that's a wedding!" he said.

"Sum kina curcus," Krista tried to say.

"I'll remember it forever. I want my wedding to be exactly like Quillie's." Katie looked moonstruck.

"I'm hungry," whispered Chelsea, and rubbing her eyes, added, "And, I'm tired."

Bob picked her up, motioning for us to follow him around back to the kitchen in order to get ahead of the frenzy. He knew crowds and free food were a dangerous combination.

Our plates were brimming over when Quillie and Dickie finally made it into the dining hall. Quillie looked absolutely radiant. They stood at the door, greeting each and every guest.

"I thought the bride and groom were the last to arrive. In fact, I thought they had to be introduced," Bob commented.

"Honey," I said, "There's absolutely nothing normal about this wedding. Quillie makes her fate. Dickie better hold on for a fast ride!"

We stayed for another hour, visiting and dancing to the wonderful sound of fifties music. Right before we left, I searched for Mom and Dad to say good-bye. I found them sitting at the back of the room. They had pushed off their shoes; leaned their heads against the wall; and fallen asleep, holding hands. Seeing this was worth the whole day.

Bob came up behind me, putting his hand in mine, "Let's go home, my bride."

The drive home was quiet. No one mentioned the dresses. Somehow they had managed to press themselves into a sensible row of pleats. Looking back at my three sleeping daughters, I felt a swell of pride and love.

I slid over, wanting to get close to Bob, "Wasn't it wonderful?"

Bob's answer came after a long pause, "Abby . . . I think there's something terribly wrong with Katie." His eyes brimmed with tears.

I fastened myself to the silence. The day of reckoning had arrived.

CHAPTER 30

Krista

The ridiculousness of Quillie's wedding was unparalleled. Somewhere in my diary was an ongoing list of things I'd never do to my children, and I updated it with this insanity as soon as I got home. In it, I detailed the pink and white polka dotted monstrosity I wore, but nothing could truly describe the horror of it.

The wedding was a spectacle, though I enjoyed it from time to time. A huge smile crept to my face to see Mom and Dad joking around and having fun. Katie, who had redefined melancholy these days, was in heaven. She danced and danced at the reception. She pulled in Chelsea, and they were both giggling out on the floor. Of course the sight of Quillie and her bevy of females dancing alongside them was like watching a bunch of pigs at the farm when their food was thrown into the air.

I was in a bad mood. The doglike animal on my skirt seemed to be taking on a life of its own. Everyone was making comments, and I was done with it. To make matters worse, I could not eat a darn thing. The outrageous outlay of food was one of the best parts of the wedding, but I wasn't about to shove anything into a syringe. The delicious smells of barbeque, chicken and dumplings, banana pudding, field peas, and various desserts were actually starting to make me nauseous from sheer desire. In the end, I was forced to sit and drink water, and Mom and Dad did the obligatory milkshake stop on the way home.

There was some good news. I was finally ready to have the wires cut, my mouth freed from liquid hell. I presented Mom with a menu before Quillie's wedding that included homemade macaroni and cheese, steak, chocolate cake, and black-eyed peas. Mom reminded me about the doctor's warning to go back on solids gradually since my body was not used to them. This was one area where I was determined to take some risks. Bring it on!

Just before we headed to my appointment, I called Alyssa. After hearing every detail of the meal I was anticipating, she seemed a bit put off. And after eight weeks of being on a liquid diet and starving a good portion of every day, I was even more confused about her self-imposed food boycott. There were times during the day the want of food pained me, and I did not know how she could turn the switch off. *Wasn't it human preservation to feel hunger and feed it? Do you somehow become numb to the need, the desire for food?*

Still, she was happy I would soon be able to talk again and get to know more kids at school. When she mentioned that, I thought of something for the first time. *Was the acceptance I had achieved at high school a direct result of people feeling sorry for me? Would I disappear all over again once I was a normal teenager with pimples and glasses? Would I have gotten into Civenettes?* I was starting to wonder if my jaw being wired shut was my gimmick, my ace card.

"Mama, do you think kids only like me because I'm, well, me or because they felt sorry for me?" I asked on our way to the doctor's office.

"Krista, people like you for you, and Alyssa knew you before, and she's your best friend," she said.

I knew she was right, but I was still nervous. My stomach's needs soon overcame my fears. I wanted to kiss the doctor when I heard the snap of the wires.

"Don't forget, young lady, slow and steady on the solids." I heard, but the imagined smell of macaroni and cheese and cake was so intense his words were fading quickly.

Mom, true to her belief food could cure all, made every item exactly as I requested. Dad complained it was way too much food, but I was too deep in my own euphoria to care. Every bite tasted like liquid gold. I take that back, just gold! I had never enjoyed a meal so much in my life. I knew I wouldn't share this ecstasy with Alyssa or Katie because I wasn't going to feel guilty about it for one second.

This meal's leftovers were served two more nights and one lunch, a real treat. Mom was up to her neck in school, and big meals were becoming a rarity. Too often it was McDonald's, pizza, or sandwiches. Normally, this would be a teenager's dream, but I had gone way too long without real food. Nobody else cared as much as I did; or maybe, they were too busy walking on eggshells around Mom. My mouth was always two steps ahead of my heart and brain, and Mom was certainly in no mood to hear about my food desires.

"Listen, for eight weeks I catered to every food need you had, and quite frankly, I've catered to everything for this family for too long. School is important to me, and I need just a little support!" she yelled one day when I groaned about another sandwich. After that, I kept my food opinions to myself.

Making tensions worse, Dad was traveling a lot more often. I heard Mom muttering to herself about how she had had shit detail (her words, not mine, of course) for all these years; and now that she needed him, he was always gone. The stress levels in the house were off the charts. It did seem odd Dad would be gone so much, especially with everything going on.

One Saturday afternoon, I was talking on the phone with Alyssa when I heard Mom's frantic cries for all of us to hurry up and get downstairs. I wanted to finish the call and took my time.

"Krista, seriously," she said walking into my room, "we have to get your dad to the airport, do you want to go or don't you?"

"I'm coming, I'm coming!" I yelled and then exchanged a hurried good-bye with Alyssa. On our way home, I was planning to remind Mom about my driver's education classes starting next week. After six weeks of classes and supervised driving, I'd get my permit. In North Carolina, God love the lawmakers here, you could get your permit anytime after you turned fifteen and had completed a driver's course. Since Mom and Dad had drug their feet signing me up, I bet it wasn't on the top of their to-do list.

We piled into the car and had barely shut our doors when Dad was reversing down the driveway with a very frustrated and pained look on his face. Suddenly the car jolted, and it was clear we had hit something. Dad, not sure what had happened, instantly pulled forward to see what he hit, and the car jolted again. He slammed the car into park, and threw open his door at the same time Mom did. The next thing I heard was a wail from Mom, and then I saw Dad fly by the car back to the house.

I opened my door to see what was wrong, and I saw Mom cradling our dog, Sugar. She was crying.

"Mama, what can I do?" I asked in a panic.

"It's okay, please just sit in the car, Krista," she said as she rocked back and forth with Sugar.

Dad barreled out of the house at this point, handed Mom a blanket, and then helped her get into the front seat with Sugar. Katie and Chelsea were both crying, and our neighbors came running across to help. My dad stood talking to Jeff's parents, and suddenly his mom was getting into the driver's seat, and my dad was getting his suitcases out of the trunk.

"I'm going to go with Mr. Manning to the airport, and Mrs. Manning is going to take you guys to the vet, okay?" Dad said and gave each of us a quick kiss. He then opened Mom's door and asked if she was sure he should go. She simply shook her head and offered her cheek. He kissed her and was off.

"Girls, I don't want you to be upset. We are going to take Sugar to the vet. Krista, I'm going to go in with your mom, and I would like you to keep the girls calm in the car," Mrs. Manning said. I was worried about Sugar and wanted to go with them. Mrs. Manning put a hand on my mom's shoulder and eased down the driveway and then off to the vet's.

At the animal hospital, Mrs. Manning came around and opened Mom's door, and they headed inside. At first, we sat there in stunned silence, but then Katie and Chelsea started crying again. Though I was the big sister, my nerves couldn't handle absorbing their sobs. I started a rousing game of rock, paper, and scissors. After that failed to hold their attention, we played games invented on the spot. We tried to make as many words as we could from "Animal Hospital." We helped Chelsea.

It seemed an eternity when I looked up and saw Mrs. Manning walking Mom back to the car with one arm over her shoulder and the other in front of her. It was clear she was helping Mom walk. I was afraid if she let go, Mom would go tumbling forward. Sugar was not with them. Chelsea noticed it as soon as I did.

"Mama, where's Sugar?" she asked as soon as Mom sat into the car.

"Girls, Sugar was hurt very badly, and, well, she didn't make it." Mom said. Only then did I realize Dad had run over her. The jolting we had felt had come from the tires hitting Sugar not once, but twice. "She died while we were in the office."

I was numb; I think everyone was. We walked into the house and slunk to our various rooms of comfort. I was too sad to be worried about everyone else. I sat on my bed and cried. I would miss the sounds of snores emanating from under my bed every night. Sugar had always patiently sat and let me pet her and discuss all my teenage drama. She was a soft warmth of support, never judging or trying to offer suggestions.

The next day, on the way to school, I told Alyssa. Even her brother was consoling. All day I walked around in a daze. One of my favorite teachers pulled me aside and asked what was wrong. We were standing in the hallway, so a few people heard me tell her about Sugar. I didn't know whether they stopped because a teacher was comforting a student or because the *special girl* was actually talking.

At lunch, Alyssa had a troubled look. Something was bothering her. She always lacked energy and was barely able to walk at times, but today she seemed particularly pale and slumped in the shoulders.

"Krista, I'm not sure I should even tell you this," she started. *Man,* I thought, *did conversations starting with that line EVER go well?* "I overheard Cindy with a few of her cronies saying you deserved to finally understand death after what you did to her in junior high. And that was only your stupid dog."

I was angry, but I also realized she had a point. I never wanted to see any of my wrongdoing in last year's debauchery, and I certainly didn't think any small thing I might have done deserved what I endured. Still, I knew what Sugar meant to me. Sure, I had not noticed her for days sometimes; and prior to the accident, I could not remember the last time I held her in my lap. I didn't talk to my friends about her and tell them what a great dog she was. No matter what it seemed like on the surface, Cindy must have loved her dad. And even though I hadn't meant what I said at the time, I could see why it hurt her. It was no excuse for what she did or the fact she was still harboring it. In the end, I would never offer her an apology because she did not deserve one. Still, I felt sick to my stomach to have this realization at such a sad time.

I went to the school office to call Mom and get some relief from the incredibly long day. There was no answer. So many things had changed since Mom went back to college, and I suddenly felt very alone. It was even worse when I finally got home, and nobody was there. Not even Sugar, who would usually come to greet me. When I went to the cookie jar to get a sugary welcome, I realized the sound of the jar lid would not bring her either. She had loved cookies.

Mom came rushing in the door with Chelsea, Katie, and McDonald's about an hour later. My stomach almost turned. Katie ran upstairs. I thought she might still be dealing with Sugar, too; but she looked angry, not sad.

"Sorry, Krista, we were at the doctor's office. I should have left you a note, but I was in such a hurry," Mom said.

Just then the phone rang. "What now!" Mom uttered as she picked up the phone. The whole time she was listening, she looked at me. "I'm so sorry. Let me know if there's anything we can do. I'll bring Krista to see her as soon as I can."

Mom hung up the phone, came over to me slowly, and took my hand. "Honey, that was Alyssa's mom, Alyssa's been admitted to the hospital."

"The hospital, what do you mean? What happened?" I asked. "I just saw her!"

"She collapsed right after school. They are going to be keeping her for a while."

Abby

I tossed and turned all night, unsettled by Bob's comment in the car about Katie, and the continued conversation after the girls went to bed. I was so wrapped up in the wedding festivities, I failed to notice Katie did not eat anything, but Bob did. He said he tried everything; but when he forced a piece of fried chicken on her, she ate one bite and ran to the bathroom. Not knowing what was wrong, he followed her. When she ignored his call, he opened the bathroom door and heard her vomiting. He waited until she came out to see if there was anything he could do. He said the vomiting and her lingering in the bathroom was not what frightened him. It was the cold-as-steel stare she gave him when he asked if anything was wrong. He shivered, relaying the story.

We talked a long time. Finally Bob summed it up, "Abby, we've got to focus on our family. Who knows what kind of medical treatment Katie's going to need, and there's the other two girls. Look at their issues. And, you know I'm going to be traveling a lot more, especially since my boss relocated to San Jose. I don't know how much support I'm going to be. I think," Bob paused, "Considering everything we've got going on you might want to reconsider school, for now, that is. I'm also worried about expenses. I've taken a financial hit with Krista's dental work." He put his hand on my shoulder, "I wish things were different; I truly do."

Guilt is not a friendly companion. The hot breath of disappointment licked at my consciousness, but in the process, peeled open raw wounds. For weeks, even in the midst of personal joy, I had felt a nagging voice chastising me for neglecting my family and putting my needs ahead of theirs. In response, I lowered my head, acknowledging he was right.

I also knew I had consciously been dodging the piper concerning Katie, knowing a problem was ready to erupt. The pit opened beneath me, beckoning me, but I stepped to the side and straightened my back with a determination unequal to any I had known before.

The next morning, Bob left early for work, preparing for a two-day meeting in California. He was right; the family needed me to be focused. I imagined donning Wonder Woman's powers, her Lasso of Truth and her indestructible bracelets, to fight the issues at hand. My endeavor began with breakfast.

"Wow!" Chelsea exclaimed, walking into the kitchen, "Pancakes and bacon, on a school day. Is it someone's birthday?"

Katie arrived, her fingers pinching her nose, "What stinks?"

Krista pushed Katie out of the way, blasting off in frustration, "Yr kllng me. I cnt et dis!"

"Settle down. This is a day of celebration. I'm starting early. For your last shake, I have blended up a chocolate chip malt, spiked with two tablespoons of peanut butter for an extra burst of energy. Tonight, I'll fix everything on your list; I promise." Six cups of coffee pumped gold adrenaline through my veins—the effect, exhilarating!

"Ses los it!" Krista announced, plopping down.

"Well, she'd better not hand me that stuff!" Katie fumed, crossing her arms in open defiance.

"Give me seconds!" Chelsea repeated twice, in quick succession.

"Okay, girls, while you're eating, let's run through today's schedule. We have to get everyone ready and off to school. Krista has a doctor's appointment. I have . . . well, a lot! When we get home this afternoon, we have to prepare for a celebration party." I rolled it off without a pause, flipping pancakes onto plates and hitting the blender button on high. "This is going to be a perfect day. Do you understand?"

I left the room, my last command trailing me, "Finish eating, grab your schoolwork, the train is arriving, you'll board in fifteen minutes."

"I think those women really got to Mama," Katie offered.

Chelsea responded, "Yeah! Do you think she'll start dressing funny? Maybe Grandma and the *Aunties* can come back. We can have a real party. Quillie's so much fun!"

"Dos nutcks cum, Im plying for adopton," Krista steamed.

I paused outside the doorway when I heard Katie's comment, listening to their review of what would or would not happen if Mother and her cronies came. The idea was indeed intriguing but certainly not practical. I needed to do this one myself. Acknowledgement became empowerment.

Trying to get everyone off to school ended up a major feat. Chelsea lost her homework. Krista announced at the last minute I needed to take her to school. And when I opened the trash to toss something away, I saw Katie's whole breakfast dumped. Seeing the food in the trash almost sent me ballistic, but somehow I managed to suck in the scream and back away without exploding. I had less than thirty minutes to get everyone in the car, make it to three different schools, and head downtown to withdraw from college.

Talking to my counselor proved harder than I imagined. He tried to get me to reconsider, focusing on finishing rather than making excellent grades. I held firm, thanked him and saw each professor before leaving campus.

Promptly at one o'clock, I had Krista in the car headed to her appointment to have her mouth freed from eight weeks of being wired. I could tell she was energized and ready. Krista had been a real trooper, but the last week had been trying, especially during the wedding. The abundance of food, and the fact she couldn't have one bite, was torment.

On our drive over, I had to listen hard to understand when Krista asked, "Mama, do you think kids only like me because they feel sorry for me?"

It took a minute to process her question, my mind racing down a list of things to do. "Krista," I paused, putting strings of thought in order, "I think people like you because of who you are. You have a great sense of humor. You're smart, and you're courageous. How many people would have tried out for clubs with their mouth wired? I'm amazed! Remember, Alyssa knew you before your wires, and she's your best friend. A lot of your new friends haven't had the opportunity to see the real you. It's going to make them like you more," I answered.

Krista nodded, growing quieter.

We were in and out of the doctor's office in record time. My seat belt had not fastened when Krista reminded me of her supper menu. "How could I forget? You've had it in my face for days!" I laughed.

When we got home, I sent her upstairs to do homework; I had to prepare a feast. Once I had the bulk of the food cooking, I zeroed in on the next imperative item on my list—call and get an appointment for Katie. I inquired and was given the name of a pediatrician who had an excellent reputation, diagnosing and working with children and teens with eating disorders. I dialed the number, surprised and relieved when the receptionist told me there had been a cancellation for the next morning. I closed my eyes and prayed a sincere "thank you" to God.

I was in the throes of cooking when the younger girls came home from school. "Did you thank Mrs. Lewis for the ride home?" Carpooling had been a lifesaver for me.

"This house always stinks!" Katie's voice shrilled.

"Well, hello to you, too. How was school?" I asked, watching her walk by and ignore my greeting.

"Mama, I had the best day!" Chelsea lifted up her hand, showing off the smiley face stamp on its back. "Today, I was the teacher's helper. Tomorrow, I get to hold the flag."

She turned her hand back and forth to give me the stamp's full effect. "Why don't you show it to Krista? She got her wires cut today. She'll want you to see her new look."

My second surprise of the day was when Bob called, saying he had rescheduled his trip till the next day. He decided he didn't want to miss celebrating Krista's big day.

At dinner, the table overflowed with every item of food on Krista's menu list. Twice, I offered her halfhearted cautions not to overeat but figured Maalox would fix an achy stomach. I ignored Bob's complaints about the greasy food, and Katie's attempt to mound and unmound her food instead of eating it.

After dinner, I asked Bob to linger behind for coffee and conversation. When the girls headed upstairs to do homework and baths, I began, "You opened my eyes last night, and I appreciate it. There have been a lot of things I've avoided, thinking only of myself is only one of them."

"Now, Abby, I didn't say . . ."

"I'm not accusing; I'm agreeing. Half my problems are just that; I have too many problems. I need to pare down my life, concentrate on you and the girls." I paused, feeling a knot forming in my throat, then tears gushed. "Gosh, I'm sorry. I don't know where this is coming from. I've been so resolved . . ."

I grabbed a tissue, blowing hard to release the last reserves of emotion, before resuming, "I withdrew from college this morning. It was the right thing, considering what's going on around here. When life settles down, I'll go back. While I was there, I met with a professor of mine who recommended a doctor for Katie. I called; we're going tomorrow morning."

Bob caught me off-guard. He looked bewildered, distressed, "You didn't have to rush and make a rash decision about school. I made a suggestion, not an edict. And, are you sure we need a doctor involved with Katie? What have we done to correct the situation? I'm not sure . . ."

Interrupting, I added, "I didn't rush into anything. I've had months wondering, questioning, denying, and chastising myself over her. I'm sad it took our conversation to get me to act. As a mother, I should have been proactive. Today, I did what I should have done weeks ago. About school? Yes, I'm sad. Maybe more than I realized, but I need to prioritize. School didn't win out."

Bob's eyes softened. He pulled me into his arms and said, "I love you."

The next morning, I picked up Katie from school and headed for our consultation with the pediatrician. Dr. Easley was compassionate, asking me questions and listening. The whole time I talked, she kept her eyes directed toward Katie. Her gaze, like someone peering in at caged animals in a zoo, was gentle and inquisitive.

Finally, she directed a question to Katie, "How do you feel about what your mother has told me?"

Katie sat stiff-straight, her hands folded neatly in her lap. She showed no emotion, none whatsoever. Dr. Easley did not repeat the question but sat relaxed, waiting for a response. I wanted to reach over and shake her but sensed the doctor was in total control, offering me a chance to settle back, still my nerves and observe.

Silence continued. Dr. Easley did not waver or ease her intent stare at Katie. Her face held power, control, and confidence in one small commanding smile.

Katie shifted in her chair, casting her eyes down. Shifting again, she said, "I eat what I want, when I want. No one can make me eat if I don't want to."

Dr. Easley let the words spill out and bound off the walls in quiet resolve. When Katie lifted her eyes, she met the doctor's head on. I was astounded how much control and determination a child could have, but apparently Dr. Easley wasn't.

"Katie, I don't want to force you to do anything. We will be working towards a goal over the next few weeks," she announced.

"Goal? Weeks? What are you talking about?" Katie stood up, putting her hands on her hips.

"We'll discuss it next week," the doctor said. Twisting toward me, she added, "Mrs. Simmons, I need Katie to be here weekly. You can arrange appointment times with the receptionist to fit your calendar. Each session will be one hour. Over the next six weeks, I'll see Katie alone. At the end of that period, you and your husband will join us to review the goals and progress. Do you have any questions?"

I gulped and swallowed hard, "Well, no, I guess not."

Dr. Easley abruptly got up and left the room. Katie bounded out in the opposite direction, yelling, "I'm not coming back!" I followed, admiring Dr. Easley's approach.

On the way home, Katie refused to talk or stop for lunch. Rather than return her to school, I took her home, hoping she'd have time to think and accept the next few weeks.

Bob called to check on Katie, "Wouldn't you know it, I'm stuck in Chicago! Hey, what did the doctor say about Katie?"

"Six weeks of appointments. Afterwards, we sit down and review their sessions," I summarized.

"What did you think of him?"

"He's a her. She's fine, a straight-shooter," I continued.

"Why do I have to drag information out of you? Can't you just tell me what happened? I don't have all day, they just announced the plane's ready to board."

"I'm sorry. That's all I know. I did check into the insurance, and it will pay for the prescribed time."

"Okay. I'm depending on you to keep me posted. Gotta go, love you!"

Pushing my shoes off, I sat back, reflecting on Katie's appointment. I liked Dr. Easley. Calm washed over me, knowing, without knowing why, that Katie was in good hands. The idea strengthened me, one sinew at a time.

Bob got home and had to do a turnaround back to California to confer with engineers on an environmental issue. Getting tickets at the last minute and repacking put us in a tether to get him to the airport. Bob rushed by clinching two suitcases, screaming for everyone to hurry up. In response, I yelled, too. Then the nightmare hit. In our haste, we forgot to check on the dog, Sugar, and Bob accidentally ran over her.

My neighbor, Mac and her husband rushed over when they heard the commotion, and realizing our dilemma, jumped in to help. Her husband took Bob to the airport. Mac took us to the veterinarian's office, staying with us even after Sugar died. Later, seeing how upset I was, she urged me to go upstairs and rest while she entertained the girls. She made me feel we'd been friends forever, restoring my resolve to pursue her company. I'd seen the change in Mama, how the friends in her life had opened her up; I wanted that, too.

We managed to get through the next few days of loss, grieving our dog's tragic death. Quiet settled over the house, creaks and moans, unheard before, became evident. What we strained to hear were toenails tapping against the hardwood floors or a bark at the backdoor.

In the middle of our heartache, more was delivered. Mrs. McConnell called to inform us Alyssa had been admitted to the hospital. My heart sank as I hung up the phone and went upstairs to tell Krista. She was distraught. Comforting her, my thoughts switched to Katie. Would we be dealing with the same issues down the road? Was it possible to fix a broken person?

Out of nowhere came an old nursery rhyme:

> Humpty Dumpty sat on a wall
> Humpty Dumpty had a great fall
> All the King's horses and
> All the King's men
> Could not put Humpty together again.

CHAPTER 31

Krista

Extreme queasiness enveloped me as soon as we stepped into the hospital and headed to the elevators. I hadn't been there since my jaw surgery, and then I was too nervous to take in my surroundings. We headed up to the floor to see Alyssa; along the way, I was overwhelmed by the soured smell of sickness and loud wails of pain coming from patients. *Or was it delirium and fear I sensed?* It was all very eerie. Mom had insisted Katie come with us, and Chelsea was at a neighbor's house. Even though this was a Saturday morning, Dad was still out of town.

When the elevator doors opened, we sought out the nearest nurse's station to find out Alyssa's location. We talked to her mother before coming, but she thought Alyssa might be moved to another room. Just as the desk was in sight, we ran into Mrs. McConnell.

I desperately needed to see my friend and see her alone. I wondered what led to this and how things were going with her father. I knew I couldn't get any information with her mother right there. Alyssa's condition and her difficulties with her dad were certainly related matters. My best guess—he was in combat mood now and ready to declare war.

We headed to Alyssa's room. I knocked on the door and heard her faint voice beckon us. The door felt like a hundred tons when I tried to push it open. When we entered, I noticed Alyssa had several tubes: one running to her arm from the IV, some going into her nose, and several more darting under the blankets. Some were familiar to me from my own recent surgery and hospital stay. She resembled a skeleton, especially without her deceptively baggy clothing. I fought the urge to rush over and hug her. *Could such a simple show of affection cause her pain?* I feared bones might break at the slightest touch. Wanting her to know I cared, I stepped closer, pulled a chair next to the bed, and gently took her hand.

"How are you doing?" Mom asked and took her other hand. We were flanking her bed.

"Feeling a little better," she said, painfully slow. It seemed to take every ounce of her energy just to speak.

"I'm sorry you're sick," Katie said in a barely detectable whisper. I had forgotten she followed me into the room with Mom.

I turned to Katie. She was so pale I feared she might puke. It was too much for her. Within minutes, Katie abruptly headed out the door. After a quick apology to Alyssa, Mom followed her.

"Is Katie okay?" Alyssa queried.

"I think she's scared," I said, letting the words linger. I stood up and gingerly put my arms around her. All I felt were bones. I lessened the hug. "I'm scared for you, too."

I loved her like a sister. I had never felt so connected to someone in such a short period of time, and I knew it wasn't fair to her not to be honest. But I didn't know how. It also wasn't fair for me to have to watch my friend doing this to herself.

"Alyssa, what do you need? Is there anything I can do?" I asked. I wanted her to give me a detailed list so I could run out and do what she needed to make everything better.

Her response came in the form of sobs and tears. The shudders ransacking her body pained us both. I was terrified she would pass out from the pure force of it.

"My dad is going to kill me!" she said. "Or worse yet, he's going to kill my mother. He screams at her all the time because of this."

I wanted Mom in here, somebody who could offer some real advice—someone who could tell her to focus on getting herself better and not worry about everyone else. Someone to say, "Your dad loves you and just doesn't know how to help you." I couldn't say these things because I couldn't bear to have her turn away from me. But Mom was dealing with Katie outside, and I couldn't run to her for help. As much as I loved Alyssa, the thought of seeing Katie go through this was even more heartbreaking and sickening.

I decided to give advice a try. I still had Alyssa's hand. "Alyssa, your dad loves you. I just know he does." I wasn't entirely sure because it didn't seem love should come with a salute and a heavy hand. And I'd already screwed up judging love with Cindy and her dad. *What the hell did I know!* But I needed *Alyssa* to believe it. "They want you to get better. I want you to get better. I can't survive high school without you there every day." I wanted to scream, "*I need you!*" but it seemed selfish.

She kept crying. I was afraid my attempts had only made her feel worse. Just then, her mom came into the room, and an icy chill came with her that went straight through me. Alyssa stopped crying immediately and gave her mother a death glare. Her mom acted sheepish and lethargic. She seemed to need her own hospital bed. Then, and I swear I could feel his presence even before I saw him, Colonel McConnell walked in. The air in the room got thinner, and I saw what looked like an apology on Alyssa's mom's face, but I wasn't sure if she was silently apologizing to me or to Alyssa.

"Krista, I'm glad you came, but we need to have a word with Alyssa. We saw your mother and told her you'd be heading out. Thank you," he said. I fumbled to push back my chair and squeeze out a quick hug and good-bye to Alyssa. I didn't want to leave her, especially with them, but I lacked the courage to go against a commanding officer's orders. And clearly that's what his words were.

I stepped outside the room to see Katie lying in Mom's lap. Mom steadily ran her hand from Katie's shoulder to her waist. I wanted to leave them, head straight to the nearest elevator, and find my own way home. I could only think of two viable options: run away as fast as I could or start screaming and see who might listen. Chelsea was lucky to be blissfully unaware at her friend's house. And Dad was always blissfully unaware.

I stood quietly by Mom until she noticed me. "Hey, honey," she said when she noticed me. "Why don't we head home?"

Returning to the car was laboriously slow. An older lady parked in the hall in a wheelchair was moaning, "Mama. Hungry. Mama. Hungry." She said the mantra over and over. As we walked by, she grabbed Mom's arm and gave her a pleading look, "Do you know my mama? I'm hungry." Katie and I clung to each other, and Mom softly told the lady she would find someone to help. The whole time she spoke to the woman she caressed her hand, and the woman seemed to calm down. An aggravated nurse finally came over.

"Stella, now I done already told you your Mama ain't here, and it ain't quite lunchtime, so let's just get you back to your room," the nurse said and jerked the woman's wheelchair down the hall.

Poor Stella, poor Alyssa, poor anyone that had to be in this place, I thought. Realizing Alyssa and Katie had a choice angered me.

No words were said until we picked up Chelsea and her normal little kid chitter-chatter filled the car. She and her new best friend, Christy, had drawn pictures for each other. Chelsea couldn't wait to proudly display them on our refrigerator.

Chelsea was first through the door. She ran straight to find a magnet for her coveted piece of artwork. Mom went to her desk and started listening to phone messages while Katie slinked upstairs to her room. I hung back hoping to get a few seconds with Mom. Maybe Mom could start with a hug followed by some discussion and advice. *Please, I don't know where else to turn.*

Mom's back was to me, but I could sense the tension building. I noticed one hand gripping the phone while her other one wrote down messages. There was a succession of pushed buttons followed by a muttered, "Of course."

"It looks like your father is delayed and won't be getting in until late," she said without even turning around. "Krista, can you see if we have Hamburger Helper or something in the cabinet?"

"Katie's not going to eat, anyway," I mumbled.

"I'm just sick to death of what everybody wants to eat and not eat," she said, whirling around.

"I agree!" I shouted and left the room. She wasn't going to be any help to me today.

Later, I came back downstairs, quickly forced down some dinner, and then excused myself. I wanted to go to bed and start another day, and nobody was fighting it. Tomorrow was church, and I had some praying to do. *Dad would be back, too. Maybe between God and Dad things will start looking up,* I thought.

I'm not sure what time I woke, but it was dark outside. I got up to go to the bathroom, and I could hear voices. Dad was back. I glanced at the clock, but it was blinking twelve. I knew I shouldn't listen, but I was compelled to learn what was going on. It wasn't like they were into holding family meetings lately. I tiptoed closer to see what I could make out.

"Seriously, Bob, you've got to be kidding," Mom said in muffled frustration. "I knew you didn't want me going back to school, but . . ."

"It's bad enough trying to convince the girls that I don't plan these things, Abby. Please don't tell me I have to convince you, too. You know better, so I don't know why you are being so hateful," he said. "It could even be for the best. Things haven't been great lately, and you know it."

I turned away. I couldn't listen anymore. Nobody had to tell me what they were talking about. I had asked for it, begged for it. I guess the old saying was true, "Be careful what you wish for."

Abby

For several days, Mrs. McConnell and I exchanged phone calls. During one of those calls I mentioned, "Mrs. McConnell, I'm embarrassed to ask, but I don't know your first name."

"Dear me, I hadn't noticed. Being in the military everything is so formal. Please, call me Betty."

"Then, Betty it is. Hey, I know you have a lot on your plate right now, but Krista would like to visit Alyssa. That is, when you think the time is right."

"She would love it. Alyssa is progressing rather nicely. Let me see, would Saturday work? Maybe I could run some errands. I've left her so little," she stated.

We hung up, having set a time to visit. Hoping she would be moved from the psychiatric floor, Betty recommended when we got to the hospital to check at the nurse's station to confirm Alyssa's room number.

I made arrangements for Chelsea to play with her friend, Christy, knowing nothing would make her happier. Christy's mom already had plans for a special outing for them, a movie and a stop at Swensen's Ice Cream Parlor.

I decided Katie would go. Maybe the visit would stir up questions or help her see anorexia face-to-face. My tactic, trying to talk to her, reason with her, had fallen on deaf ears. There seemed to be no answers to why she was changing from a sweet little girl to one who hid behind words as effectively as she hid her food under napkins.

When Saturday arrived, Krista could not contain her excitement, "Mom, does this mean Allysa's getting better?"

Katie interrupted, "I'm not going. I don't even like her."

"It doesn't matter; you're going. Right now Alyssa needs encouragement and caring friends."

We arrived at the hospital on time, walking down the hallway to the room number provided by the desk nurse. Betty spotted us first, her hand waving us over, "Alyssa's going to be so surprised. I didn't tell her you were coming, afraid something might happen. She doesn't need disappointment right now."

Betty nudged Krista into the room, and we followed. Alyssa's mane of dark hair fanned across the pillow. When she turned toward us, it was hard to keep from gasping. Katie's fingers dug into my arm, pulling me back into the hallway. I excused myself, letting Betty go ahead.

"What's wrong?" Before she could answer, Betty came by tapping my shoulder, announcing she'd be back shortly. I smiled, returning my gaze on Katie.

"Mama, I can't go in there. She looks scary," she pleaded.

Her alarm was justified. The room's stark fluorescent lighting cast an opaline flush to Alyssa's sunken cheeks and highlighted the fringe of downy hair along her jawline. Her pale green eyes portrayed an ethereal illusion, amplified by thick, dark lashes.

I cupped Katie's chin, lifting her face to meet mine, "I understand, but she needs our support, not rejection. She's very sick. This can happen when you have anorexia."

She trembled. For a moment, I thought about taking her downstairs to protect her, but I had done that long enough.

"Katie? We'll only be in the room a few minutes. Krista wants to spend some alone time with her, okay?"

Katie didn't say a word but led the way back to Alyssa's room. Walking up to her bed, she touched her hand, saying, "I'm sorry you're sick."

Alyssa smiled, softly responding, "Me, too. How's school going?"

Katie struggled, growing pale and shaky. Finally, I offered pleasantries and announced we would wait in the visitor's lounge and give Krista and Alyssa time to catch up.

Once she got past the door, Katie ran. When I caught up with her, she cried, "Mama, she's dying!"

I crushed her to me, pain and fear cycling from her wracking sobs to my breaking heart. Not saying a word, I edged over to a nearby set of chairs, holding her, rubbing warmth into her shaking limbs. Before long, calmness seeped in and enveloped us. I put my head against the wall and closed my eyes, breathing out to expel pent-up tension.

A strong male voice jolted me, "Mrs. Simmons?"

I looked up into the stern set face of Colonel McConnell. He stood stiff-board straight, arms pinned to his sides. His jarring voice felt more like a senior officer rendering a command, rather than a neighbor trying to get my attention.

"Yes, sir?" I said in response.

"I understand your daughter, Krista, is with Alyssa. I will dismiss her for you to take home. Mrs. McConnell and I need some privacy with our daughter."

I stared and gulped, nodding a *yes* to his dictate. That's when I noticed Betty standing behind him. She never looked my way but followed behind him as he turned to leave.

Krista met me in the hallway. We fell into a single line formation, resembling a worn-out, defeated troop. We passed other patients, most dull eyed and hopeless. One, an elderly woman in a wheelchair, grabbed my arm, begging for her mother. I leaned over, offered her a crumb of attention and got back in line to exit the hospital. The drive home wrapped us in unanswerable questions, each crushing under their own weight.

Everyone went separate ways when we got home. Midafternoon, Bob called to say he'd be home shortly. The silence in the house was corollary. I decided to take advantage of it and soak in a hot tub, scan a few magazines, and hopefully drift into a dreamless sleep.

Half-dozing, I heard the bedroom door open. Figuring it was Bob, I called out, "Hey, I'm in the tub. I'll be out in a minute."

"No problem. I'm busted. Take your time."

By now, the misty lull of sleep had ended. I toweled off, applied lotion, and slipped into a nightgown. When I entered the bedroom, Bob offered a slight smile, "I missed you."

I folded the duvet down and slid under the cool sheets. Once comfortable, I answered, "Missed you, too. How was your day?" I grabbed a magazine off the nightstand and began flipping through the pages. When I didn't get a response, I looked up. He had one hand on the door frame, the other in his pocket, pensively studying me.

"What's up?" I asked, patting the bed and sliding over.

Sitting down next to me, he stated, "Abby, I've been transferred." He paused; but when he saw my reaction, he took my hand in his, "I didn't tell you about it earlier. I had hoped to buy some time or get my manager to change his mind, but I couldn't."

I pulled my hand away, trying hard to process what he had said. Another move?

He continued, "I know the timing is terrible. I've struggled with how to handle it. You can't imagine how many scenarios have run through my mind, ways I could avoid this, but none are viable. I'm not willing for us to be separated or do a commute job. I'm sorry."

I sat quietly. For once in my life, I did not go into a tirade. I wanted to, but stopped when I saw the grief-stricken look on his face. "I don't know what to say. Rather than say something wrong, I need to think about what this means, how it's going to affect us." I paused, taking his hand back in mine, "I'm not mad at you, but I'm mad as hell to think our lives will get uprooted again. I don't think I can . . ." I stopped.

Tears fell. There was no controlling their flow or the crushing feeling of loss. Bob took me in his arms. He undressed me for the first time in years, grazing, slow, and seductive. My lips found his, letting hurt and ache rise, wanting nothing more than to forget everything, but this.

The next morning we got up late. Everyone was already downstairs watching television when we stumbled into the den. Chelsea got up and ran to us.

"Krista said we had to be quiet 'cause Daddy came in late. I'm starving!" She said, dragging me toward the kitchen.

I followed. Before my mind went into a million fractured pieces, I needed the normalcy of routine to ground me. "Let's fix something fast. Something you can help with. How about French toast and sausage links?"

Chelsea was already getting eggs out of the refrigerator. Watching her reminded me how much she had grown up this past year. A pang of sadness hit me.

"Can I crack the eggs, Mama?" she asked.

I studied her for a moment, "Do you think you can?"

"Sure. Krista let me crack one the other day when we made brownies. I only got one piece of shell in the batter. Krista didn't even scream at me," she remarked, proudly.

During breakfast, Bob and I kept looking at one another with a knowledge we did not want to voice. Krista acted very curious about Bob's trip, which made me wonder if she had overheard something, but on second thought, realized it had been too late.

Early afternoon, I searched for and found Bob working in the garage. "Hey, I wanted to say . . . we're a family, and we do what a family has to do."

Bob dropped the oil-filled rag he was holding and grabbed me tight. "Thank you," he whispered.

When he released me, I added, "Let's wait and get more details before we tell the girls. I want Katie to have more time with Dr. Easley. Who knows we might have a breakthrough."

Bob nodded, "You're right. My guess is they'll have me at the new job sometime after the first of the year. We'll make it work."

Walking back into the house, a veil of dread draped over me, but I refused to tumble into depression. When Quillie called later that day; her voice was like a ray of sunshine, "Abby? Is that you?"

"Quillie! What a surprise! How're you and Dickie doing? We're still reeling from that beautiful wedding."

There was a long pause. Its lull made me queasy. "Quillie, are you still there?"

"Oh, sweetie," she moaned, "I just don't know how to tell you this. I've got some real, bad news."

Panic wrapped its cellophane hands around my throat; breaths came in raspy, struggling strokes. The floor trembled, and I clutched the counter.

In a soft, anguished voice, she said, "Abby, honey . . . your daddy just died."

Chapter 32

Krista

I first heard of *Scared Straight* in school. Teachers made us watch the videos about kids sent to a mock prison as a hard-love approach to putting their lives back together. The videos were harrowing. It amazed me how violently teenagers held on to the need to control their lives even if it meant their own destruction.

The trip to see Alyssa scared Katie straight, at least for the time being. She had cried for hours in her room once we came home from the hospital. Mom and Dad had barricaded themselves with her for a good session of soul-cleansing crying and loving. The house breathed with a new sense of hope. Katie and I did not talk, and Mom and Dad were not sharing about what happened in specifics; but she started eating more. Thanks to one overheard conversation—the only way I seemed to get any information these days—I knew Katie was also participating more in her doctor's sessions.

If only the hospital stay had impacted Alyssa in the same way. I visited her several times, and each conversation became more and more about her father and what he wanted and what she was absolutely determined not to give him. Mom dropped me off to spend time with her a couple of times, only to have Mrs. McConnell bring me home early because the Colonel had stepped in and wanted me gone. One time, Alyssa begged him to let me stay.

"Dad, please don't make her leave," she pleaded one Saturday morning.

"You are not living up to your end of the bargain, Alyssa," he said. He stood by her bed with his arms crossed. I was standing in the corner. Mrs. McConnell came over to stand beside me. She had a hand on my shoulder for her support as much as my own.

Hell blazed in Alyssa's eyes. Fear gripped my heart. I wanted to yell at her dad, tell him I was helping. She needed me, and I needed her. I wanted Alyssa to get better because I loved her; I really did. My days at school were hollow with her absence. I was functioning, and certainly it was nothing like the Cindy and Kerry days of last year. Still, part of me was always at this hospital, always worried.

"You cannot take everything!" Alyssa screamed. She sat up in bed and pulled at her tubes. One broke free, and alarms went off.

Her father took two dramatic steps—and I swore the room moved with each one—to Alyssa and pinned her down. Alyssa was fighting him with everything she had left. The nurse rushed into the room and gave Alyssa a syringe full of something. Colonel McConnell never eased his grip, and Mrs. McConnell ushered me out of the room.

"I'll wait, please don't take me home," I pleaded as she gathered her purse and pushed me toward the exit.

"Honey, you need to go," she said.

I pulled on her arm to force her to stop for a second. I stepped in front of her and said, "Alyssa needs me."

She hugged me. She put a hand on each of my arms and looked at me. "I know she does. You mean the world to her, and I need you not to give up on her. I don't know what she'd do without you." She paused, tears glistening in her eyes. "It is such a blessing that she has you. Just give this some time. Keep coming, keep doing what you're doing, and know it's helping."

I started to cry, and she never took her touch away. She held me the whole way to the car, and she kept her hand on my leg the whole ride home.

She dropped me off and hurried back to the hospital. When I walked in, Mom could tell I had been crying.

"Krista, honey, are you okay?" she asked, running over. "I would have come to get you. I hate that Betty had to leave to do that."

I told her what happened. It all poured out, the look on Alyssa's face, how her dad restrained her, the tears in Mrs. McConnell's eyes, and in my own. "Mom, I want to help her, and I'm not doing a good job."

"Honey, you are there for her, and I think she needs that as much as anything. Don't give up on her; that would be devastating."

It hit me how selfish I had been lately. One year of stupid girls and I begged Dad to move us. From what I heard the other night, it sounded like I'd get my wish. I let my own weariness with life cloud everything, and the payment would be abandoning Alyssa. She had given my sister back to our family, and I repaid her this way. I was disgusted with myself. I wanted to come right out and ask Mom if the move was really happening, but the phone rang, and she went to get it.

I waited in the living room for her to return, acting scenes out in my head on how to confront her about the move. After a good twenty minutes, I wondered what was taking so long and went looking for her. She was not there. The door to the garage was wide open, so I stepped outside.

Dad was holding Mom. She was not crying, but she was leaning into him for support. Dad saw me and simply shook his head. I turned around and left them.

I walked upstairs to find Katie and Chelsea each in their separate rooms. They followed me to mine.

"How's Alyssa?" Katie asked.

"She's doing better," I lied. If I thought Katie needed encouragement to stay on her path, I would have given her the gory details, but she seemed to be doing well. I would save the story for a day she was not.

"Girls," Mom said from the doorway. I jumped a bit at the sound of her voice. None of us heard her approach. Dad was at her side with his arm around her shoulders. Mom continued, "I just got a call from home," she stopped for a moment and let out a breath, "My daddy died."

We ran to her, clung to her, and enveloped her with our love. Dad joined in, and our family became a mesh of bodies breathing together.

We pulled up to the church in silence. Other than an impressive steeple, it was an ordinary place of worship. Nothing to distinguish it from the ten other churches we passed since coming into town. We filed in quietly and were escorted to the reserved pews in front.

Mom, Dad, Katie, Chelsea, and I all sat behind my grandmother, her best friend Quillie, Mom's sister, Mandy, and Mandy's son, Connor. Uncle Teddy and his family were in a pew behind us. We all exchanged hugs before we sat down. Mom and Teddy had put together a video montage of their father's achievements, and it was playing now. My grandfather had played baseball in the farm leagues and Minors. The Yankees signed him but he threw his pitching arm out before much playtime. He had served in the navy and loved archery and golf. There was not much about him with his kids, but there were several great pictures of him and my grandmother together.

My aunt's husband—I would say my uncle, except this was her third marriage, and he was old and not interested in the title—sang a horrible rendition of "How Great Thou Art." Whoever asked him to sing had clearly not heard him sing before.

My grandmother was crying most of the service, and Quillie's arm completely wrapped her. Connor sat on her other side; and several times, she leaned her head over and laid it on his shoulder.

I was so taken with people's faces and emotions I did not hear much of the actual service. Well, I heard the one song, but no one could have missed that.

Mom didn't shed one tear during the service. I know she loved her dad, so I didn't understand why she seemed in such a dry cloud. I thought of Cindy, and again felt remorse about judging her feelings and relationship with her father solely based on how I thought she should have acted when he died.

When the service concluded, Mom and Dad whispered a minute before Dad said, "Girls, your mom needs to stay with your grandmother for a couple of days to help her through this. We're going to head back home, so that you guys don't miss anymore school." He began leading us out.

Chelsea ran back to Mom and threw her arms around her legs. Katie and I exchanged looks. We didn't want to leave Mom either, but we were too old to get away with throwing ourselves on her.

We watched Mom bend down to one knee, hug Chelsea, and hold on before Dad walked over and took Chelsea away. He turned us toward the car and nudged us in. As he pulled away, we strained our necks looking out the back window, waving until we could no longer see Mom waving back.

We were halfway to our house when Dad broke the silence, "Girls, your mom is very upset. We need to support her right now."

We can't support her if we aren't with her, I screamed in my head. Moving wasn't going to be very supportive of Mom right now, either, but now didn't seem the time to discuss it with Dad.

Abby

Waves of emotion flooded my heart. How could Daddy be dead? He had just played golf and attended Quillie's wedding. I know he had been tired, but death?

A vivid picture of Dad surfaced. Tall, six foot four, handsome, and quiet. He never shared much about himself, always standing behind the scenes, observing. Mama pretty much ran the show, taking care of everything and everybody. But there was an underlying strength that drew you to him when you needed a hug or refuge. Often I would find him and sit near, never really talking, but wanting to draw comfort from his presence. I know Mama did that, too. Sometimes I would find them in the kitchen, and he would be holding her in his arms, quiet and gentlelike. It gave me a feeling of safety as if the world was in his embrace.

I was a junior in high school when I found out Dad had played minor league baseball. I asked him about it, and he shrugged, "Yeah, I guess." Then he walked over to the television and turned up the volume. That was it, discussion over.

So I went to Mama. "Did Dad play baseball after he finished high school?" I asked.

Mama looked up, her face changing to softer tones as she explained, "Your dad never finished high school. He got his grandpa to sign him into the navy at seventeen. A scout for one of the minor leagues saw him pitching somewhere down in South America and spent the next three months getting him out of the service and under contract. He played all over. Once we got married, I traveled with him. That's how come you and Teddy stayed with Mother so much. Life on the road wasn't a place for children."

Mama picked up the hamper and asked me to follow her to her bedroom. I trailed behind, picking up Teddy's toys strewn all over the floor. By the time we got there, my upturned shirttail brimmed over with baseballs, a sword, and two Johnny Reb cannon balls.

Balancing the clothes basket on her hip, she pointed to the closet, "Look in there, up high on the right-hand side and pull down that brown box. It has Albert written on it."

She never looked my way while I struggled to get the large box down from the shelf. When it tipped forward, I caught the box as it hit the bridge of my nose. Between tears and a throbbing headache, I managed to lug it over to Mama's bed.

"What's in here?"

"All the newspaper clippings and flyers about your Daddy when he played ball. I think there's some pictures of us in there, too. Oh, there's probably a bunch of match covers we collected at bars we went to after the games. We used to keep them in a fish bowl." She expelled a blast of air, "That seems like a lifetime ago." Mama's lips turned up into a crooked smile, her eyes glistening.

I spent the next hour going through one scrapbook after another, learning more about Dad than he would have ever wanted. What impressed me the most were the pictures of him in his uniform. He looked so assured and handsome. I only had preconceived notions about him. It was as if he began the moment I did. For me, these precious things fleshed him out, making him more real than the man I passed daily in the hallway.

I pivoted on a downward spiral, trying to escape from the reels of pictures projected on the back wall of my mind. Right now, I had to shut out the pain and concentrate on the things that needed to be done. But first, I needed to see Bob.

He was right where I had left him, stooped over the fender of the car, peering at the engine. Before I got to him, he heard me, greeting me with his brilliant smile. Seeing my troubled face, he was by my side in two strides, "Abby, what's wrong?"

I leaned into him, "Just hold me." Time stretched into a tight band before wavering into a broken prism of excruciating pain. A numbing took up residence, amortizing speech and movement.

Bob kneaded my back with his hands, pressing in tighter, releasing pain to puddle at his touch. "Abby, please tell me what's going on." Gently, he pushed me away and lifted my chin, "Is something wrong with you or the girls?" he asked.

"Daddy's dead."

Disbelief washed his face, crumbling his features, "No!"

Bob stared deeper into my eyes until the truth ripped off the scab of perceived reality, sealing grief in the lines around his eyes, "I can't believe it. I thought Al was doing better."

Clearing my throat to push the sound forward, I stated, "I thought he was too, but Quillie just called." With a sudden shock of revelation, a scream burst forth, "Oh my God, Mama needs me.

I struggled to get out of his hold, but Bob yanked my arm to get me to listen, "Wait, there's nothing you can do right now. Let's go in and talk. We have to tell the girls and make arrangements. I'll call and get my schedule cleared for a few days."

"No! I've got to talk to Mama first. I have to hear her voice and know she's all right. Surely Quillie will stay with her until I get there. Maybe Mandy's there or Teddy . . ."

Bob nodded, "You're right. Let's call Nadine first, and then we'll manage the rest."

I dialed the house three times before Quillie picked up. She tried to allay my fears about Mama, telling me people were pouring in with offers to help, giving hugs and expressing words of sympathy. Mama would not be alone; Quillie would make sure of it.

When she put Mama on the phone, her pain crushed me, "Abby, Quillie's here. She won't leave, but I need you to come home." Her voice filled with muffled tears, "Your daddy's gone. My sweet Al is gone."

Quillie talked to me for several minutes after Mama dropped the phone. By the time I hung up, the shaking had ceased, and I was ready to formulate a task plan for my absence.

Together we told the girls about Dad. Each of them mourned in their own right, whether for him or for me, but their hugs and tears started a healing. Afterward, Bob and I spent a couple of hours talking about the next few days and the things he would have to coordinate. We decided I'd go home the next morning, and Bob and the girls would follow once the final funeral arrangements were made.

When I arrived at Mama's house, cars were parked two blocks around it. Fortuitously, the car blocking the driveway pulled away as I drove up, giving me the opportunity to drive right up to the back door. The house was packed. Tobaccie spotted me from the cookstove, laid down the spoon, and rushed to envelope me in her open arms. Sweet aroma of cinnamon and cloves filled the air around her.

Thoughts of planting myself in her bosom faded when I heard voices in the next room, "Where's Mama?"

Tobaccie motioned toward the living room. The doorway was jammed with gray suits and flowered dresses. Finally, I said "excuse me" enough to get to where Mama was sitting. When she saw me, she tried to get up, but folded back into the chair with a thud.

Panicking, I rushed over, "Mama, are you okay?"

With a squeaky voice, she answered, "I've been drugged, Abby." Her voice, slurring, thick. "They took my Al, and now they're trying to take me," she said, alarmed, her eyes wide with disbelief.

Quillie came up behind me, wrapping her thick arms around my body, "She'll be fine. I'm gonna get her to bed now, and shoo this mess of folks outta here. The arrangements are made; the funeral will be the day after tomorrow."

An incredulous look crossed my face before I could confine it, but Quillie immediately squelched it by saying, "Al and Nadine made all the preparations when Al had his heart attack. Al didn't want Nadine to worry about anything."

Quillie was a drill sergeant, directing people out the front and back door, even Tobaccie, spoon and all. She picked Mama up out of the chair and tenderly carried her to the bedroom, undressing her and putting her in her pajamas. Quillie even took the Pond's cold cream and washed her face, the same way Mama did every morning and night. I stood there, watching her croon and hum, her cadence a soothing balm.

Next, she turned her attention to me, "Go to bed. I'm gonna go check on my man, but I'll be back tomorrow. There's enough food in the fridge to feed an army, so don't cook." She reached over and smeared a kiss on my face, then left.

I talked to Bob. We decided they would come right before the funeral. I thought it would be easier for all of them, and Bob agreed.

The next day, Mama barely managed without a steady supply of pills from the brown bottles lining the kitchen window sill. Quillie doled out the prescribed dosages while I watched, hoping she would spill one container down my throat to ease the rawness.

When we arrived at Biggs Funeral Home, the clouds were pigeon feather marbled, veins running white to gray. Lanky mums in scant bushes peeked an occasional yellow. Next to them, flanking both sides of the walkway, were scattered cigarette butts and empty Mountain Dew cans. When we opened the double doors of the funeral home, I heard a car engine rev up at Roberson's Service Station. Two boys were laughing and squirting each other with the water hose, pretending to wash and shine a beat up Ford Fairlane. The world ticked by as sure as the Roanoke River flowed and Shaw's Barbeque smoked pork. But I was in another world, a world of permafrost, tundra bound, moving only through conscious commands.

Mama walked ahead while I held the door open for her. We were greeted by the same man in the same navy suit who had attended us for all our family's deaths. He never spoke, but took Mama's hand and led her down the hallway into a large room. Dad's casket stood in place. Flowers in a multitude of colors, shapes, and sizes rowed the walls, depicting an outpouring for a life having touched lives.

Quillie put her hand on my shoulder, "Sorry I'm late. Dang hose wouldn't stay up, so I ditched them for socks. You reckon people will notice?" she said, hiking up her long striped skirt, exposing folded white socks stuffed into loafers.

I shook my head at the sight but calmly replied, "You're fine. No one's going to be looking at your feet. I think they'll be more interested in your bright orange shirt." The comment slipped out before I could restrain it.

Quillie opened her arms wide and proudly announced, "Your daddy made fun of this shirt at last year's church bazaar. I'm wearing it to taunt him. You know he's shaking his head right now, laughing at me, don't you?" she said, rubbing my back.

From her seat near the casket, Mama raised her head to look at us. I had given her the two pills Quillie had laid in my hand the night before, and her eyes were slowly shifting right to left, unable to focus.

"Hey, girls, I need to go pee," she loudly broadcast.

Quillie reached over, pulling her to her feet, "You're a troublesome woman, Nadine, whining constantly. What am I going to do with you?"

Mama laughed, continuing a scribbled conversation no one could understand the whole way down the hall. When I turned around, the man in the navy suit was standing behind me, his arms primly angled behind his back.

"Is the room acceptable?" he asked, his face serene and tender.

"Oh, yes, I think so." I glanced around nervously, "Thank you for everything."

"You are the only clients here tonight. You may use any of the opened rooms for family or friends. If you need anything, I will be in my office at the end of the hallway," he said, rotating on his heels to leave.

Before I could respond, I saw Teddy and Samantha walk in with their two boys, Tyler and Matthew. Mandy and Connor were right behind them. While we were hugging, the door opened to a line of well-wishers. Mama and Quillie came back into the room, and we got into the customary receiving line. Quillie dragged a chair over next to

Mama and the next two hours we shook hands, closed eyes for offerings of prayer and listened to memory-march stories about Daddy.

Every few minutes, I would glance at Mama. She shook hands and nodded, barely able to sit straight or focus on words of sympathy and sorrow. Often she would spread a goofy smile at someone who was overflowing with emotion. Many times I heard people walk by, exclaiming, "Bless her heart."

I wanted this to be over. I lowered my head to rub my aching temples, and when I looked up I saw her. She was staring at me. Our eyes locked, for an instant, and I saw a ripple of hate, but it quickly slid off her face. I was struggling to put face and name together, then I saw Drake. The woman was his wife, Dora.

Aging had not been kind to her. Deep wrinkled furors lined her face, running in vertical valleys. Her wispy brown hair was as disheveled as she appeared to be. She wore a brown polyester shirtwaist, dress far too large and too long, giving her the appearance of a child in dress-up clothes. My heart ached for her.

Drake was talking to the two men in front of him. It took everything I could muster to listen to the next few people who took my hand and offered words I did not hear. My pulse heightened; my hands grew cold and damp. As he drew near, I wiped them down the side of my skirt, reaching up to take his.

Our eyes met. On impulse, I started to reach up and hug him tight, but his eyes held me at bay, at a decent distance. Realizing my near indiscretion, I stiffened, saying, "Thank you so much for coming."

Without hesitation, I swung my attention to Dora, "Thank you for coming, too. It's been a long time since I've seen you. How are the kids?"

Struggle was not a new word for me, but this emotion was. She lifted her eyes, first to Drake and then to me, "We're well, thank you," came a gauged offering.

"Abby," Drake answered, "we're real sorry for your loss. Al was a good man. I know you'll miss him. If Nadine needs anything, let us know."

He turned to Teddy with more condolences. He never looked back, not one glance. I grappled with control and emotions I had no right to feel.

Dora took my hands in hers and pulled me close, whispering, "Can I talk to you? I'll wait in the room on the other side of the office. Please?" She straightened, slid her hands out of mine and into my brother's, greeting him with a somber voice.

Hands reached out, voices proclaimed sadness, but I heard none. I had to get some fresh air. Leaning over, I told Teddy I had to take a break. I slipped past him and edged through the crowd down the hallway to meet Dora.

She was sitting in a chair, her feet swinging, barely clearing the floor. I walked up to her, staring, waiting for a response. When she stood up, I noticed she only came to shoulder height.

"I feel brazen asking you to talk to me like this, but I had to," she said.

Shifting, she continued, "I know about you and Drake meeting at R&C. Ida Louise told me, but I'm not mad or anything. I know Drake still loves you. Probably loves you more than me, but he's faithful. I know that."

I opened my mouth to speak, but she kept going, "Me and Drake are tight, you know what I mean? If it weren't for him, I would probably be out in the street. He married me when Fred Rogers got me pregnant. I weren't but fifteen. Drake hit him in the nose when he caught him beating me up behind the Tastee Freeze one night. Drake's been there for me and my youngins ever since. He's protected and loved us like a fierce lion."

I stepped back not wanting to feel so close to her, smell her floral perfume, or look at her watery, blue eyes. Eyes that appeared innocent, until you looked closer, realizing her intentions were not so noble, "I'm sorry about the R&C. It was an accidental meeting."

Dora's face held no emotion, "Don't matter none, truly. I just wanted you to know that I'm not mad. I've always been lookin' for the kinda hope tucked in other people's back pockets, but it hain't worked. Drake put that kinda hope in my heart when he talked to me. He say love has restraints in some ways and is deeper than an ocean in others. The kids and me, well, he say we're the ocean. That was a mouthful, you know?" She reached over and put her arms around my waist and hugged me. "You's pretty," she said, releasing me and walked away.

My knees were so weak I had to will my feet to move to the nearest chair to sit down. I trembled, feeling as if I were dangling over a bottomless chasm. Tears fell in unwanted streams, causing my shoulders to round, my chest to heave. Pulling on pain like a worn-out sweater, I released, crying until spent.

That's the way Quillie found me. She did not say a word but pulled me in tight, encouraging me to let it all go. The next thing I knew, Dickie had his arms around me, ushering me to his car. We did not speak driving back to Mama's. Looking out the window, I realized the landscape looked the same; I was the only one altered.

Bob arrived minutes before we were scheduled to leave for the church. Unlike him, he looked mismanaged and out of sorts. When I reached over to hug him, he could not contain himself, "Abby, I know you don't believe this, but I got up at the crack of dawn, and should have been here way before now, but . . . Christ, I almost lost it! Not one of those girls offered any assistance, dragging their feet at every suggestion. I'm totally wiped out."

I put my finger to his mouth, "Shhh, it's all right. You're here, that's all that matters." I hugged him again, feeling his body soften, "I've missed you." I reached over and gave each girl a firm hug, "I really missed you, too."

Quillie burst in, announcing the limo was waiting. She rushed by, heading toward Mama's room, but LeRoux and Salem met her halfway with Mama firmly held between them.

I reached out to touch her arm but could see she was still reeling from the morning pills. Quillie offered, "She'll be better in a little while. We aren't going to do anymore pills unless we have to. By the time we get through the service, she'll be better."

Updating Bob, I told him, "I'm not going in the limo with Mom and the *Aunties*. I'll follow you in Mama's car. We'll need a way home after the service. I've decided to stay a few days or at least till I know she can manage. Is this going to be okay?" Bob gulped before agreeing.

The service blurred; I kept my eyes pinned on Mama. As the pills wore off, grief strafed her heart, causing her to cry out many times. With each cry, a stabbing ache wracked me, irregular and unstoppable, until I sat motionless, willing my mind into a void.

When the service ended, Bob took my arm and guided me on the short walk over to the gravesite. Connor walked close to Mama, his hand cupping her arm for support. Quillie stayed close behind the family with Dickie at her side. The girls remained beside me, touching or reaching for my hand.

Approaching the tent placed over the open grave area, I heard Mama scream. I ran, grabbing her as she toppled to the ground, crying. I held her, rocking, not stopping until heaven's liquid exhausted itself.

CHAPTER 33

Krista

Our house without Mom was incomplete—a structure of wood and plaster without a heart. There was no warmth or joy; we just went through the motions. With each passing day, the air became dense and stale. Dad tried his jokes, and we would gurgle out a pathetic attempt at laughter, but we needed Mom back. Pure and simple.

Melancholy settled over me. I found myself sitting in my room and playing Karen Carpenter over and over again. The words were constantly echoing in my head.

Love, look at the two of us
Strangers in many ways
We've got a lifetime to share
So much to say
And as we go
From day to day
I'll feel you close to me
But time alone will tell
Let's take a lifetime to say
I knew you well.

When Karen Carpenter sang, I wanted to cry—her soul seemed to travel through the depth of her voice. I knew I should turn on a fun song like "Footloose" and attempt to bring happiness back to my corner of the world. I couldn't. I needed the sadness because I could feel tears, but they would not come. I hated these moments, being a teenager and a female. The last few weeks encased me in moodiness. My brain was still functioning because my grades were strong, and my schoolwork was as easy as ever, but it could not force logic and calmness to overpower the fog.

"Krista, phone!" Dad yelled. It was our first weekend at home since we left Mom in Williamston. I walked into my parent's room to pick up the phone, and I could smell Mom's perfume. I wanted to go into her bathroom, find the perfume, and spray it on myself to see if it would provide some comfort.

When I answered, Mrs. McConnell's heavy voice said, "Hi, Krista. How's your mom?"

"She's having a hard time. She stayed behind in Williamston, and I'm not sure when she'll be back," I started. I knew Mrs. McConnell was asking as a formality, but she sounded miles away. "How's Alyssa?" I asked.

"She'd love to see you," she said. "Do you think your dad would mind if I come get you so you can visit with her?"

"I'm sure he won't, but let me go ask," I said. I put the receiver gently down and ran downstairs to find Dad.

"Hey," I said when I found him in the garage. I also noticed Katie and Chelsea playing in the backyard. "Can Mrs. McConnell come get me today so I can go see Alyssa?" I asked.

He looked weary and continued fiddling with things on his workbench. He did not look up but said, "Sure, just be back in time for dinner. If I'm not here when you get back, I've just run to get some food."

"Don't worry, I won't be gone long, Dad," I threw over my shoulder, heading back inside, "Mr. McConnell won't let me stay."

I picked up the phone in the kitchen to avoid a run up the stairs, "Mrs. McConnell, Dad says it's fine. When do you want to come get me?" We went over the details, and she pulled into my driveway within twenty minutes. I had spent the time hanging out with Katie and Chelsea in our clubhouse.

Katie noticed Mrs. McConnell's car, "Krista, are you going to see Alyssa?"

"Yes, do you want to come? I'm sure Dad won't mind," I said. I genuinely wanted her company.

"No," she said. "Well, you know, Dad needs me to help with Chelsea," she added. "But, I do hope she gets better. Can you tell her I said that?"

I nodded and then waved to Dad before hurrying to the car. I stopped dead in my tracks when I noticed Colonel McConnell in the front seat. I looked nervously back to Dad; and for once, he sensed my worry and walked over to the car.

"Colonel McConnell, how are you, sir?" Dad asked and extended his hand through the car window. They shook hands and exchanged a few pleasantries.

"Mr. Simmons, you have quite a remarkable young lady here, I hope you know that," Colonel McConnell said as I got in the backseat. "She has been a great friend to our daughter."

"I'm glad to hear that," Dad said. Even he appeared intimidated by the colonel's presence and booming voice.

"We were sorry to hear about your loss," Mrs. McConnell said over her husband.

A few more pleasantries and we were on our way. What was normally a fifteen-minute drive seemed an eternity. Between the silence and my own beating heart, I nearly jumped out of my seat when the colonel spoke.

"Krista, Alyssa's Mom, and I have a favor to ask you," he said as he turned into the hospital. He pulled the car to a stop in the parking garage and then turned to face me, "Alyssa is getting better, and we think they will release her soon. Her mother and I cannot be with her at school. Since you are in most of her classes, we need you to let us know if she is throwing away her lunch food, not eating at lunch, or other concerns. Can you do that?"

I was terrified, caught off-guard and completely insulted all at the same time. I was not going to be some kind of spy on my friend, and I could not be her babysitter either.

"Clearly you are concerned, Krista, and I understand. But a unit is only as strong as each team member. We would ask Jack, but he has such a different schedule."

I realized why Alyssa fought against him so hard—I was not a member of his beloved military, and I was not blindly following him into any kind of war zone. "Colonel," I said. I wanted to choose my words carefully. My hands were trembling as I continued, "I truly want to help Alyssa, but she's smart, and she would figure out if I was watching her and acting strangely. I care about her too much to do that. I want her to get better, truly I do, but I want her to know I'm her friend, and I'm there for her."

He stared at me before his booming voice said, "Young lady, if you are not part of the solution, you are part of the problem." Mrs. McConnell put her hand on his shoulder, and he shrugged it away.

"We understand," she said, "we know you care about Alyssa. We just want to help her." She opened her door and got out of the car, so I followed suit. The colonel was not following.

"Is he coming?" I asked.

"He'll be in soon," she hesitated for a moment. "Please don't tell Alyssa about our request. I'm afraid she'll get upset."

"Don't worry Mrs. McConnell, I won't," I said. In my book, the sooner this whole awkward situation was over the better.

When we walked into Alyssa's room, I saw she had a tray of food in front of her. It actually looked good. Roast beef with mashed potatoes and vegetables. My stomach started to growl. It was the first time I had seen real food in the same room as her since she came to the hospital.

Some color had come back to her skin, and she smiled with a slight sparkle in her eyes when I came in.

"Good news, Mom and Krista!" Alyssa said. "They said I can get out of here next week." She turned to me, and I noticed she was almost shaking with enthusiasm, "I can't wait to get back to school. I've actually kept up with my work, and I just want to hang out and feel normal again."

My greatest hope was this enthusiasm would change her behavior. It was hard to imagine she would ever want to end up here again. The nurse and her mom stepped out, presumably to discuss more details about Alyssa's possible parole. From the conversation earlier, I knew this would indeed only be a parole. There were too many conditions on her release. I was afraid—if Alyssa sensed any interference by her father, she would never really try to get better. I wish there was some way to tell him to try some reverse psychology with her—a term I picked up from the help-me-help-my-teenager classes Mom used to take.

The nurse snuck me in an order of fries, and Alyssa and I chatted like old times while watching reruns of *Happy Days* on her small hospital television.

"Krista," the booming voice of the colonel ended my joyful moment with Alyssa, "Betty will take you home now." He gave me a quick pat on the back, something he had never done, and then he looked me dead in the eyes. The warning to keep our conversation private was unmistakable.

After a hug for Alyssa and a very silent and quick ride home, I found myself in an empty house. Dad had left a note about heading out to pick up dinner. I sat down on the couch and was about to wield the remote when the phone rang.

"Hello," I said exhausted.

"Krista, sweetie, is that you?" Mom asked. I thought I was exhausted, but Mom sounded completely wasted.

"Oh, Mom, I'm so glad it's you," I said. I was about to pour out everything about Alyssa, but I stopped myself. Mom needed me to be strong right now, and I did not want to add to her burdens. "We miss you, Mama," I said simply.

"I miss you girls so much," she said and then excused herself for a moment before continuing, "Honey, I do not have long to chat. Your grandmother is having a hard time. Is your dad home?"

"No, I just got back from seeing Alyssa, and there was a note that he would be home soon with dinner," I said.

"Please have him call me, and please continue to be my big girl and help him out, okay?" she asked.

"Of course, Mom. When will you be home?" I could not help asking. I understood my grandmother needed her, but nobody deserved her like we did.

"Soon, I promise. Give my love to your sisters. Bye," Mom said and she was gone.

Within minutes, I heard the door open. Dad walked in with several white take-out bags, followed closely by Katie and Chelsea. We sat down to eat, but with little enthusiasm. We were starving, but for Mom's cooking and Mom's presence. Fast food was quickly becoming unappetizing. It was like having my jaw wired all over again. At first, the idea of junk every night was appealing, but it quickly lost its luster.

After dinner and a quick returned call to Mom, Dad sent Katie upstairs to give Chelsea a bath. Katie was thrilled because this was usually a job reserved for me. I knew something was up.

"Krista, I think I should tell you something," Dad said.

Crap, I thought, *please don't let this be about the move.* With everything going on, I hoped things had changed. "What is it, Dad?"

"Your mom and I wanted to tell you girls together, but with everything going on, well, I just thought it would be better if I broke the news. When she gets home, there will be a lot to do," Dad said. I sat motionless and refused to help him out by breaking the silence, so he continued, "I know we have not been here that long, but a great opportunity has come up in Madisonville, Kentucky. I will be the plant manager there."

I could not find my quick-tongue or my angry rebuttals. I wanted to rail against him about so many things, but I knew I had asked for this. I also knew once he told us it was a done deal, inevitable.

Abby

I put the car in park and turned off the ignition. Threads of tepid rain stitched an eyelet fringed gray sky to the earth, matching our mood. Neither Mama nor I moved. Solitude filled the car and gave birth to connected breath tones. We chose to forfeit words, welcoming the reprieve to focus on life outside rather than inside.

Mama made the first move, climbing out of the car, still clinging to the Peace Lily she had confiscated from the gravesite. With effort, she struggled to her feet, "Abby, guess we'd better go on in."

I grabbed a bag out of the trunk and got in step with Mama. We marched up the hill to the house. A burgeoning weight filled my soul, one of Death's many offerings.

Mama opened the front door to a veiled depth of blackness. Running her hand along the interior wall, she found the light switch and flipped it on, peeling darkness into thin shadow ribbons on the ceiling. I stood behind her, propping the door open, while she struggled to get the plant inside.

"Abby, you know you don't have to stay," she said, out of breath.

Mama wrestled with the plant, knocking off two large elongated white blooms. When she tried to rearrange her grip on the pot, she dropped it. Dirt hit the storm door threshold and bounced up onto our pantyhose.

Her voice cracked, "Now look what I've gone and done."

"Mama, please, why don't you go get on something comfortable? I'll clean up this mess and make sure the plant survives," I offered. "And yes, I am staying."

She didn't say a word as she walked toward her bedroom. Her shoulders were hunched over, her gait slow and shaky. The blue suit she had chosen for the funeral was a hand-me-down from Salem. The sleeves had to be rolled up, and the waistband turned over twice to be presentable. Right now, one sleeve was hanging below her wrist, the other shoved up above her elbow. The hem, hiked up on one side, made her appear tipsy and off-balanced. That's when I noticed her hair had not been combed in the back. Ridge lines created from tossing and turning were still there.

Instantly, an image of Dad sprung to memory. He had been standing at the front door waiting for Mama to go to the bank with him. When she came into view, his face went into a full-blown scowl, "Woman? You don't mean to leave this house looking like that, do you?"

Mama didn't say a word. She grabbed her bag off the end table and huffed, shaking her head back and forth. Her high step toward the door was in open defiance. He held the door wide open for her, exclaiming, "Go right on, Ms. Nadine. Ain't nobody gonna be lookin' but me. I've seen worse!" He left the house laughing so hard he almost missed the second step down. Ted and I stood with our mouths gapped, wondering if she would slug him with her bag. When she didn't, we exhaled and went on about our business.

Yep, I bet he was laughing now, too. I thought. *But then again, maybe he was crying. I hoped one day I'd get to ask him.*

Looking down, I saw we had dropped everything on the stoop. I pulled my overnight bag and pocketbook over the plant and wedged them beside the foundling chair. Mama said she found the chair in 1946 in a trash heap behind some minister's house. All my life, we had called it this. Dad offered many times to buy her a new chair or have it reupholstered. She refused. We begged her to throw it away. She refused. After years of coaxing, we gave up. That's when we started piling things around it, never on it. It was a filthy eye sore she also refused to hide away in other rooms.

It took about thirty minutes to clean things up, settle in, and make a pot of coffee. Mama still had not come out of the bedroom, so I began rummaging through the kitchen cabinets for ingredients to fix something sweet.

My eyes went straight to the marshmallows; the peanut butter jar stood right beside them. With one quick swing on the refrigerator door, I found the saltine crackers. Mama always said crackers stayed crisper in the fridge. Now I had the perfect treat to soothe our souls, Depression Cookies. Come to think about it, they had graced many tables from my grandmother's to Mama's and mine. Many hands had held them over discussion, laughter and tears. Remembering made my stomach toss in unsettling flips. I rubbed my eyes, drowning out questions both Mama and I had to face.

I put the cookies together on a baking sheet and turned on the oven. Still waiting, I busied myself, memory walking around the living room.

Mom and Dad's house was a typical 1940 mill-constructed, clapboard home with four rooms plus a bathroom. We were lucky the bathroom was on the inside and not on the back porch or outside. The living room, dining room, and kitchen encompassed a fourteen-by-fourteen foot area of knotty pine paneling. For the first time in my life, it felt like a dungeon, claustrophobic. *How in the world did we all live in here,* I pondered.

Bypassing Dad's chair, I went and sat down on the edge of the couch. The steady ticking of the anniversary clock was unbearable. It took all my reserve to sit there and not get up and run. I started to call Bob but realized they were still journeying back home from the funeral. I crossed one leg over the other, undid them, and recrossed in the other direction, pumping my foot up and down. When that didn't work, I anchored both feet to the floor. Before I knew it my legs were piston bobbing, even faster, doing the Saint Vitus' Dance, again.

Putting more pressure on my legs, I closed my eyes. Images of the day flashed by in rapid succession: the funeral, the people filing by, the kind words and silent nods, faces I knew and didn't, and Mama's loud cry of "No' when they lowered Dad into the grave. It was stifling. More disconcerting was the fact that I hadn't cried. I felt numb, like a thousand wasps had stung me. I could see everything, but could not move or react. Bob kept his arm around my shoulder ready to catch me if I staggered. He took my staring and pallor as grief. It was none of those things. The reality of it was disturbing. It made me feel something was terribly wrong with me.

When I finally exhaled and opened my eyes, Mama was standing in front of me. We stared for several seconds, two women not knowing where to begin.

Mama finally broke the stillness, "Do I smell coffee?"

Jumping up, I rushed past her, conscious to give a wide berth and not brush up against her. "Yeah, I started a pot a few minutes ago. I put some cookies together, too. I'll slide those in the oven; I wanted them to be warm." Moving in overdrive from one end of the kitchen to the other, I talked so fast I could barely keep up with myself. "And I got the plant back together. I put it over by the heater. Figured it'd be okay, for now, since we aren't running the heat," I rambled on and on, filled with excess nervous energy. Realizing Mama had not followed me into the kitchen, I stopped.

"Mama?" folded out of my mouth in quiet unsettledness.

Pivoting I saw her, small and helpless, wispy as a broomstick, looking all broken and unsure. Fear gripped her face, and she shivered. A soft pant escaped her parted lips, and I wanted to rush in, draw her close, and protect her, but an edge of resistance stopped me. There she was, standing in front of me in those bright yellow silk pajamas, the ones

Daddy loved, and I could not find my way to her. Could not will my legs to move; so, we stood staring.

She finally broke the tension, words spoken so soft I barely heard them, "Think I'll have that cup of coffee now."

She walked over, sat down in the nearest chair, and locked her eyes on her hands, lying limp in her lap. Diverting my stare, I rushed to get two cups, saucers, and dessert plates. I poured coffee in the cups and grabbed the evaporated milk out of the refrigerator, adding it to each. I slid the hot steamy liquid toward her. Then I took the cookies out of the oven, placing two on each plate and offered her one.

She reached out to take it and spied the cookies, "This is fitting, isn't it?

"It seems to be the only consistency in our life, doesn't it?"

We sat sipping our coffee and crunching on cookies, refusing to let our eyes meet. When I finished my cup, I got up to refill both. She held out hers in midair, a normal reaction to a normal action. But today was not normal. Death would never feel a part of this world. It was uncomfortable to even say the word *death*. Our society had come to the point of making up substitution words for it, refusing to face its certainty while begging for its final promise. Euphemisms such as "he passed on," "he met his end," or "he perished" did not make death easier to digest. The one saying totally beyond my understanding was—he bought the farm. How many times did I hear that one today? I wanted to scream: Death is final! No one was coming back to the farm or from the journey. It was the end of all ends. Then again, who shouts such things?

Mama got up and put her used dishes in the sink. "Abby, when you finish your coffee, let's talk. I'm going to go put my feet up." Without looking at me, she left. I heard her sit down in the chair, expelling a deep, long sigh. Its weight made me heave and sink deeper into helplessness.

Cleaning the kitchen gave me another break from the evening ahead, from the talk. I didn't know what to expect; and until this very moment, I realized I could not remember the last time I had spent solitary time with her, and never in grief. This fact startled me.

Becoming more uneasy, I rushed past her announcing, "I'm going to put on something comfortable. I'll be right back."

I went into the bathroom and dropped the toilet lid to sit down. Lowering my head into my hands, I kept thinking I could not do this. I could not sit in this house another moment. I did not want to engage in morbid conversations about her losing Dad or what she would do now that he was dead. Maybe it would be better to leave Mama to grieve on her own, work things in her own time frame.

Teddy sure hadn't offered to stay. It was embarrassing how quickly he loaded his family up and said good-bye. Bob stood around shifting from one foot to the other. I could tell he was struggling, not wanting to leave and knowing he should. The girls hovered, too. Chelsea cried when Bob finally picked her up and said, "We need to go." The older girls looked wide-eyed, not knowing what to say. I willed my hand to wave good-bye, internally screaming for them to stay, but all I could do was sigh.

Sighing had become the only way I could judge one episodic moment from another. Time appeared to collapse on itself between each inhalation and exhalation, leaving behind only labored breathing to mark its duration.

Putting my hand on the sink, I pulled myself up and looked squarely into the medicine cabinet mirror. I whispered to the ashen image, staring at me, "Abigail Bower Simmons, you will do this. Your mother needs you. You're all she's got, pitiful as you are."

I quickly put on my pajamas and walked back into the living room. Mama's head was drooped, her chin doubled up on her chest. I could hear her breathing, soft and rhythmic. Relief washed through me like a tonic, a soothing reprieve.

Deciding to give her a few minutes to rest, I sat down in the chair across from her. A sharp-edged spring pierced my fanny, causing me to grasp my mouth with both hands to extinguish a yelp. *Damn*, I thought, remembering I hated this chair, too. It was a reject from some aunt who died shortly after Mama got it, and she refused to get rid of it, too. It was horrid. A small winged back chair ringed with a multitude of stains. The only saving grace was the large orange flowers that masked some of the dirt.

One thing about Mama no one could deny—she didn't care if things matched or if they looked clean. As long as it was presentable to her, it worked.

Perusing the room, I noted most things were hand-me-downs from various relatives, friends, yard sales, or roadside throwaways. I did not detect one new thing, except for the pictures. Suddenly everything looked tired and used up.

"Abby?" Mama brought me back to the present.

"Hey, there, I thought you were asleep." My voice stalled, then quick-started. "How about we get you to bed? You've had a long, rough day, and you need to get some rest." I got up, and walked over to her; but she threw her hand up, motioning for me to sit back down.

"Yes, I'm tired, but we have to talk."

Easing myself back into the chair, my left butt cheek elevated to avoid the spring, I asked, "What about?"

Mama looked down, then up, and away. The glass wall emerged and separated us. It was the very one we had used for years. It was patterned to resurrect itself without being commanded. I was afraid to say anything. I knew how she could be; she needed time to process things. It was all I could do not to go back into that Saint Vitus' Dance again. My nerves bristled.

Finally, she looked me in the eye and began, "Abby, I don't know quite how to begin . . . there are things I need to share with you."

I interrupted her, "Mama, don't worry about the next few days. I'll be here. We can tackle the insurance stuff together and go over to the funeral home to make sure everything's finalized."

"Abby," she said, raising her hand, "I need for you to shut up and let me get through this, okay?"

Stunned, I lowered my left cheek without thinking and yelled, "Shit!"

She pointed at me, "There's a loose spring in that chair. Why don't you save yourself and sit in your daddy's chair. It's real comfortable." She motioned for me to move, "I used to sit in it during the day when your dad was at work. It kinda rolls up around you. Always reminded me of your daddy's arms when he knew I was hurt or confused. I bet you never realized that, did you?"

I shook my head no, not wanting to admit seeing them in one another's arms.

"Well, there's a lot of things you don't know. Wait here. I'm going to get something." She got up and walked the short distance to her bedroom.

I sat motionless, my heart rate elevated. Sweat beads lined the bridge of my nose, collecting until they slid, falling onto my pajama top. I grabbed a Kleenex to blot my face, my mind racing. *What was all this mysterious talk about? Shouldn't we be reminiscing about Dad? Talking about all the fun times we had or what a great provider and father he was or make up things to make us feel better?*

Hearing Mama's footsteps stopped the introspection. At once, I was stunned over how negative and angry I felt. Suddenly the house, me in it and a supernatural feeling of wrongness, drove a cold chill up my spine. I shivered, pushing deeper into the chair.

Mama returned, huffing as she settled back into the chair, then she began, "I do not want you to disrupt me while I'm talking, even if you encounter long pauses. I may sound disjointed and rattled at times, but let me go on. In the end, it'll all connect."

That's when I spied a rectangular box in her lap. It was wrapped in an old brown paper grocery bag—A&P stamped on its front. There was no hint as to its contents. Curiosity killed the cat, my grandmother would say, but I held my tongue, honoring her request.

She took the box and laid it on the floor beside her chair. "Well, I've rehearsed this for many years. I've dreamed about it, had nightmares about it, and here I am not knowing what to say or how to begin."

She stalled, looking down at the floor. Her struggle was evident. "Abby, I've always wanted to apologize for not being the mother I should have been—to you, Teddy, and Mandy. There were so many things you didn't know and wouldn't have understood. Things I still don't understand. Guess that's why I kept quiet. I didn't know how to share them with you. I'm thankful I got to share it with Al before he died. He was so gracious and loving, like always."

She cast a look over to her left and sniffed. "Abby, it's not just a story about me. The story begins with my mother." She stopped, grabbed a Kleenex, and blew her nose, continuing, "I didn't know about it for a long time, couldn't have back then. Families kept secrets close to their heart or gun; one or both were enough to keep silence, for most people, that is. Mind you, I'm not expecting you to understand. I just want it told."

"My mother wasn't a loving woman. Oh, I saw her be loving with you and Teddy, but she wasn't with me and my brother. She was different. I always felt cheated. And, I guess I was also a little jealous of her. She could be a charmer to customers who came into the store. Oh, you may not have known, but mother worked in her father's grocery store. She was smart as a whip, not formally educated, but could out figure most men. Grandpa had her manage the books. Then, they lost it all in the Great Depression."

Taking a deep breath, she continued, "I was left alone, a lot. Even when she was home, I felt alone. When she wasn't working, she'd sit in the parlor staring or reading. Strange, until now, I never put it together . . . my love of reading and hers. I remember feeling books were her family, a way to hide from us.

"Anyway, one day, I was bored beyond my young girl limit. Knowing Mother wouldn't be back from work for a while, and my brother was across the street at a friend's house playing, I decided to venture up to the attic. In our household, this was the biggest crime you could commit. Mother said no one was to go up there. She said it was full of private things, old things that had belonged to her parents."

Mama motioned for a glass of water. I got up and attended to her. Funny, how we had mental telepathy about certain things and none about others. I rushed back to my seat, totally intrigued. I had no clue where she was going with this.

She took a deep swallow of water, wiped her mouth off with the cuff of her sleeve, and proceeded, "Like I said, I was bored. I had just finished reading *Anna Karenina*, which I devoured in one day. The librarian always said I was an enigma. I think she was actually surprised I knew what it meant. Guess that's what reading can offer you, a homemade education."

She paused, "Wait, where was I? Oh, yes. At first, I tried to talk myself out of going to the attic for fear someone would come in and find me, but eventually curiosity got the best of me.

"The attic was floored and partially walled. It housed most of my grandparent's furniture and paintings, that is, the stuff Mother refused to get rid of. In a far away corner, I spied a tri-fold screen with a dull painted scene on it. Thinking it was somewhat intriguing, I edged over to where it stood, watching the placement of each foot.

"I almost tripped several times, but finally got to my destination. The picture on the screen was indeed nondescript, dark and brooding. I could make out very little in the late afternoon shadows, filtering through the broken gable louvers. As I started to leave, thinking I had wasted my time, I spied an old trunk strapped with a thick leather binding. Upon further investigation, I saw the lock had been broken.

To secure it, a ragged piece of cotton fabric had been looped through it and tied in two anchor knots. The fabric was yellowed with age, one corner a dark rust color.

"I should have left. I know that now, but I didn't. I managed to move the screen over to the side and kneel down beside the trunk. A thick carpet of dust lay on top and bugs clung to it, even in death. It was hard to pry it open. My fourteen-year-old hands were not strong nor were they used to hard work. The effort was exhausting."

Mama took another drink of water and excused herself. I had been spellbound by her story and hated she got up to leave. I feared the weariness wrapping around me or was it the chair or was it . . .

She settled back down, "Sorry, my bladder has a mind of its own. Sometimes even when I obey the call, it still manages to wreak havoc, and always at the wrong place and time. But that's another story, isn't it?

"Let's see. Oh yes, I somehow managed to get the trunk open. I rummaged through the contents, nothing special: a wedding dress, lots of papers, a couple of hats, some dried flowers, but nothing of interest, and I closed it. No, I mean, it slammed. The weight of it was unbelievable. It slipped right out of my hands. But, when I studied it to make sure it wasn't damaged, I realized the lid was cocked up on one side; so, I opened it again. The liner had fallen out. I worked and worked trying to get it back in place, hoping no one would notice, but every time I pushed it up, it pulled down further. On my third attempt, it fell down completely and with it came this box."

She sat for a short spell, collecting her thoughts, "Well, as you can guess, I was terrified. I knew Mother would kill me, but I didn't know what to do. My hands shook so hard it took an enormous amount of effort to push and shove to get things back in place. Then, I carefully shut the lid. Tears burned, and it hurt to breathe. I kept thinking, *what have I done?*

"I hurried downstairs. For days, all I could think about was the box. It filled my daytime and nighttime thoughts like an obsession.

"After a week, I couldn't stand it anymore. I had to have that box. So, I carefully planned my next trip to the attic, retrieved it, and worked carefully to make the trunk look the way I had found it.

"I had prepared a place to stow it, a carefully carved out section of wallboard in my closet. I slid the box inside, between the studs, and closed the makeshift wall-hatch. I covered my crime with a calendar Grandpa had brought me from the store."

Mama started withdrawing, her eyes veiled in shadows. She looked around the room, frantically casting her eyes from one side to the other. Fear mounted its steed. Whatever Mama was trying to tell, this part must be painful.

I kept quiet, honoring her request. I wrung my hands and pushed my toes into the floor, choking down the words I wanted to say to break the tension that was spreading like ink on an oilskin tablecloth.

Finally she looked up, but not at me, beyond me. "I kept the box hidden away for days. The urge to open it gnawed at me like my monthlies did—deep down in my belly. But I had to plan it right. So the next Sunday when Mother took her nap, and my brother was out with his friends, I locked my door, closed the window, and drew the curtains. Then I got the box out of its hiding place, handling it like an expensive porcelain statue, so afraid I'd drop and break it.

"I gently peeled back the wrapping paper, so nervous I could feel my heart thumping in my throat. I took the top off and peered inside. In it, beneath layers of scrap material, lay a black Bible tied up in mildewed tobacco twine. The name on the bottom right hand corner was my mother's, Mary LaDora Grier.

"I held it in my hand feeling disappointment rise like dark, bitter bile. This was what I had risked so much for? What I had dreamed about? I could not believe it. I kept shaking my head over and over again. Finally, I decided to finish the deed, thinking it might contain something of interest."

Mama stopped, sipping more water. "The twine frayed as soon as I touched it. Shreds fell all over my lap. For a second, it was all I could think of, the mess. Mother hated messes. She would have blistered me over it, but I pushed the thought aside. Once I had removed the cording, I tenderly opened the Bible. I wondered, as I lifted the edge, how many times my mother had put her finger in this exact spot.

"I was shocked and bewildered to see the entire innards of the Bible cut out, and a small journal tucked inside. For a moment, I chastised myself thinking it might be personal, but curiosity took over again. I pulled the faded pink fabric book out and opened it."

Mama lifted the box off the floor, saying, "This changed my life. I want to share parts of it with you." She opened the journal and turned it toward me. The script was small and flawless. I leaned closer, seeing beautiful swirls in the flowing penmanship. She turned it back around and began thumbing through the pages.

CHAPTER 34

Krista

Maybe I should call Mom and suggest Grandmother come live with us, I thought. Anything to get her home. What in the world could be taking so long? Surely Grandmother had a group of friends waiting in the wings to swoop in and take care of her once Mom left. Mom was probably sitting around out of obligation not accomplishing anything anyway.

I wanted to call and beg her to come home and tell her about Dad's bombshell, but I was afraid knowing what she would come home to would guarantee her continued absence. At least with the last two moves, Dad and Mom had prepared beautiful speeches on the cities about to become our new homes. We had all kinds of outdoor fun in Tennessee, and the ocean and beautiful weather in North Carolina. If Mom knew Dad was going to officially tell us, and actually he had only told me, she would have found a selling point for Madisonville.

In addition to laying the move news on me, Dad tried to inspire me to help him around the house by reminding me how much he and Mom leaned on me as the oldest. Our financial arrangement to make sure I would come directly home every day after school, watch the kids, and prepare dinner didn't hurt.

One thing I was determined not to do was take on the burden of telling Chelsea and Katie about the move. But I knew I had to tell Alyssa. A few people at school might care a little and wish me well, but Alyssa would be devastated. I considered calling her parents and breaking the news so they could help me, but I imagined her dad calling in the troops to barricade our departure. Or worse, he'd make her think it was another punishment. Nothing would shock me when it came to that man. Still, I wasn't quite ready to tell her.

After returning from school, Katie and I sat at the kitchen table and did homework. We even gave Chelsea a very important coloring project, a "welcome home" card for Mom. Then the girls helped me with an amazing dinner from a box.

Dad walked in and seemed surprised for a moment before he gave us the news we had been waiting days for, "Your mom called me at the office today. Your grandmother is doing better, and she should be home in a couple of days."

"Yeah, yeah, yeah," Chelsea said. "I made her a card today!"

"She's going to be very happy to see that, Chelsea. I think we should do our best to help Mom out. She's still going to be sad. She also said she wants to get working on your Christmas lists," Dad said. He shot me a glance.

I get it, I thought, *don't attack her right away with my issues over the move and Alyssa.* I had been bottling up so much forever now; and instead of getting credit for it, I was put on notice to keep it up.

I collected dishes and headed into the kitchen. Katie offered to help, but I asked her to take Chelsea up for a bath. She ran off with a huge smile, the power and responsibilities of late were so getting to her head.

I expected Dad to come in and engage but quickly realized what a silly notion it was. Mom had once told us the reason Dad had been so successful in his career. He was known for getting cooler heads to prevail in labor disputes and, particularly, at disgruntled plants. He had a gift. "Sometimes your father has to tell his employees they're going to hell, but he has the charm to inspire them for the trip," Mom described.

So funny, I thought. *All these hellacious moves, and Mom had always been the motivator.* I guess Dad used up all his skills at work.

My thoughts turned back to Alyssa. As much as I truly hated to leave her and to tell her about the move, there was still a decent part of me not completely upset at this turn of events. I was not going to throw Dad a bone just yet, but a fresh start would not be awful. Granted I would be moving with little left in my sophomore year, but I could not imagine Kentucky would have many challenges I had not already faced.

Mom would be home soon, and she would help me talk to Alyssa. She would know what words to say.

Mom returned. She didn't talk about Grandmother much or why she was gone so long. I gave her several days without any of my nonsense, as Dad would say. We were all minding our p's and q's because we could breathe again. Our house was a home once more.

Alyssa's phone call on Saturday morning changed my resolution to give Mom space, "Hey, Krista, I'm coming home today, and I'll be back at school on Monday." The words were sweet, and I was so happy for her, but the bitterness of knowing I must tell her tainted it.

"You timed that one well. In two weeks, we have a week off for Thanksgiving," I said, trying to relieve the pressure.

"You're not going away for Thanksgiving, are you?" she asked, panicked.

"No, I don't think so. If anything, my grandmother will come here to get away from it all. But she's also got these crazy old lady friends she might be doing stuff with," I said.

"Good, I couldn't face a week at home with my parents without the escape of having you over or going to your house," she said.

My heart started pounding. I was afraid if I looked down, I would actually see my chest heaving with anxiety.

"You know I'm always here for you, Alyssa, no matter what," I said. It was true and all I could think to say. "Call me tomorrow when you get settled."

I hung up and sat for a while. I was hoping the answers would come from the heavens and fall into my lap. Nothing came. I knew I had to talk to Mom.

"Hey, Mom, where are you?" I called from my bedroom.

"In the kitchen, hon," she answered.

I went downstairs slowly, conjuring up all my questions and needs. "Where's Dad? The girls?" I asked.

"He's doing God-knows-what in the garage again, and I think your sisters are outside playing," she said and continued to grab ingredients.

I was sidetracked momentarily, "What are you making?"

"Katie actually asked for fried chicken with gravy and rice," she said.

"Ugh! You know I like it with mashed potatoes," I said.

"Krista, let's give Katie this one. She is doing so much better," Mom said. Immediately, I felt like a complete loser. Katie really had come a long way, and all I could think about was me.

"Anyway, that was Alyssa on the phone, and she's coming home today. She'll be back at school on Monday," I said.

"That's great. I'll have to give her mom a call. She sent me the nicest card when I lost Daddy," she said. "But, Krista, I really do think Alyssa has a long way to go. I don't want you think this is the end of her recovery."

"I don't, Mom, and that's what I wanted to talk to you about," I said. *Here we go*, I thought, *no turning back*. "Dad told me about the move."

"He did, did he?" she asked putting down all the treasures she had been holding. "I was not aware of that."

"He didn't want you to have to deal with it when you got back," I said. I was not sure why I was defending him, but I continued, "I don't think he's told Katie and Chelsea, yet, though."

"We'll need to have a family meeting, I guess, and get this all out in the open. Maybe over Thanksgiving when everyone is not running this way and that," she said.

What? Go around the table and say what we are thankful for, like always, and then have one of you guys throw in a tidbit about how life will be upended again. That should be fun.

"My biggest concern is Alyssa, Mom. I don't know how I'm going to tell her. She needs me. I'm afraid of what will happen if I go," I said.

Mom didn't say anything right away. Instead she pulled me in and squeezed, "Honey, we will tackle this move as a family like we always do. And you and I will talk to Alyssa. You can write and call her from Kentucky and still be there for her. Actually, I will call Mrs. McConnell at the beginning of the week and give her a heads-up. You never know. It could bring them all closer together."

Hope glimmered as I sank into the embrace. Anticipating a move would be a new and unfamiliar territory.

Abby

It was getting late, but we were at a juncture. Mama sat straight up with a new sense of urgency. Feeding off her renewed energy, I perched on the edge of my chair, eager to listen.

"I know most of this by heart," she said, thumbing through the journal. "Let me see . . . oh, here it is. Do you mind if I read it?" She started, then stopped, laying the journal down and retracting her question, "Rather than read, I need to give you some background information. Trust me; I learned more about my family reading this than I ever did living in the house with them. Mother worked hard at keeping everyone out. I mean, everyone."

I settled back into the chair, realizing how important this time was for her.

"Before we moved to North Carolina, my mother's people migrated from Piedmont, South Carolina, to West Point, Georgia. West Point thrived as a small southern mill town situated right on the Chattahoochie River. My grandfather, Mother's stepfather, owned a grocery store, and she worked in it from the time she could pick up a bag of sugar till her parents lost the business in 1930. I have no recollection of my grandmother. All I know of her is written in the journal. After my grandparents lost their business, Mother stayed home with us for a while. Then she worked in retail until she retired. Our home was quiet, too quiet. You did what was expected, asked few questions, and kept to yourself. Three times a week, a black woman came in to clean, wash, and cook. She talked to me and gave me more attention than Mother ever did."

Up until now, Mama spoke clearly, almost detached. Now, she fidgeted, "I'm going to summarize some of the journal. To be honest, it's painful for me to glimpse into mother's life when I can't ask her any questions. She seemed so full of life in the words on the pages, a fullness I never saw or realized."

She paused, collecting the story, "When Mother was fourteen, she was smitten with her minister, Rev. Jeremiah Winston. He was married and had an infant son. There seemed to be something about him that drew her to him, attention for the most part, I imagine. But then, there were the small touches here and there, singling her out from the other young girls at church, and feeding her loneliness. At the same time, she filled her idle time reading romance novels her friend's aunt brought down from New York. I'm sure the girls begged, and she probably thought it dicey to kinda accommodate them. Mother wrote about how the steamy stories tempted her in ways she knew were not pleasing to God or her mother. This was a pivotal time for her—a time of search and discovery, a time of weakness and poor choices.

"I'm just going to highlight some of the parts, those that affected my life and later our family. One day, when I'm dead and gone, you can read it and decide for yourself how you feel her life tangled into ours. Now, remember our pact, no interrupting."

June 7, 1925

    I had a pleasant week going about my work. On Sunday, Reverend Winston delivered a superb sermon. After church, he touched my hand lightly, inquiring about my health. I blushed red. Kathleen saw it all. She's sure to make trouble for me. Lately, she watches everything I do.

    We, younger girls, gathered on the grounds, whispering and laughing about silly things. We were especially excited to hear Miss Anna Faye Brigham was due in on the train at three o'clock. Mary Louise said she had a surprise for us. I hope it's Zane Grey's latest book. During the conversation, I overhead Mary Louise say the Reverend took tea at the library on Mondays. I planted it in my mind to go there to study, hoping to catch a glimpse of him. I know I'm wrong to have feelings for him, but I do.

    I am amazed at how my body is changing. Feelings I never had pour out like cooling rain on a hot summer day, "a sinuous silver ribbon of sin wrapping itself around me". I read the line in a book and memorized it, laboring over its delicate meaning. I continue to ask God to forgive my evil thoughts and deeds.

October 10, 1925,

    I saw Jeremiah in the prayer garden behind the church and joined him. Leaves of varying colors rained on us, giving a soft, warm feeling to a somewhat mellow fall day. He read the Bible to me, telling me his name, Jeremiah, meant the exalted of God. His voice soothes me, making me want to touch his arm, but I don't for fear of what would happen.

    My prayers continue. I pray my feelings would transfer to James. He is sweet and cares for me, wanting to daily carry my books home. At times, I don't know which way to turn—toward the silvery smooth voice of a man or the high pitched tone of a boy.

Mama dog-eared the page and closed the book. "I wanted you to hear her words. I've given you a little background on Rev. Winston. James was a neighborhood boy she grew up with. Their families were very close. I learned from the readings that everyone kinda assumed they'd get married. Also, Mother's real father died when she was young. When she refers to Papa, she's talking about her stepfather. Bear with me as I read two more entries."

November 8, 1925

I received two Bibles for my birthday. Mother and Papa each gave me one. When I opened hers, he flamed in anger, accusing her of sabotaging his gift. He wanted to know why she did not confer with him about such an important decision. Then he slapped her, hard and mean, saying she embarrassed him. It wasn't the first time. When she saw I was looking, she averted her eyes and ran out of the room. I hate him. He can make me call him Papa, but he will never be my father, never!

I took Mother's Bible back to my room and hulled it out with my brother's knife. I put my diary in its center, at the heart, for safekeeping. It would always be a reminder of his anger and my mother's broken heart. I tied it up in twine and hid it in a safe place. I told her I would always cherish her precious gift, but to keep peace and keep her out of harms way, I would only take the Bible he gave me to church.

My deceitful ways continue. I think I am headed straight to Hell. Reverend Winston rages on every Sunday about Hell's fire and damnation until my mind feels blistered. I fear

I might become like Hester Prynne, marked with an A, scorned and scoffed for life. I don't know how to stop thinking and dreaming about Jeremiah. God, please forgive me.

November 10, 1925

Dear God, I beg you again for forgiveness. I have sinned against Thee. I have not even tried to keep away from Jeremiah. Mary Louise grabbed my arm at school yesterday and told me I was an idiot to carry on so. She reminded me he was twenty and married. I know this. But when I am with him, I am full. I don't feel so lost and lonely.

Today, Jeremiah called me the angel of his dreams. He leaned over and kissed my forehead. It was a tender, sweet kiss. I could feel the heat and fullness of his lips and swooned. Before he left, he whispered, begging me to meet him at the church that evening, and I did.

After supper, shortly after Mother left to help deliver the Skinner baby, I slipped out of the house. Fear tightened its bands around my chest, yet I trembled with excitement.

When I got to the church, he was waiting near the old, giant oak. A brilliant full moon haloed the tree, casting an ethereal light to the edges of the top branches. The night felt surreal and charged with energy. As he took my hand and drew me toward the church, I felt I was floating on clouds.

Once inside, my eyes fell on a simple white cotton shift draped across the back of the pew. Jeremiah reached over and handed it to me, lead me to the prayer room at the back of the church and asked me to slip it on. Then, he excused himself. Candles lined the communion table at the back of the room. Shadows danced and played off the walls in tones of soft, buttery yellow.

I took off my clothes and slipped the shift over my head. As it brushed over my breast, I closed my eyes, delighting in the coolness. It made me shiver the same way I did when I read about Heathcliff and Catherine's unresolved passion, love as immovable as the rocks around their home.

I sensed I was in a dream state. I barely felt my feet skim the wooden floor as I moved toward the small cot in the corner of the room. Lying down, I stretched out, feeling the world had slowed down for this moment. Then, Jeremiah entered the room. He had on a similar shift, resembling a groom come to claim his bride. My heart raced as he came closer, knelt down beside me and took my hand. His voice was melodic, whispering words of pure love. Love for always and forever.

I listened as he recited a familiar Bible verse, "The victims commit themselves to you; you are the helper of the fatherless." It was true. He was talking about me. I remembered Mama telling him how father had died. Heavenly Savior, forgive me. You know I didn't mean to get stuck in the tree. I didn't know father would climb up to get me and then fall and break his neck. Mother never forgave me. Oh how different her life would have been had father lived. Now she looks at me with such contempt.

Jeremiah's words continued, sounding like God's approval to my prayer. I knew God had sent him to me, to love me, save and show me his divine forgiveness. His chant continued, and I fell under its spell as his fingers lightly stroked my face, neck and arms. But, when they moved under my shift, I nudged him, saying 'no.' This touch felt wrong. I pushed, gently at first, repeating 'no.' The more I pushed, the more persistent he became. I shoved with all my might, one more time, and he slapped me, hard. I started swinging my arms and kicking my *legs, trying to make him stop, but he ripped my shift, tearing off an end piece and shoving it into my mouth.*

I tried to scream, but there was no sound, only the raspy exhales of nasal breath. Vomit rose, scorching my throat with bitter bile. Panic set in, and breathing became unbearable. Then, a miracle happened. Peace washed over me. I stopped fighting the nightmare. Light beamed bright, colors erupted into beautiful patterns, and I began to float toward them.

When I opened my eyes, I was alone. The room was dark, except for the faint moonlight stealing around a small curtained window. I tried to stand, but waves of nausea sank me to my knees. My hands drew to the pain in my abdomen, and I was repulsed to feel my nakedness. Stretching my hands forward in search of my clothes, I finally found them wadded up at the end of the cot. I dressed, fell and eventually crawled until I managed to stand and stumble out of the church.

I don't remember how I got home, but was thankful Mother was still out. I drew water and bathed, scrubbing hard, wanting to cleanse his foul scent from my body. Back in my bedroom, I grabbed my Bible and turned to the scriptures, reading the rest of the passage Jeremiah left out, "You, Lord hear the desire of the afflicted; you encourage them, and you listen to their cry, defending the fatherless and the oppressed, so that mere earthly mortals will never again strike terror" Never again, I cried!

We sat for a time. Me, soaking in my grandmother's pain, and Mama, resting, her head pressed into the back of the chair.

"I have another entry I wanted to read, but I can't do it right now. I'll try telling some of her story." Mama straightened up, gathered her thoughts, and continued, "Later on, Mother found out she was pregnant. Back then, an out-of-wedlock pregnancy was treated severely.

Often young girls were driven away from home. So, she duped James, her sweet friend, into marrying her. It's all in the journal. Come to think of it, I do want to share this with you."

Mother pushed pages back and forth until she found the right one, then read:

> *Forgive me God. I do not want to hurt James, but I had to make things right. I want the nightmare to end. Father, I am so frightened. I know I have no right to ask anything of you. I am the worst of sinners, but please help me. Help me love James and be a good wife to him. Lord, protect the life growing inside me.*
>
> *I know there is no comparison, Lord, but I've wondered about your mother, Mary. She had to have been a strong woman and Joseph an understanding young man to deal with her pregnancy. You alone provided that. Would you grant us some of that strength?*

"Mama," I said, breaking my promise to be quiet, "I'm moved by this story and sorry for grandmother. She suffered a lot, but she sounds like a woman who did what she had to do and didn't whine or feel sorry for herself. I'm touched you wanted to share this with me, but I don't see why. Look, we've had a long emotional day. I think we both need to go to bed. Let's finish this tomorrow." I was beyond tired. For me, the story was finished.

"Abby, please stay with me. Right before I found Mother's diary, I was seeing a young boy in my church, Nathaniel Winston. Mother did not know about it. In fact, she had forbidden me to have anything to do with the family when they moved back to town. She said they were no good. Secretly, Nathaniel and I planned to elope the next year. We were intimate, but decided, after a time, to refrain, wanting things to be different, better than for most young people around us. Then I found Mother's journal. I immediately ended the relationship." She sat up, clutching the chair arms, shaking, eyes bulging, cheeks growing a blotchy red, "Do you understand what I'm trying to say, Abby?"

My head pounded, spinning wildly. "You're scaring me. I don't..."

"Abby," she said through clinched teeth, "I . . . I got pregnant." Her eyes softened as she leaned back in the chair. "We were young and in love. I still see him—hair the color of a soft red sunset. And his eyes, a beautiful bark, cinnamon color, the kind you want to melt into. But for God's sake, he was my half brother. His dad, Jeremiah Winston, was the minister who raped my mother, and later abandoned his family. I'm his daughter! I'm a grief birth, born of Satan's handicraft."

Horror purged my senses, filling them with putrid blackness. Reality spiraled, faltered, then plummeted into an abyss of violated promises, echoing unbearable hurt and humiliation. Death stood tall with his scythe, ushering in a dark season of life. The crack widened, and the story flew at me like a poisoned dart.

Mama stopped talking. From the look on her face, I knew she was gauging my reaction, and at the same time, dealing with her confession, pain, shame, and embarrassment.

I struggled to resurrect a comment, something to ease her hurt, but only managed a pitiful, "I'm so sorry."

She lifted her hand, signaling me to be quiet and listen. Taking a deep breath, she continued, "When I read Mother's diary and put two and two together, I went to bed and stayed there for two days. I never told Nathaniel. I roamed around the house not able to eat or concentrate. When Mother asked me what was wrong, I lied. One day, Nathaniel showed up at the door, and Mother told him to stay away and not return. I never sought him out or talked to him again."

She stopped, wringing her hands in a nervous fidget. "I began to think of ways to kill myself, rid myself of the sin growing within me. In desperation, I remembered the elderly black lady, Jessup, who used to work for Mother. She did our laundry, but it was common knowledge she cured folks of their ailments, white and black. Black folks said she was tetched with a special gift.

"The next day, I got on my bicycle and pedaled ten miles out in the country, stopping at several shacks before I found her. She remembered me, but didn't once ask why I was there. She touched my arm, and her eyes fluttered shut, then blasted open, white and milky. It almost scared the life out of me, but when she uprighted herself, she promised to fix everything.

"I clung to her in relief. She cradled me as I wept, pain and fear pouring out in a torrent. I don't know how long I stayed in her arms, but when the tears stopped, she gently took me into the kitchen, had me strip from the waist down and lay across the table. I heard pots clanging, water boiling, material ripping and foul odors that made my nose run.

"Jessup came and stood over me. She put her hand on my forehead, closed her eyes, and silently mouthed words whose silence only frightened me further. She took my hands and stroked them before gently lashing them to the side of the table with a black leather strap, crooning and singing 'In the Sweet Bye and Bye.'"

The air in the room was stifling. Mama sank deeper into her chair, staring at the floor, barely talking above a whisper, "She put a thick paste in my mouth and lifted my head long enough to sip water and swallow. When she touched my private area, I winced and yelped. Jessup kept singing, stroking my leg. Above me, the ceiling slanted, waving back and forth like grain in a gentle breeze. I felt a sharp pain below my navel, and my legs shook like fish flopped down on the ground, out of control and helpless.

"I must'a passed out. Lifting my head, I saw blood everywhere. Jessup was cleaning me up. I tried to talk but my mouth was thick, my jaws tight. I watched as she packed an herb pouch in a rag, placed it between my legs and belted it around my waist. Then she slipped the cotton pantaloons up and over the thick wad. Within minutes, she had me sipping a strong drink, humming as she finished cleaning up the floor, but no amount of scrubbing could clean my heart.

"Awhile later, she had a black boy put me in the front of a wagon, my bike in the back and took me home. Jessup sat between us, patting my hand. I just prayed to die. At the edge of the drive, Jessup pulled a small vial of liquid from her apron and pressed it into my hand, saying, 'Take it three times a day till gone.' I got in the house unnoticed and went straight to bed. Mother threatened to take me to Doc Eagan, suspecting anemia because I was so ghost white. But, I refused."

I got up and walked over to Mama. I knelt down on the floor in front of her, pulling her in close. Her shoulders heaved, and she kept saying, "I wanted to die. I just wanted to die."

I hugged her hard, pressing my body into hers, trying to absorb some of the pain. I don't know how long we remained this way, gently rocking until her sobs became deep strokes of breath. At some point, she lifted her head off my shoulder and looked into my eyes, "The story isn't finished, Abby."

I nodded and sat back down on the floor, keeping my hand on her thigh. I didn't want to break the connection. For the first time in my life a bonding energy coalesced, blurring the past, wrapping hardened feelings and hard times into a soft fringed shawl of understanding.

She looked me straight on and heaved a deep sigh, saying, "I married your father years later. He pursued me like a hunting hound. We'd only known one another about six weeks. Everyone said we were crazy, but I wanted to be crazy." She stopped and laughed out loud, hitting her knee, "That man could make me roar with laughter." Reminders softened her, and the glass wall started to drop.

She sat quiet. I knew she wanted to hide behind the words, but she continued, "I've done so much to commit my life to hell. For some reason, I thought I could start over and create a normal life. But I couldn't. I tried, but it didn't work. Misery ate at me even when

I should have had joy. Your Dad tried to love it all out of me, but after so many tries, I think a part of him gave up. And, I couldn't seem to reach out to any of you. I felt I didn't deserve you. I could never . . . Oh Abby, I killed my first baby. I always felt unfit to love . . . to be loved."

Mama's face went blank. She pushed me away from her, and started screaming. It caught me so off-guard I stumbled backward. I struggled to my feet, trying to comfort her, but she kept swinging her arms wildly about her frame. She screamed and screamed until her voice refused another sound. Spent, she started shaking violently. I got up to call the ambulance, terrified she was having a seizure or stroke; but she grabbed my arm, pleading for me not to leave her.

"Mama, please, I have to get some help. You're scaring the hell out of me. Please, let me call the ambulance or if nothing else, I'll take you to the hospital."

Still shaking, she sat up, "When I'm finished, you can take me anywhere you like. Right now, call Quillie. I need her to come over here."

I went straight to the phone. Quillie's number was on an index card taped to the paneled wall. With shaky hands, I picked up the receiver and dialed. She answered after the first ring, "Hello?"

"Quillie, this is Abby . . ."

"I'm on my way." She hung up the phone before I could finish.

"Mama, she's on her way." Then it dawned on me, "Mama, does Quillie know about Grandmother or Nathaniel?"

"No, Abby, she doesn't. But Quillie has special gifts." She stopped, bracing against a surge of shivers, "Honey, grab me a blanket out of the cedar chest in my bedroom?"

When I got back, Quillie was coming through the front door. She went straight to Mama, enveloping her in a hug. Reaching up, she took the blanket from my hands and tenderly tucked Mama inside.

"Nadine, are you okay? Honey, Quillie's here. It's gonna be all right. I'll take care of you, sweetie," she continued to croon, rubbing Mama's arms and brushing hair from her face. I sat down two feet from the scene, wrapped my arms around my knees, and remained quiet. Quillie was working her magic; and I was glad to remove myself, if only for a few minutes. A dead-on hit by a freight train would not have left me anymore traumatized than this day.

Quillie tapped my shoulder, and I lifted my head, realizing I had dozed off, "Oh, Quillie, I'm so sorry. I didn't mean to go to sleep." I twisted around and realized Mama was not in the room. "Where's Mama?" I asked, panicky.

"Nadine went to the toilet, girl. She's fine. How are you, baby? This has been some day." Quillie had both sides of my face cradled in her large hands. Warmth and reassurance filled my empty reserve. "You're going to be fine, child. When Quillie gets a call, you can rest assured that she will take care of it! Right?" she asked, looking deep into my eyes.

Mama entered the room, looking pale, but much better than she had minutes before, "Mama, are you okay?"

"Well, I don't know about okay, but I'm here, thanks to the good Lord, you, and Quillie. Quillie sit in that chair over there." Then she quickly added, "No, not that one, honey. Sit in Al's chair. Abby, you stay here beside me. Now, let's finish. We all need to go to bed." Mama sat on the edge of her chair and looked intent at me, "Abby, I'm sorry for tonight. Now, that's the last time I'm going to apologize for my past. I've been stuck there too long."

She patted my arm, and I nodded an understanding. "You know how life used to be when you lived at home. Well, nothing changed after you left. Not until your Daddy had his heart attack, that is. Suddenly, I wanted to breathe again. I felt Death at my backside, waiting and laughing. He had finally come to call and get his payment. That's when I ran into Quillie at the hospital. She became my lifeline.

I can't tell you how my life has changed because of her. You see, Abby, I realized that I was not the only one with a past, the only one trying to hide from something. But I was the only one dying under it. I knew Quillie's past, and Piney's, and Tobaccie's . . . all of them. I shunned them on the street for years for their improprieties; felt I was too good for them, more educated, more self-righteous.

"But that day, in the hospital, marked a new beginning for me. Quillie knew I was broken, flawed and fragile, but she just swooped me up and loved me anyway—loved me just the way I was. She didn't try to change me or act disappointed like everyone else did. She introduced me to other women, each with their own raw story, and they accepted me, too. Those precious women ignored the fact that in years past I wouldn't have nodded to them in the grocery store. Now they were laughing and loving pain right out of me. At the same time, I knew it was wrong to leave your Daddy so much while he was recuperating, but I needed these women. When I was with them, it was the only time I truly felt whole. Didn't you feel it . . . didn't you see the difference in me when we came to your home?"

Mama reached out, wanting me to recall, understand, and accept. I nodded, a lump forming in my throat. Here I was looking at a face I had known for over thirty-five years, and yet it was strangely new.

She sat up, "Okay, I'm done. I'm through harboring a lifetime of pain. I hope you can forgive me for not being there for you. I've let a lot of people down; I know it, but I want to start over. I want to be happy. I'm broken up about losing my Al. But, before he died, we talked, and I told him everything. He held my hand and said it didn't matter. He said he had always loved me. He loved the broken me, Abby. Don't you see . . . Quillie started the healing, but it was your Daddy who completed it. It was then and there that Dorothy's words from *The Wizard of Oz* made perfect sense—'If I ever go looking for my heart's desire again, I won't look any further than my own backyard.' Al was my true love. Nathaniel was just a sweet childhood crush. The realness of life was always Al. But, I spent too much time, and wasted a lot of years, pining after a lover-ghost and a dead baby. Al's forgiveness healed me. It made me see I'm okay. Love edged me back out of my own planted heartbreak."

Tears streamed down my face, "Oh, Mama, I'm so sorry." I reached over and hugged her. Quillie came, wrapping her arms around both of us.

The night ended soon after that. Quillie left and went home, but not before a lot more hugs and tears. I walked Mama to her bedroom and tucked her into bed with a promise to have a nice, warm breakfast ready for her when she got up.

All of a sudden it hit me, and I turned to ask, "The foundling chair? Mama, what about the foundling chair . . . was it . . ." I halted, unable to go further.

At first she looked puzzled, then smiled, "I found that chair in the trash outside Mrs. Winston's home. It was 1946. Nathaniel had been killed in action in France the year before, and his mother, feeble and unable to care for herself was being sent to the county home. It was right before Al and I got married. I never understood why I needed it so desperately. No one ever knew where I got it, not even Mother. It was another one of my secrets. But over time, the chair took on a much more complex meaning. It reminded me of my sweet Nathaniel, our love and lost baby, but more importantly, the way your father took me in and loved me—kind of adopted me. I was the foundling, never the chair. But to get rid of it would be like getting rid of myself, broken, blemished, and used. I always thought as long as I kept it safe, I wouldn't drown in my own sin."

Mama's eyes were starting to close. I kissed her cheek and said, "We'll talk tomorrow."

Opening her eyes slightly, she whispered, "There's no more questions. All the answers are dead." She closed her eyes, let out a sigh, and turned away from me.

I left the room quietly, crossing the hall to my old bedroom. I felt a sudden rush of longing. I kicked off my shoes and climbed up into the bed my grandfather had made and laid on top of Grandmother's crochet quilt. I looked up at the ceiling. *Nothing will ever cover those stains. Well, nothing but the right application,* I thought.

In midwhirl, I willed my brain to halt and go into nothingness. I stretched my limbs corner to corner and breathed in deeply, closing my eyes with the exhalation. The night-black shutter of my eyelid exploded into ribbons of color. And there she was, the girl from my childhood dreams. I reached out to her, wanting to join her, be with her again.

* * *

Ringlets of blonde-mahogany hair bounced as she ran through lush green fields. In her hand she wielded a stick, waving it back and forth granting wishes to the unseen. She danced and bobbed, swirled and curtsied under a seamless Alice-blue sky. Then she stopped abruptly and picked up a "wishie." Holding it up to chin level, she pinched her lips together and blew out a steady stream of air. The spherical clock of the dandelion burst, sending free hair seeds, floating wind-fed and free.

PART III

Back to the Table

Summer 2000

Abby

I splashed water on my face, running my fingers through my hair to fluff it up. I wanted to go downstairs looking refreshed. In the space of a few hours, the past had come slamming back. Why now? I thought I had stopped jumbled yesterdays from controlling my emotions, but I guess I had not learned to control my dreams. But it was strange they were about my teen years; I'd almost forgotten the incident with Henry, my paper doll family, and fears of the future. Maybe it was a gentle reminder how much those forces shaped my life. "Or tormented it," I said aloud.

Over the last decade, I had worked hard uncovering the woman I knew lived inside me. I'd get glimpses of her, beg her to come out, cheer her on when she tried, and forgive her when she refused. One day, she appeared, strong, lusting to create her world, rather than have it created for her. With a lot of effort and a lot of encouragement from my family, I finally got a degree, Finishing magna cum laude. The dream had taken a long time to come to fruition; but it had, in its own time, and was probably more appreciated. In the end, it wasn't the degree I was most proud of, it was having my daughters witness me rise and meet challenges, sometimes two steps back, but always pushing for that one step forward. I hoped my efforts would be a reminder and a boost for their confidence when their lives were stretched.

Once downstairs, I headed to the kitchen to make a pot of coffee, heat the kettle for hot chocolate, and pull out the ingredients for Depression Cookies. They had become a salve: healing us, bringing us together, joining lives around the table.

Hearing everyone laughing surged my excitement. "What's going on out there?" Not waiting for an answer, I ventured into the dining room. Krista, Katie, Chelsea, and Bob were sitting around the table surrounded by photo albums.

"How long have you guys been down here?" I asked.

Chelsea waved me over, "Mom, this is hysterical. Please, tell me it isn't Dad. When did he ever have a crew cut? I mean it's so fifties. Did you guys live in a time warp or something?"

Bob grabbed the picture from Chelsea's hand, "I think I look damn good here." He strained to adjust his bifocals, but once in view, exclaimed, "Wow, I'd forgotten I had that much hair."

Katie popped up, squealing, "Oh my God! Look at Mama's hair! Someone would have to climb on a ladder to untangle that mess." The picture was passed around, and even Bob rolled with laughter.

Joking, I reminded them, "Hey, beehives were the style back then. Google it, you'll see more hideous hairstyles than that!"

Krista got up to check on Janie, sleeping in the next room. She returned with a large rectangular box in her hand.

"Mama, I didn't mean to be nosey, but I found this earlier with the photo albums. Grandma's name on it. Do you mind if I open it?"

My heart skipped a beat, "It was Mama's. I found it after she died, tucked away in the closet. I hadn't seen it in years. She wanted me to have it." Bittersweet sorrow filled me when I thought about the contents, a box of sunrises and sunsets, a telling of life measured out without wasting one ounce of joy or grief.

"If this is too painful, I'll put it back in the cabinet," Krista responded.

Wiping away tears, I said, "No, I knew there would come a time when I would want you to see it. This is the perfect timing."

"Abby, do you want me to leave?" Bob had learned over the years women felt more comfortable discussing certain issues between themselves. He never felt threatened by it. Rather, I think he saw it as a time to escape the full range of emotions involved.

I looked at him and smiled, "Not for the world. We need you here for support."

The kettle whistled, signaling a shift in attention. I excused myself, announcing I'd be back with dessert. Rising from the chair, my legs wobbled, and a shot of searing pain drove downward to my calf. Groans erupted before I could stall them, "Sorry, it's getting harder to get up if I sit too long. Aging is not what it's cracked up to be!"

Krista helped, and we returned with steaming mugs of hot chocolate, coffee, and a tray of Depression Cookies. "Eat them while they're hot," I said, passing around the tray.

We sat quietly, sipping strong brew and the velvety smooth flavor of hot chocolate. From the umms and ahhs, I knew it was perfect, especially the cookies. There was nothing more addictive than its salty-sweet taste.

Chelsea, her mouth chock-full, sighed, "These are sinful."

"It's fitting, isn't it?" I said, hearing Mama's voice say the same thing.

When we finished, I picked up the box off the table, "Krista, I want you to open it. Katie, you and Chelsea sit close. You, too, Bob."

Krista looked nervous, but I smiled and nodded. When she opened the top, she pulled out two photo albums and a black Bible bound tightly with tobacco twine.

"Who is Mary LaDora Grier?" she asked.

"That's my grandmother."

Krista laid the Bible down, picking up one of the albums. Flipping through the pictures, she asked questions about various people, passing them around for everyone to see. Bob got a thrill looking at the old cars in the background, pointing out ones like his family had owned. The second album contained faces they recognized: their grandparents, Teddy, Mandy, and me. The last few pages were devoted to pictures of the grandchildren from birth on. Mama had put little sayings by some, and on others dates, signifying certain times or special events. It grew quiet when they got to pictures of Mama, Quillie, Clara, Salem, Piney Jodelle, Annias, LeRoux, and Tobaccie on their many trips, and last, but not least, the special section entitled, Quillie's Wedding.

Katie pointed to one of the photos, showing Quillie and Dickie surrounded by friends and family, "I dreamed of this for years, this type of wedding, and yet, when Luke and I got married I settled for traditional and dull. How could I have forgotten?" she wailed, softly.

Chelsea picked up a picture that had fallen on the floor. She sat for a long time looking at it. When she turned to me, tears clung to her lashes, "Do you remember this one, Mama?"

I reached over, taking the picture from her hand. It was a picture of Mama and me taken at the funeral home after Dad died. Our faces were ashen, sadness scribbled in unhealthy lines across our brow and false smiles. I pulled it to my chest, closed my eyes, and rocked.

Bob came behind me, his strong arms circling my shoulder for support. Shutter speed images flashed across my memory sill: Mama handing me a beater full of frosting to lick, sewing my junior and senior prom formal, mad as hell at me over countless silly things, rocking and crooning my first baby, begging me to come home the years I lived away, her trials during Dad's sickness and death, finding a new beginning with the *Aunties*, our talk after Dad died, and her last words to me before she died.

When I opened my eyes, I saw a room full of love. Bob reached down and kissed my neck. Chelsea reached over for my hand. The older girls nodded. We remained like this until we heard rustling and whimpering coming from the next room. Not wanting to break sacred quiet, I raised my hand to signal that I would get Janie.

She smiled when she saw me. "Hey there, precious," I said, picking her up and pressing my nose into the fold of her neck. "Umm, Umm, you smell good!" I laid her on the bed and changed her diaper, both of us talking in our own separate language. I spied the rocker in the corner and Chelsea's old bear, the one she had given to Janie, and decided to spend a moment rocking her. The creaking of the rocker sang its soothing lullaby, peeling away sadness into soft memories.

Janie watched intently, first me and then the bear. "You know, my mama would have loved you," I said softly. Janie giggled, looking past me, extending her chubby hand as if to grab an imaginary gift. Warmth swept my cheek, faint and brief. Janie dropped her hand, pushed back, and smiled at me.

When we got back to the dining room, everyone squealed. Janie, capitalizing on the attention, danced, bouncing to the rhythm of high pitched voices and laughter. Krista came over to retrieve her, giving her a big hug and kiss. Then she put her hand on my arm, saying, "Mama, please go sit back down."

I lifted Janie's hand to kiss it; and when I did, I saw the Bible, opened, its hull exposed. The contents, a faded pink fabric journal, stood open to the place marked years before.

Returning to my chair, I asked, "Did you read it?"

Krista sat down with Janie and looked at Katie. Katie swung around to Chelsea, and she to Bob.

Bob spoke first, "Abby, honey, we don't know quite what to make of this, but we know it's special."

I shifted, "Yes, it's very special. I'd like to share it with you, but it would create a lot of questions. You're welcome, at any time, to read its memoirs. But when you finish, there are to be no questions; the answers are all dead." I recalled Mama repeating it twice to me—once, when she revealed its sorrow and the next with her dying breath.

We watched Katie reach over, pick up the journal, slip it back in its hallowed home, and tie the twine back around it. She retrieved the box, adding the Bible and the two albums before securing it, end over end for a tight seal. Once completed, she walked to the cabinet, shoving it toward the back.

Returning, she had our full attention, "Mama, some things don't need to be answered. Things happen in life that change the way we think, the way we view the hunger to survive, and even the desire to forge ahead in the middle of pain and disappointment. But we can grow and usually are better for it. The destruction comes in not being allowed to grow beyond our mistakes in order to become whole again. It's easy to lose the way. It's family that sits you back on the wall and heals the breaks with love. Trust me, I know."

We were quiet.

In the solitude, I remembered the note Mama handed me two weeks before she died. "Abby, I want you to have this," she said, before pointing to the dark lettering written at the top—BLACK BADGE OF COURAGE.

"I don't remember where I got this from," Mama said, "but, it describes how I see life. Keeping it posted and reading it often, gives me the strength to tolerate life's droughts and stretches of cold. You know, those lonesome valleys of life."

"Why use black for courage? Isn't that negative?"

"Life is mysterious and unknown. You have to have a lot of strength to get through the perspective and depth of each day, whether it's joy in birth or grief in death. For me, black symbolized the kind of courage that snubs its nose at defeat and keeps on trying."

I smiled until the image faded, "By the way, there is one answer I can share with you."

I walked over to the refrigerator and came back with the note Mama had given me, taped on a three-by-five index card. Since it had left her hand, its message had never left my heart.

Black Badge of Courage

> Courage doesn't always roar. Sometimes courage is the quiet voice at the end of the day, saying, "I will try again tomorrow."
>
> Author Unknown

Krista

Though I enjoyed every second of it, I never realized how long it would take to convince a tired child to nap. She wanted me to lie down with her after reading four books—one of which I read three times. I simply could not deny her when she turned back to page one every time I finished. When I thought she was settled, I edged off the bed and put pillows around her. She mumbled a bit, so I stayed a few more minutes.

I realized I was freezing. Mom kept the house at a frigid sixty-seven degrees lately, so I opened up the closet to see if I could find one of her old cardigans. I opened it very slowly, wanting to avoid any creaking Janie might hear. As I rummaged, I noticed a huge moving box in the corner. The number of different colored moving company stickers caught my eye and made me chuckle. Curiosity forced me to pull it out.

On top were several of my own personal memories. I remembered Mom wanted me to go through some of my old stuff and decide what I wanted to keep. These items, she insisted, needed to be stored in my own house because she and Dad were considering downsizing.

I got to work. I pulled out yearbooks and simply laid them aside. No great memories there. Not sure I needed to keep them. I did not have a freshman yearbook because I was still at a junior high, and my sophomore one had been mailed to me in Kentucky. There was not one signature in it. *Great, my kids will pull it out someday and think I was a complete loser.*

My junior and senior ones were about as pathetic. I was brand new for my junior year. Discouraged by the scene in Kentucky and by my lack of friends, I begged Dad to let me take a job at the local Roses Department Store. They loved me there and gave me more and more responsibility until I became their nighttime customer service desk specialist. My entire focus was getting into the University of Kentucky and getting on with life. I met my future husband within two days of college starting . . . and the rest was, as we say, history.

Still, the sophomore yearbook called to me, so I opened it. When I did, a piece of newspaper floated out. I laid the yearbook aside and picked up the piece. Tears started right away. It felt just as raw as the first time I read it.

> Alyssa Marie McConnell died on August 15, 1990. She was a loving daughter and friend. She will be greatly missed and is survived by her father, Colonel Thomas McConnell, Camp LeJeune, Jacksonville, NC; her mother, Betty Alexander, Wilmington, NC; and her brother John Thomas, student at the Naval Academy in Annapolis, MD. Services will be held . . .

The obituary was sent to us by our old neighbor, Mrs. Manning. I was home on Labor Day weekend—my first break as a freshman at the University of Kentucky in Lexington, Kentucky. I arrived with the great news of a guy I had just met, the one I would marry, and Mom shared the obituary with me. I cried and cried. But once the tears were gone, deep anger was left. There was one face, and one face only, fixated in my mind . . . Cindy. Truly, I ran away because of one girl, and it had cost me the life of my dearest friend. But in all honesty, I was angry at Alyssa, too. The waste of a good life.

I remembered the day so well . . .

"Mom, this is so unfair," I yelled.

"I know, sweetie," she said as she stroked my hair. Katie and Chelsea were in the room with us. "I got this a week ago, but I couldn't bear the thought of telling you on the phone."

Once I realized where my anger was turning, I asked Mom if she still had our old Wilmington phone book. She always saved them in case she had to call back for some reason.

"Yes, of course, in your dad's office," she said a bit perplexed. "You want me to get it for you?"

I nodded and she brought it to me. I turned to the *S*s and found Sheppard. Nancy Sheppard, there it was. My gut told me they had not moved, and as soon as I saw the number it came back from the depths of my memory.

"Mom, do you mind if I use the phone in your bedroom?" I asked. "I'm going to call Cindy."

"Cindy?" Mom asked.

"I think it's high-time I give her a piece of my mind," I said and started toward her bedroom.

"Krista, do you really think that's a good idea?" Katie asked concerned.

"I think it should have happened many years ago," I said with determination on my way out.

I sat down on Mom's bed, my hand already sweating so badly I almost dropped the phone when I tried to pick it up. I attempted deep breathing to steady my heartbeat and nerves, but it didn't help. I dialed the number.

"Hello," Cindy said. I recognized her voice immediately.

"Cindy, this is Krista, do you remember me?" I asked.

"Of course, of course! Hey, how are you, girlfriend?" she asked with a heightened tone to her voice.

How are you, girlfriend? You almost single-handedly ruined my high school experience and my self-esteem, and you are going to ask me that, I screamed in my head.

The conversation went into an unexpected alternate reality where we were still friends. We talked about going to college, how we both had boyfriends, our plans for the future, and even the weather.

Cindy was the one to bring up Alyssa and how sad the community had been when she died, "Krista, I really am sorry about Alyssa. I know it must be hard on you. I hoped someone would make sure you knew."

She actually seemed to care; nothing from our past had affected her at all. A sudden rush of overwhelming nauseousness hit me. I said good-bye and rushed to the bathroom and puked. I puked up the wasted years of caring what she thought or if she had understood what she had done to me. I continued to vomit until it voided from my entire being.

I was free.

The past could bring back such current sadness and pain. Losing Alyssa has been so hard. She had given me back my sister, and I had failed to save her back. The day I found out, Mom reminded me of the reality of Alyssa's condition. She was in a downward spiral when I met her; and though I gave her a safety net to enjoy for a while, I could not have stopped her fall. It was too much for any one person.

As I continued through the memories, I found letters from Cindy, most of which were from our best friend days. Then I found a letter from Helen detailing Cindy getting horrible grades at the community college and still living with her mom. *Guess popularity wasn't the end all, be all.* I also found old artwork from us girls and even letters I had written to Mom from college. It was all jumbled up . . . wedding mementos from my wedding and Katie's were right after our first grade projects. Almost at the bottom, I pulled out a box. When I opened it, I saw a Bible. I knew I wanted to give it more attention, so I laid it aside and started pulling out pictures of Quillie's wedding and even older pictures as far back as Mom and Dad as teenagers and before.

Knowing this would make for great fun, I tiptoed out of the bedroom after glancing over to see my daughter's beautiful face. I joined everyone back in the living room.

"Finally got her down, huh?" Katie asked.

"I did, and then I started rummaging through this box of old memories and decided share it. Come over and see," I said.

We were going through everything and having a ball. Still, the box I left behind never left my thoughts. The draw was demanding, so I decided to go back and retrieve it. Janie was still sleeping away. I slipped back out careful to not waken her.

As soon as I presented the box to Mom, I could see a veil of pain. It was obviously significant; but since a great amount of anguish also seemed attached to it, I was hesitant to explore it further.

After some discussion, Mom gave her blessing for all of us to read it. We hesitated. It was easier to dive into the memories captured in color and paper than to engage in ones clearly encased in pain. Katie lamented her traditional wedding after years of promising herself an extravaganza like Quillie's, Dad bemoaned his clothing and hairstyle over the years, and Mom found pictures of her mom and dad who were both gone now.

We were having such a good time, so we slid the box to the background. I felt its presence, but I wanted to enjoy the moment. Mom heard Janie's rustlings before I did and ran to get her.

"Mom sure loves that little girl, doesn't she?" Katie said.

"I imagine she sees you girls when she has Janie around," Dad said. "We always say it went too fast. Well, in your case Chelsea, I sure wish it'd go a little faster with the college years, so your Mom and I can get on the fast track to retirement."

"Whatever, Dad," she moaned. She was only in her sophomore year and still draining Dad's finances.

Mom came in cuddling Janie. I reached for my daughter, but clearly she was enjoying Mom's smothering. "I'm chopped liver here," I announced. "Then I go home, and it's all about Daddy. Go figure!"

Mom touched my arm, "Speaking of your hubby, I'm glad he'll be here for Thanksgiving. Luke, too, Katie," she added. "We have so much to be thankful for."

I agreed. I got a leftover depression cookie off the tray. *How fitting.* These cookies were always around when we gathered. It was our bond, a delectable delight born of rough times.

Katie looked a little sheepish as she offered, "Well, actually, Luke and I were going to tell you together, but I can't help myself," Katie said, a broad smile erupting. "We will have something else to be thankful for soon." We all looked at her and waited through her planned pause.

"Yup! Should be here by early March!" she said beaming. We jumped up squealing and hugging, just like old times. Mom cried, clutching Janie even closer. Even Dad was teary.

The rest of the day rounded out with more memories and hope for the future. When I went back into the guestroom to put Janie down for the night, I carried my great-grandmother's diary with me and put it carefully into my suitcase. I noticed some pages appeared to be falling out, so I reached down to tuck them back in and noticed more loose papers. I pulled them out . . . I recognized my grandmother's writing almost immediately and realized she had started her own diary entries and included them.

The desire to read it burned. I wanted to understand more how choices shape and affect our lives. I saw how emotional it had been for Mom to see it and to talk about it, and I knew it was very special. *Who were these women whose genetic makeup lined mine? Could peeking into their lives refract some hidden knowledge?*

I would read it one day. But for now, I refused to touch this day of happiness with even one drop of sadness.

Not one drop.